DNA makes RNA makes Protein

DNA makes RNA makes PROTEIN

Edited by
Tim Hunt, Steve Prentis, John Tooze

1983

ELSEVIER BIOMEDICAL PRESS
Amsterdam – New York – Oxford

© 1983 Elsevier Biomedical Press

ISBN: 0 444 80491 9

Published by
Elsevier Biomedical Press BV
P.O. Box 1527
1000 BM Amsterdam
The Netherlands

Sole distributors worldwide except for the U.S.A. and Canada
Elsevier Biomedical Press (Cambridge)
68 Hills Road
Cambridge CB2 1LA
United Kingdom

Sole distributors for the U.S.A. and Canada
Elsevier Science Publishing Company Inc.
52 Vanderbilt Avenue
New York, NY 10017, U.S.A.

Preface

In the past three years *Trends in Biochemical Sciences* has published nearly one hundred articles in the areas covered by this book – genome structure and organization, transcription, translation and post-translational events. To keep the size and cost of this volume within reasonable limits, this number had to be whittled down to less than half. Some articles were omitted simply because we had available more up-to-date reviews on the same topic; others were excluded because they explored fascinating by-ways – entirely appropriate in a monthly magazine such as *TIBS*, but not in a book which traces the main themes of this rapidly advancing field of science. One of the prime aims of *TIBS* is to present the latest advances in a way that is comprehensible to all biochemists and molecular biologists, and we hope that this collection of 37 reviews and six Journal Club articles will prove valuable to scientists with a wide variety of specialist interests.

Unlike a magazine, the contents of a book must be frozen in time. Had we chosen to delay the publication of this volume it would have been possible to include several important reviews which will appear in the next few issues of *TIBS*, but this is true whatever 'cut-off' date one chooses. As it is, the oldest of these articles was first published in September 1979; the most recent, in November 1982. All but ten first appeared within the last 2 years and eight of the older ones have been supplied with addenda up-dating them to July 1982.

We should like to thank all the authors for giving permission for their work to be published in this book and all the present and past members of the *TIBS* editorial board who have commissioned so many excellent articles.

STEVE PRENTIS

Introduction

The General Idea
Tim Hunt

'Jim, you might say, had it first. DNA makes RNA makes protein. That became then the general idea.'[1] Thus did Francis Crick explain to Horace Judson years later, long after he had written with such clarity and force on the subject of Protein Synthesis in the 1958 Symposium on 'The Biological Replication of Macromolecules'[2]. This article is celebrated for its prediction of the existence of tRNA (though by the time the article appeared in print, tRNA had been discovered), but it is chiefly worth reading and re-reading, even today, for its enunciation of the two principles that together constitute the 'General Idea'. The first principle is the Sequence Hypothesis; the idea that the sequence of amino acids in proteins is specified by the sequence of bases in DNA and RNA. The second principle is the famous 'Central Dogma'; not DNA makes RNA makes Protein, but the assertion that 'Once information has passed into protein it cannot get out again.' It isn't completely clear why one is a hypothesis and the other a dogma and the two together an idea. The Dogma stuck in some throats, mainly because it was called a dogma, with the heavy religious overtones.

Crick explains that calling it a dogma was a misunderstanding on his part; he thought the word stood for 'an idea for which there was no reasonable evidence', blaming his 'curious religious upbringing' for the error. But it probably wasn't that much of a mistake after all, for the Oxford Dictionary allows dogma to mean simply a principle, though the alternative 'Arrogant declaration of opinion' is probably how most who were not molecular biologists took it, considering its never-modest author. That is probably how they were meant to take it, too. It was the most important article of faith among the circle of biologists centred on Watson and Crick, and remained so for quite a long time, until the mechanism of protein synthesis became clear. Crick said that if you did not subscribe to the sequence hypothesis and the central dogma 'you generally ended in the wilderness', though he did not offer alternative scenarios for public consumption, even though they probably played an important part in convincing him of the dogmatic status of the General Idea's second component.

In any case, the General Idea has proved completely correct (at least as stated in the 1958 article), and it is now difficult to recapture the atmosphere in 1970 which caused such glee in certain quarters following the discovery of reverse transcriptase. The importance of this enzyme was not that it 'demolishes the hallowed idea of the molecular biologist that RNA cannot synthesize DNA'[3] (as we have seen, this wasn't the idea at all, let alone hallowed) nor yet that it solved the cancer problem (which it still hasn't), but because it opened the way for recombinant DNA technology to be applied to higher eukaryotes. It was a crucial tool, to be sure, but only a tool. The trouble was that Watson was very emphatic that RNA did not specify DNA synthesis in his justly famous and very influential textbook[4]. Even the 3rd edition says that 'DNA makes RNA makes protein' *is* the Central Dogma, and adds that 'RNA never acts as a template for DNA', though a couple of footnotes admit the existence of RNA tumour viruses and reverse transcriptase. It is curious and amusing to look back on this flurry; words like 'hallowed', 'shibboleth', 'heresy' and 'inspiration' joined 'dogma' in the pages of *Nature*[5]. They were the last shots in the battle, as it turned out. Crick wrote a measured rejoinder[6], and from then on it was possible to be a respectable Molecular Biologist. It is still difficult to understand why the General Idea aroused so much resentment among the more conservative biologists.

It is crucial to understand the General Idea. Together with the principle of Natural Selection it forms virtually the only solid ground biologists have to stand on. Some of the opposition to the Idea may have come from the constraints that it may appear to place on living systems; the discovery that they contain a highly structured logical core was perhaps surprising, and somewhat alien to biological thought, which tends to emphasize the variety, heterogeneity and general illogicality of living systems. The strength of the Idea lies much more in what it forbids than what it permits. The Central Dogma necessarily eliminates Lamarckism in any shape or form, and the idea of DNA as a self-replicating store of information for the synthesis of proteins is wholly in accord with the Darwinian conception of life. As Howard writes, the DNA molecule itself 'embodies in a pure chemical substance the three Darwinian conditions for evolution: heredity from the accuracy of self-replication, multiplication from the fact of self-replication and variation from the rare inaccuracies of replication'[7].

The General Idea has its limitations too, which in a way forms the subject matter of this book. The statement that DNA makes RNA makes Protein is first and foremost logical, and stresses the one-way flow of information. The chemical and structural content of the Idea is low, and its physiological content zero. It fails to specify what proteins are for, and fails to mention the important fact that the regulation of the information flow requires proteins to interact with DNA. But the Central Dogma was strictly concerned with protein synthesis, not with the control of gene activity. Indeed, when the Central Dogma says that 'information cannot escape from proteins' it sounds close to a denial of the very essence of living systems, for proteins are the very stuff of life. We wouldn't get very far if all we had was DNA. But it is at the level of proteins that simple logic fails, as the linear strings of information are converted into three-dimensional structures according to defined but complicated rules (which are still incompletely understood). And these structures have specific functional properties, which are quite impossible to deduce by simple examination of the structure.

While it is a challenging problem to predict the structure of a protein from its amino acid sequence, it would be impossible to tell what a protein *did* in a cell, even if every atom in its structure had been mapped with complete precision. It might be possible to guess, based on analogies with known enzymes, but given the sheer variety of potential substrates, and of the variety of possible transformations that might be performed on each of the substrates, it seems to be a task bordering on the logically impossible to say what a particular gene, and hence protein is for

based solely on its base (= amino acid) sequence. Likewise, it would be difficult to design an enzyme from scratch, if you were a real genetic engineer. What we actually do at present is to analyse. A gene is revealed by the effects of its loss or alteration. An enzyme can be discovered, purified and sequenced. Its gene can be located, purified and cloned. But it is difficult, if not impossible, to deduce the function of a given DNA sequence. It may be possible to rule out the possibility that it specifies a protein owing to the lack of extended open reading frames (though the existence of introns make this difficult in the genomes of higher organisms). It is even more difficult to recognize a regulatory region of DNA, for these sequences are exempt entirely from the rules of the Central Dogma or the Sequence Hypothesis. They are presumably constrained only by the requirements of uniqueness (so that they can be addressed unambiguously) and of recognizability by certain proteins. We will return to this theme later.

One gene, one enzyme

At about the same time that the 'transforming principle' was being identified as DNA, the slogan 'One gene, One enzyme' was coined by the fungal geneticists. They did not proclaim it as a dogma, or even really label it a hypothesis; it seems to have served principally as a useful definition of a gene. It is very interesting that this definition did not arouse the same kind of bridling hostility among the intellectual community of the day that the Central Dogma provoked. Was it because the implications were less explicit, because nobody cared, or because the slogan was couched in physiological terms?

A gene was a heritable entity which had detectable effects in organisms and an enzyme likewise something you could assay by its function. At that time, there was no general acceptance that genes were made solely of DNA, so to translate 'One Gene, One Enzyme' into the current equivalent 'DNA makes Protein' would not have been possible in any case. Even if it had, there is a curious disparity between the logico-chemical language of the Central Dogma and the functional description of One Gene, One Enzyme; a great deal is changed by the translation. What is gained in rigour is lost in physiological descriptiveness. It is an old conflict, between the observational biologists, the physiologists and the biochemists which is by no means over, as witnessed by Penman's current insistence on the importance of the cytoskeleton, for example[8]. Actually, according to Francois Jacob, even Monod had difficulty at first accepting that repression acted at the DNA level, that it could be as simple as that. His objections, which Jacob overcame in part by reference to toy trains, make interesting reading[9].

In fact, neither of these catch-phrases really deal with our current concerns. We now take for granted that the General Idea is correct (though the detailed mechanisms remain in many cases to be worked out). What is needed is an extension, like this:

DNA → RNA → Protein → Cells → Organism

Why not? Is this not what we all believe? Today's Dogma? The trouble is, such a statement of faith (for such it surely is) acts to keep the vitalists away, but it does not serve any other profound or useful function. It does not define limits, or exclude possibilities. As soon as the sequence of amino acids in proteins is achieved, they fold and gain properties which are no simple function of that sequence and, of course, the proteins interact with each other and with the structures whose building they catalyse to generate a system endowed with life. The pleasing simplicity and rigour of the General Idea is left far behind as soon as this messy and illogical realm is entered, where the laws of chemistry and physics are obeyed all right, but the system is more or less unfathomable because of the complexity of its interactions.

What is even worse is that the role of DNA goes well beyond simply specifying amino acid sequences. The instructions in the genetic material include what might be called analogue information as well as the digital. It isn't enough to specify the structure of proteins; the amount and timing of their synthesis must be specified as well. This specification is not made by a simple digital code based on the sequence of bases in DNA; rather, DNA is recognized by proteins in much the same way that they recognize any other substrate, and proteins control the expression of the information contained in the linear strings of bases. Proteins do not read DNA as ribosomes read mRNA, using simple nucleic acid recognition rules. Proteins 'feel' the shapes they find in the grooves, or look for sequence determined, but not rigorously sequence defined alterations in the backbone, like Z-DNA, for example. This is a very active area of research, from which exciting results are just beginning to emerge (look at recent work on the two lambda repressor proteins, cI and cro[10,11]). Evidently, proteins recognize the right sequence when they see it. At which point, the General Idea has to give up, because the proteins' trapped information is being used to control the flow of information from DNA. There is no more logic in that than in the recognition of glucose-6-phosphate by glucose-6-phosphate dehydrogenase. Certainly nothing to be dogmatic about.

A considerable oversimplification

The articles in this book all deal with the General Idea in one way or another (hence the title of the collection), beginning with various aspects of DNA and ending, after the mechanism of protein synthesis, with a set of pieces on what might be called 'topological' aspects of protein synthesis and assembly. It is a pleasure to have this opportunity of giving these articles a second airing.

The collection does not pretend to be complete. Some holes will probably still be there in time for the next edition, if there is one, because of the difficulty of the subjects. It should be possible to write a general article on the organization of chromosomes, for instance. After all, centromeres and telomeres have been isolated[12]. Genes and gene families have been cloned and sequenced. Great walks along hundreds of thousands of base pairs of Drosophila DNA have been taken[13]. Chromatin domains have been mapped[14]. Vast tracts of repetitive sequences occur in certain locations. Above all, comparisons between the simplest viral genomes and those of the highest eukaryotes are possible, but nobody is yet prepared to try to make simple sense of what is known, to account for the various features in structural and functional terms. It is admittedly a hard challenge. However, good examples of what can be done abound within the following pages.

The very first two articles succeed in summarizing an enormous amount of information in an easily accessible, interesting and thought-provoking way. Both deal with the question of how proteins find the right place in DNA. Both are confined to the prokaryotic realm, but such evidence from eukaryotes as is beginning to emerge suggests that similar principles apply to protein–DNA recognition in eukaryotes, though additional factors such as the accessibility of chromatin are likely to complicate the picture. On the other hand, it would be quite wrong to suppose that Jacob and Monod had the only word on the regulation of gene expression in bacteria. The article by M. Watson describes how premature termination of transcription is used to regulate the synthesis of mRNA from several operons which specify the enzymes for amino acid biosynthesis. In these cases it is thought that ribosomes play an important role, by determining the secondary and tertiary structure of the nascent transcript. Such a mechanism is very unlikely to play any part in eukaryotes, where the ribosomes are excluded from access to RNA by the nuclear membrane. However, the fact that such unexpected mechanisms are used ought to serve as a warning not to be too blinkered by existing orthodoxies. A further example of the variety of modes of regulation gene expression is found in the article by Nomura and his colleagues towards the end of this collection, where they discuss translational regulation, something which wasn't really supposed to happen in prokaryotes!

There should be much more to say about the topic of the regulation of gene expression in higher organisms, but despite a huge amount of work, extremely little is known. There are isolated examples of regulatory proteins which bind to specific eukaryotic DNA sequences like the (admittedly viral) SV40 T-antigen. The existence of consensus sequences at the 5′ end of structural genes suggest that at least promoters exist in eukaryotes. The RNA polymerases which recognize these sequences are exceedingly complex molecules, and most of the work on them has been devoted to purification and molecular characterization, as Marvin Paule describes in his article. Rather similar uncertainties surround DNA replication in higher eukaryotes, as the articles by Harland and by Scovassi *et al.* make plain. By contrast, the topology of DNA molecules and the enzymes that catalyse the various knottings, windings and weavings are surprisingly well understood, and well covered here in the articles by Wang, by Abdel-Monem and Hoffman-Berling, and by Watson. The reason they are so well understood, apart from the undoubted quality of scientists in this field, is that simple technological advances (for example, the gel system for analysing super-helicity) have enabled biochemical and genetic analysis to proceed hand in hand, while at the same time the underlying topological theory has supplied a firm conceptual basis.

Equally impressive is the progress in understanding the variety of ways that DNA can be rearranged. In 'Models of DNA transposition' A. I. Bukari gives a splendid account of various mechanisms and models whereby transposable elements of one sort and another are thought to move around their hosts, illustrated largely with examples from bacteriophage Mu and work from the author's laboratory at Cold Spring Harbor. It appears that gene rearrangement is also quite common in higher organisms; examples include the yeast mating type locus, the surface glycoproteins of trypanosomes (the article by A. Bernards) and the immunoglobulin genes of mammals (by N. Gough).

One of the important questions about the genomes of higher organisms is the extent to which rearrangements accompany differentiation. According to the classical view, supported by the famous experiments of Gurdon and colleagues using nuclear transplantation in frogs[15], there is neither loss (which would necessarily be irreversible) nor irreversible modification of the genetic material during the differentiation of specialized cells from the zygote (red cells apart, of course). However, as the various examples mentioned above clearly show, there are genes in eukaryotes whose expression depends on their environment, that is, precisely where they are in the chromosome. In the case of the immunoglobulin genes of course, the mature genes have to be assembled by various recombinational events. Are these highly specialized cases examples where a high degree of variety is called for, or are they representative of processes which are commoner than we now suppose? The fact is that we have at present no inkling how the diversity of cell types is generated, working from a common store of genomic information, and if there are lessons to be drawn from such examples as sexuality in yeast or the variation in the African trypanosome, it is surely that there are many different solutions to these problems. Every case has to be worked out for itself.

Rearrangements lead naturally here to mitochondrial genomes, thanks to the work of Bernardi and his colleagues. In the space of 3 years, the petite mutation of yeast has been almost fully elucidated, shedding unexpected light in all directions, including insights into the replication of mitochondrial genomes, their transcription and the mechanism of methotrexate-resistance in mammalian cells. A more straightforward view of the mammalian mitochondrion's genetic apparatus is provided by Attardi, whose beautiful work on this system over the years came to full flower with the elucidation of the complete DNA sequence. It is a pity perhaps that there is no first-hand account of that work from Sanger's laboratory. What is instructive is to realize how important both approaches are; sequence information by itself is pretty dry stuff, but once both the sequences and the functional maps can be compared, an avalanche of understanding is released. There is however that most important and intriguing aspect of mitochondria to be elucidated; how the nuclear and mitochondrial genes are co-ordinated. C. Leaver touches briefly on this point in his article on male sterility in maize, which turns out to be all about plant mitochondria. A nice story which combines genetics, plant breeding and agriculture with molecular biology.

Whereas it is possible to comprehend the entire mitochondrial genome, comparable studies of chromosomes are confined to rather more restricted regions by the sheer magnitude of the problem (the X-ray film alone for sequencing the entire human genome would cost several million pounds). C. Bostock discusses the mysterious satellite DNA, E. Ullu the only somewhat less abundant but equally inscrutable Alu family of sequences. 'Real' genes are represented by Collagen and the corticotropin–beta lipotropin precursor (older articles on globin[16] and ovalbumin[17] were cruelly discarded). Collagen is astonishing for its large number of relatively enormous intervening sequences, whereas the ACTH–beta-LPH precursor has a relatively simple gene but a complex pattern of processing to yield a multiplicity of biologically active peptides. Not all eukaryotic genes are this complicated. For example, there are histone and interferon genes which have no intervening sequences.

The question that always springs to mind is: why is the concentration of coding sequences so low in the chromosomes of higher organisms? Not only do the intervening sequences usually exceed the exons by far within a given transcription unit, but the spaces between adjacent genes are enormous in the cases (e.g. the globin family) that have been studied. This is not true of the mitochondrial genome, however, which Attardi calls 'A lesson in economy'. How does the nucleus tolerate such prolixity, and why is the organelle so spare? At present there do not seem to be any simple answers to these questions.

The discovery of split genes and the existence of splicing was made during studies of gene expression in adenovirus, an organism that L. Philipson used to call the 'lambda of eukaryotes', not without reason. Two articles, the first by Philipson and Petterson, the second by Broker and Chow, give a bird's-eye view of the organization and pattern of gene expression in this complex virus which has consistently given leads ahead of time as to what to find in the expression of regular cellular genes. What does not perhaps emerge from these articles is the incredible amount of work that lies behind these elegant and complex stories, though the acknowledgements section in Broker and Chow's piece, regretting that lack of space did not permit more thorough citations of the extensive literature, alludes to this in quite a heartfelt way. There are still a good many good stories in that virus. Apart from anything else, how it transforms cells is still quite unclear; and there is the ironic matter of VA RNA, one of the first RNA molecules to have its complete sequence determined, very abundant in adeno-infected cells, still looking for a certain role! This is of course an excellent example of how difficult it is to deduce anything from a nucleic acid sequence which does not specify a protein, and the following article by Dynan and Tjian likewise describes the 'enhancer sequences' in that other favourite object of study, SV40, which appear to exert a long range effect and enhance transcription from neighbouring genes. It is not clear how they work, and they were only detected as a result of empirical observations; that is, their existence was not deduced as a result of knowing the complete DNA sequence of SV40. Even in such a relatively simple case, the information that can be obtained from studying the sequence of bases in the genome is quite limited at present.

RNA splicing seems to be a potential site of control of gene expression which cells do make use of during differentiation, as Wall describes in his article on the switch between IgM and IgG synthesis during lymphocyte maturation; adenovirus thought of it first, of course. At the moment, however, the precise mechanism of splicing is still unknown, except in the case of the tRNA precursors in yeast that Abelson and colleagues have elucidated. It is doubtful whether the same mechanism applies to the splicing of mRNA, and at present any hope of determining how it may be regulated seems remote. There is no shortage of ideas, though they have not received much of an airing in the columns of *TIBS*. While the notion that the small nuclear RNAs play some sort of a template role is attractive[18], evidence that such is actually the case is elusive, and meanwhile, the electrifying observation that some RNA molecules require nothing besides GTP to catalyse their own processing has set people thinking afresh[19]. Perhaps this is a special case, as is probably true of the 'flu story, equally astonishing when it broke. It had been known for a long time that influenza virus required an active nucleus in order to replicate, even though it was a negative strand RNA virus and contained its own polymerase. Nobody dreamed that it required fragments of capped host messages to act as primers until Krug and Plotch followed their noses and arrived at this startling conclusion, which they proceeded to document in convincing detail, as they tell here. Whether or not this mechanism has any implications for the operation of normal cellular processes, it is a fascinating story. While it is not inconsistent in any way with the General Idea, it scarcely flows naturally from it. The whole splicing story rests uneasily with the notion of colinearity, in fact. Once it is allowed to split genes into pieces, there is no logical necessity to put the pieces together in the order in which they were transcribed. The fact that they are, at least in every case that is known so far, says more about the mechanism of splicing than anything else. In other words, knowing what we now know of gene organization in higher eukaryotes, it looks like the sequence hypothesis is a bit lucky to have survived intact. That isn't quite fair, of course, because at the level of mRNA *translation* it still stands as a great and fundamental truth.

An extremely difficult matter

It is as true today as it was in 1957, as Crick said, 'an extremely difficult matter to present current ideas about protein synthesis in a stimulating form'[2]. Best then to draw a veil over the articles by Clark, Caskey and myself, which attempt to give a straightforward account of the process of protein synthesis, and move on to two articles which do present ideas in a stimulating form, by Fersht, and by Schimmel and colleagues.

The exciting logic of the triplet code brushed aside the problem of how codon–anticodon recognition actually worked, a problem which still remains. Fersht's work has mainly dealt with tRNA charging, a step in protein synthesis which is crucial in maintaining a high overall accuracy. More recently he has turned his attention to the process of DNA synthesis, and he compares the various mechanisms which together determine the overall accuracy of information transfer from DNA to protein. Schimmel, Putney and Starzyk discuss complementary aspects of a part of the same problem; how proteins accurately recognize RNA molecules. But they also tell how in the case of the alanyl tRNA synthetase the same protein recognizes not only its substrate, but also its very own gene, and regulates its own expression. So much for information not getting out of proteins! This is a very remarkable situation, but then, 'In the protein molecule, Nature has devised a unique instrument in which an underlying simplicity is used to express great subtlety and versatility; it is impossible to see molecular biology in proper perspective until this peculiar combination of virtues has been clearly grasped[2].'

The following article in this section, by Nomura, Dean and Yates, exemplifies this dictum in yet another way. Studying the genes for the ribosomal proteins of *E. coli*, they found that the synthesis of ribosomal proteins is regulated at the translational level; certain proteins bind to their own messages it seems, and the structure of these messages mimics the structure of the regions of ribosomal RNA to which these proteins normally bind. The story of the regulation of expression of these complex operons is a good deal more complicated, however, and doubtless further lessons will be learned from their study. It is hard to resist a wry smile, however, as bacteria were once supposed to regulate all gene expression at the level of transcription, whereas now it seems that all the best-documented cases of translational control come from prokaryotic systems. This section of the book concludes with a piece by Garrett, Douthwaite and Noller about the role of 5S RNA in protein synthesis. It shows how difficult it is to find out how ribosomes are constructed and how they work despite a battery of elegant probes and clever ideas.

Cells are not the empty containers that the name implies, and the final section of the book deals with the question of how proteins are delivered to the correct places. The general idea of the signal hypothesis is over 10 years old now, and Leader summarizes the classical story. However, the last couple of years has seen a fresh burst of information, with attention shifting from the N-terminus of the nascent peptide, the signal sequence, to the components of the endoplasmic reticulum with which it interacts. A. M. Tartakoff and D. I. Meyer describe the recent evidence, though a most intriguing aspect was discovered too recently to receive a mention. It is that the signal recognition protein (SRP) contains 7S RNA[20], a 300 nucleotide cytoplasmic RNA whose sequence is conserved among different vertebrates, and which shows homology with the Alu family. Treatment of SRP with RNase abolishes its activity, so presumably it plays a role in secretion. Whether the RNA serves to hold together the several components of SRP in the same way that ribosomal RNA is presumed to act as the backbone of the ribosome, or whether there may be some hydrogen bonding between the RNA of the ribosomes and 7S RNA à la Shine and Dalgarno is too early to say.

There is an interesting comparison to be made between the almost exclusively biochemical approach to the analysis of secretion which has been so successful in animal cells with the almost wholly genetic approach described by Schekman in his article on secretion in yeast. A large number of mutants defective in secretion have been isolated, from which it can be deduced that there are a large number of components whose correct functioning is essential for this process. The order in which they act can also be deduced, that is, a pathway is defined. However, the actual nature of the pathway is poorly defined in chemical and structural terms at present. Still, mutations are potentially extremely useful tools. Somebody should write a book called Great Mutations, with chapters on $r1789$ and $c1857$ and so forth.

We don't really know how proteins are transported into mitochondria, as Neupert and Schatz say, but the affinities with the secretory pathway suggest that their excellent article belongs in the final section rather than up with the organelles. It is still a considerable mystery how most of the cellular components find their rightful places in cells. This looks like a messy problem, one that has to be solved for each component in turn, though admittedly the idea of detachable signals, or at least optionally detachable signals has proven a surprisingly general mechanism to explain intracellular traffic. One hopes that similar kinds of attractive simplifying rules will emerge to enlighten some of the other dark corners of cellular life.

The recent history of molecular biology has shown over and over again how strange are the ways of Nature, even when you understand the Central Dogma, the General Idea, whatever you call it. Not that it's a 'considerable oversimplification'. All the Idea ever sought to explain was how proteins were made, to account for the role of the genetic material in this process. As the articles in this book show, even fleshing out the General Idea with the specific details of its implementation has taken a great many people a great many years, and has revealed a wealth of unexpected beauty in the search for the Actual[21]. The search goes on, of course; it is still an open question whether development has a logic to it as simple as that of the lac operon. Brenner puts it as clearly and wittily as anyone[22], asking whether there is such a thing as the '"Make a hand" program' encoded in the genome. This is high level stuff. We do not even know if there is a button in the genome which when pressed (metaphorically speaking) will give a red cell. It is going to be a lot of fun finding out. In effect, the hypothesis to be explored and tested was stated boldly (looking at it now, almost foolhardily) by Crick in the famous article, right at the beginning. It seems just that he should have the last word. 'I shall also argue that the main function of the genetic material is to control (not necessarily directly) the synthesis of proteins. There is a little direct evidence to support this, but to my mind the psychological drive behind this hypothesis is at the moment independent of such evidence. Once the central and unique role of proteins is admitted there seems little point in genes doing anything else.'

References

1 Judson, H. F. (1979) *The Eighth day of creation*, Jonathan Cape, London

2 Crick, F. H. C. (1958) On protein synthesis. *Symp. Soc. Expl. Biol.* 12, 138–163
3 Anonymous (1970) Aprés Temin, le Deluge. *Nature* 227, 998–999
4 Watson, J. D. (1976) *Molecular Biology of the Gene* (3rd edn), W. A. Benjamin, Menlo Park, California
5 Anonymous (1970) Central Dogma Reversed. *Nature* 226, 1198–1199
6 Crick, F. H. C. (1970) Central Dogma of Molecular Biology. *Nature* 227, 561–563
7 Howard, J. C. (1982) *Darwin*, Oxford University Press
8 Cervera, M., Dreyfuss, G. and Penman, S. (1981) Messenger RNA is translated when associated with the cytoskeletal framework in normal and VSV-infected HeLa cells. *Cell* 23, 113–120
9 Jacob, F. (1979) The Switch. In *Origins of Molecular Biology* (Lwoff, A. and Ullman, A., eds), Academic Press, New York
10 Pabo, Carl O., Krovatin, W., Jeffrey, A. and Sauer, R. T. (1982) The N-terminal arms of lambda repressor wrap around the operator DNA. *Nature* 298, 441–443 *and* Pabo, C. O. and Lewis, M. (1982) The operator-binding domain of lambda repressor: structure and DNA recognition. *Nature* 298, 443–447
11 Ohlendorf, D. H., Anderson, W. F., Fisher, R. G., Takeda, Y. and Matthews, B. W. (1982) The molecular basis of DNA-protein recognition inferred from the structure of *cro* repressor. *Nature* 298, 718–723 *and* Matthews, B. W., Ohlendorf, Anderson, R. G. and Takeda, Y. How does *cro* repressor recognize its DNA site? *Trends Biochem. Sci.* (to appear early 1983)
12 Szostak, J. W. and Blackburn, E. H. (1982) Cloning yeast telomeres on linear plasmid vectors. *Cell* 29, 245–255
13 I have heard Welcome Bender speak about his and Pierre Spierer's 'walks' along the Drosophila genome between rosy and bithorax, but it seems that nothing exists in print. Examples of a couple of strolls are in Snyder, M., Hirsh, J. and Davidson, N. (1981) *Cell* 25, 165–167, or in McGinnis, W., Farrell, J. and Beckendorf, S. K. (1980) *Proc. Natl Acad. Sci. U.S.A.* 77, 7367–7371
14 Elgin, S. C. R. (1981) DNAase I-hypersensitive sites of chromatin. *Cell* 27, 413–415
15 Gurdon, J. B. (1974) *The control of gene expression in animal development*, Oxford University Press
16 Kourilsky, P. and Chambon, P. (1978) The ovalbumin gene: an amazing gene in eight pieces. *Trends Biochem. Sci.* 3, 244–247
17 Flavell, R., Glover D. M. and Jeffreys, A. J. (1978) Discontinuous genes. *Trends Biochem. Sci.* 3, 241–244
18 Lerner, M. R. and Steitz, J. A. (1981) Snurps and Scyrps. *Cell* 25, 298–300
19 Cech, T. R., Zaug, A. J. and Graboi, P. J. (1981) *In vitro* splicing of the ribosomal RNA precursor of tetrahymena: involvement of a guanosine nucleotide in the excision of the intervening sequence. *Cell* 27, 487–498
20 Walter, P. and Blobel, G. (1982) Signal Recognition Particle contains a 7S RNA essential for protein translocation across the endoplasmic reticulum, *Nature* 299, 691–698
21 Jacob, F. (1982) *The Possible and the Actual*, Pantheon Press, New York
22 Brenner, S. (1981) Genes and Development. In *Cellular Controls in Differentiation* (Lloyd, C. W. and Rees, D. A., eds), Academic Press, London

Tim Hunt is at the Department of Biochemistry, University of Cambridge, Tennis Court Road, Cambridge CB2 1QW, U.K.

Contents

Preface   v
Introduction, The General Idea, *Tim Hunt*   vii

DNA and DNA–protein interactions

How do genome-regulatory proteins locate their DNA target sites?,
    Otto G. Berg, Robert B. Winter and Peter H. von Hippel   3

The interaction of *E. coli* RNA polymerase with promoters, *Hermann Bujard*   10

Attenuation: translational control of transcription termination, *M. D. Watson*   18

Initiation of DNA replication in eukaryotic chromosomes, *Richard Harland*   23

Eukaryotic DNA polymerases, *Anna I. Scovassi, Paolo Plevani and
    Umberto Bertazzoni*   30

Superhelical DNA, *James C. Wang*   37

DNA unwinding enzymes, *M. Abdel-Monem and H. Hoffmann-Berling*   43

The role of DNA topoisomerase I in transcription and transposition,
    M. D. Watson   49

Nucleosome structure, *A. D. Mirzabekov*   51

Models of DNA transposition, *A. I. Bukhari*   56

Mitochondrial genomes

The petite mutation in yeast, *Giorgio Bernardi*   67

The origins of replication of the mitochondrial genome of yeast,
    Giorgio Bernardi   76

Organization and expression of the mammalian mitochondrial genome – a lesson
    in economy: I, *Giuseppe Attardi*   85

Organization and expression of the mammalian mitochondrial genome – a lesson
    in economy: II, *Giuseppe Attardi*   92

Mitochondrial genes and male sterility in plants, *Christopher J. Leaver*   99

Genome organization and reorganization

A function for satellite DNA?, *Christopher Bostock*   111

The human *Alu* family of repeated DNA sequences, *Elisabetta Ullu*   119

Structure and regulation of a collagen gene, *Benoit de Crombrugghe and
    Ira Pastan*   125

Corticotropin-β-lipotropin precursor – a multi-hormone precursor – and its gene,
    Shosaku Numa and Shigetada Nakanishi   130

Gene amplification during development, *R. Chisholm*   137

The rearrangements of immunoglobulin genes, *Nicholas Gough*   139

Gene rearrangement can extinguish as well as activate and diversify
    immunoglobulin genes, *Nicholas Gough*   145

Surprising complexity in the gene locus encoding mouse λ light-chain
    immunoglobulins, *Nicholas Gough*   151

RNA synthesis and processing

Comparative subunit composition of the eukaryotic nuclear RNA polymerases, *Marvin R. Paule* — 155
Control of adenovirus gene expression, *Lennart Philipson and Ulf Pettersson* — 162
Patterns and consequences of adenoviral RNA splicing *Thomas R. Broker and Louise T. Chow* — 171
Transcription enhancer sequences: a novel regulatory element, *W. Dynan and R. Tjian* — 181
Immunoglobulin RNA processing, *Randolph Wall* — 183
The mechanism of tRNA splicing, *Richard C. Ogden, Gayle Knapp, Craig L. Peebles, Jerry Johnson and John Abelson* — 188
The 5′ ends of influenza viral messenger RNAs are donated by capped cellular RNAs, *Robert M. Krug, Michele Bouloy and Stephen J. Plotch* — 196

Protein synthesis

The initiation of protein synthesis, *Tim Hunt* — 205
The elongation step of protein biosynthesis, *Brian Clark* — 213
Peptide chain termination, *C. Th. Caskey* — 220
Enzymic editing mechanisms in protein synthesis and DNA replication, *Alan R. Fersht* — 226
RNA and DNA sequence recognition and structure-function of aminoacyl tRNA synthetases, *Paul Schimmel, Scott Putney and Ruth Starzyk* — 233
Feedback regulation of ribosomal protein synthesis in *Escherichia coli*, *Masayasu Nomura, Dennis Dean and John L. Yates* — 240
Structure and role of 5S RNA–protein complexes in protein biosynthesis, *R. A. Garrett, S. Douthwaite and H. F. Noller* — 246
The organization of 16S ribosomal RNA, *R. A. Garrett* — 252

Putting proteins in their place

Protein biosynthesis on membrane-bound ribosomes, *David P. Leader* — 257
The one and only missing link in protein secretion?, *A. M. Tartakoff* — 265
How proteins are transported into mitochondria, *Walter Neupert and Gottfried Schatz* — 266
The signal hypothesis – a working model, *David I. Meyer* — 274
The secretory pathway in yeast, *R. Schekman* — 278

Index — 285

DNA and DNA-protein interactions

How do genome-regulatory proteins locate their DNA target sites?

Otto G. Berg, Robert B. Winter and Peter H. von Hippel

The Escherichia coli lac *repressor (R) locates its DNA target site, the operator (O), by a two-step mechanism. In the first step a repressor-DNA (RD) complex is formed at a non-operator site in an ordinary diffusion-controlled reaction. This non-specifically bound repressor then 'slides' along the DNA in a one-dimensional diffusion process. A series of intramolecular (within the 'domain' of the DNA molecule) dissociation–association steps, interspersed by sliding, eventually results in target location. Similar mechanisms may be used by other genome regulatory proteins in locating specific DNA binding sites.*

Site-specific genome-regulatory proteins, such as the polymerases, repressors and activators involved in transcription, initiate their action by binding to specific base pair sequences (operators, promoters etc.). However, before specific binding can occur, the proteins must *locate* and *recognize* these targets.

The magnitude of this location and recognition problem becomes clear when we consider the *E. coli* cell. The *E. coli* genome usually contains single operator and promoter sites for each operon. It also contains ~ 10⁷ base pairs of DNA. The number of copies per cell of any particular site-specific genome regulatory protein is *much* smaller than this (for example, ~ 10¹ *lac* repressor molecules, ~ 5 x 10³ RNA polymerase molecules etc.). Therefore, in specific binding interactions the target sites are each in competition with ~ 10⁷ potential (overlapping) non-specific binding sites.

How do genome-regulatory proteins recognize their specific target sites?

This problem has recently been analyzed in detail[1]. Generalizing from the properties and behavior of *E. coli lac* repressor (the specific protein–DNA interaction about which we know the most[2]), the situation can be summarized as follows. (i) Genome regulatory proteins show affinity for non-specific sites. This binding plays a central role in both equilibrium and kinetic aspects of the interaction of these proteins with their specific target sites. (ii) This non-specific binding is largely 'electrostatic' in nature; that is, binding involves charge–charge interactions between DNA phosphates and the basic residues of the protein. The free energy driving this process results from the entropy of dilution of the small cationic counterions (K^+, Na^+, Mg^{2+}) released from the DNA when the protein binds[3]. (iii) Specific binding involves the interaction of a matrix of DNA hydrogen bond donors and acceptors, located in the grooves of the double-helix, with the sterically-complementary acceptors and donors of the protein binding site. (iv) In order for specific binding to be favored at equilibrium, the product of K and [D] must be greater for specific than for non-specific binding, where K is the (specific or non-specific) binding constant and [D] is the concentration of each type of DNA binding site. (v) Genome regulatory proteins may bind specifically to DNA sites involving a few 'wrong' base pairs (i.e. partially mutated sites); however, too many wrong base pairs result in too many unfavorable mispairings of protein and nucleic acid hydrogen bonding groups and the protein is 'pushed' into a non-specific DNA binding conformation.

With these principles in mind we turn to the problem of the rate at which genome-regulatory targets are located by specific binding proteins, illustrating our discussion with recent theoretical and experimental studies of the *E. coli lac* repressor–operator system.

Theoretical background

A bimolecular interaction between small molecules in solution generally proceeds via an initial diffusional encounter. One can calculate a separate *maximum* rate for the formation of the initial encounter complex, treating the reaction as 'diffusion-controlled'. A direct (one-step) reaction between repressor (R) and operator (O) can be written:

$$R + O \underset{k_d}{\overset{k_a}{\rightleftharpoons}} RO \qquad (1)$$

and the *maximum* value of k_a for this process can be estimated[4,5] to be $\sim 10^7$–10^8 $M^{-1} s^{-1}$.

In contrast, *measured* values of k_a, determined *in vitro* in very dilute solutions of repressor and operator[5-7], have been found to *exceed* 10^{10} $M^{-1} s^{-1}$, under some conditions. Since a bimolecular process cannot, by definition, be faster than diffusion controlled, this means that location of operator by repressor *must* involve mechanisms which effectively reduce either the dimensionality or the volume of the search process.

Therefore, the actual kinetic pathway must comprise (at least) a two-step process:

$$R + D + O \underset{k_{-1}}{\overset{k_1}{\rightleftharpoons}} RD + O \underset{k_{-2}}{\overset{k_2}{\rightleftharpoons}} RO + D \qquad (2)$$

in which the first step of the overall reaction (eqn. 2) is the diffusion controlled formation of a complex between repressor and a non-specific DNA site (D), followed by a *facilitated* transfer process involving further non-specific RD complexes, before the final RO complex is formed (see Fig. 1). Since the overall reaction must be speeded by binding of repressor to non-specific sites, such binding must *increase* the rate of target location, rather than slowing it further as would be expected if RD complexes act as competitive 'sinks' for the repressor.

Proposed mechanisms for facilitated transfer of non-specifically bound repressor

Two mechanisms have been proposed for the facilitated transfer of repressor in the non-specifically bound state. These are illustrated in Fig. 1. One is called 'sliding', and is defined as the diffusion of repressor, *while non-specifically bound*, in a one-dimensional random walk along the DNA molecule[4,6,8,9]. This reduces the dimensionality of the operator-search process, and could, under some conditions, lead to an accelerated rate of target location. The other proposed mechanism is called 'intersegment transfer'[4,10], and involves the rapid and direct transfer of repressor from one segment of a DNA molecule to another as a consequence of the relative diffusion of these segments within the 'domain' of the molecule. This second mechanism postulates the transfer of repressor by a series of 'ring-closure' events in which the repressor is transiently bound between two DNA segments (current views of repressor structure suggest that the repressor tetramer may have at least two DNA binding sites[1]). When the segments diffuse apart one of the DNA–protein contacts will break and, if the two binding interactions are equally tight, the repressor will have a 50% chance of being transferred to a new and distant site on the DNA molecule. This process may be very fast because it circumvents the large activation barrier involved in the dissociation of repressor into solution.

Both mechanisms are plausible, given the known properties of *lac* repressor (see below), and both can, in principle, lead to the desired result. Fortunately, these mechanisms are experimentally distinguishable because sliding is correlated with position *along* the DNA molecule, while intersegment transfer is not.

Experiments and analysis

The binding of repressor to operator-containing DNA was examined using a filter-binding technique[11]. This method takes advantage of the fact that protein binds tightly to nitrocellulose filters, while free DNA passes through. Radioactively-labelled DNA will therefore only be retained on the filter if it is tightly complexed to protein. Conditions are fixed so that weak complexes, such as those due to non-specific binding (RD interactions), are *not* retained, while RO complex formation holds the labeled DNA on the filter. Both equilibrium and kinetic measurements can be made this way; because of their bimolecular character even very fast association reactions can be brought into the experimental time range by dilution.

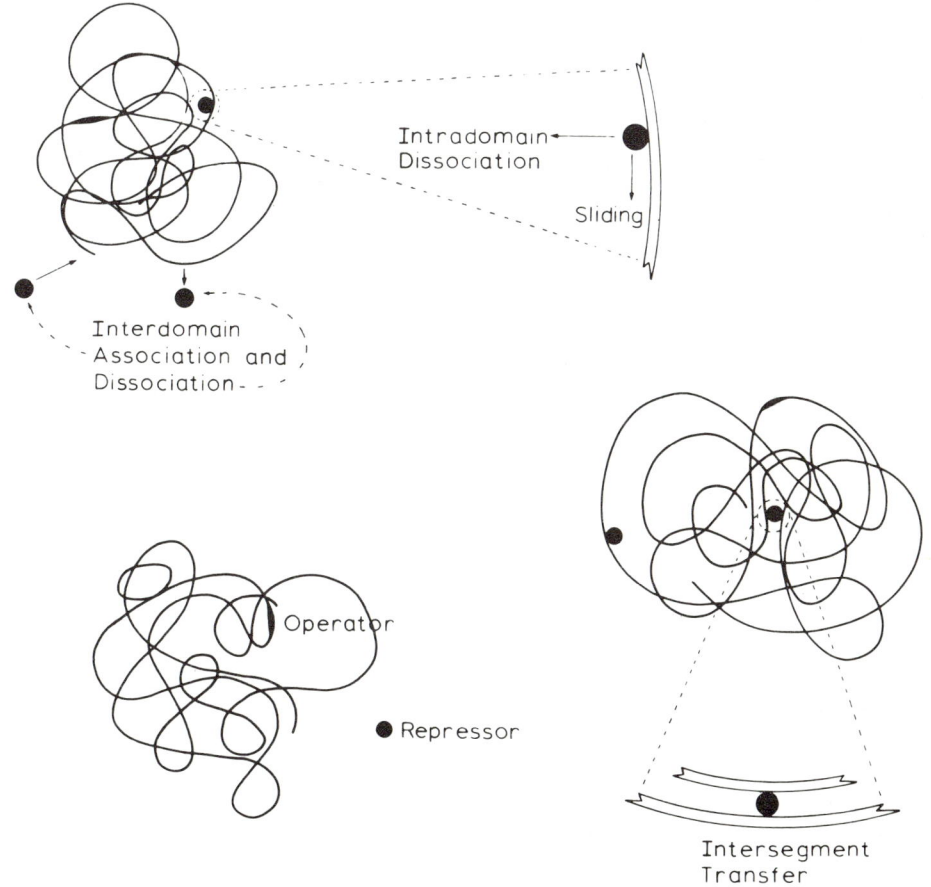

Fig. 1. Schematic view of lac repressor (R) interacting with large operator-containing (O) DNA molecules in dilute solution. The DNA molecules are well-separated into 'domains' under these conditions. The (upper right) expanded view shows repressor bound to a segment of non-operator DNA, on which it can either 'slide' or engage in intradomain dissociation–association processes in seeking its specific (operator) target site. The (lower right) expanded view shows a repressor molecule doubly bound to two DNA segments; this corresponds to the (hypothetical) intermediate state in the intersegment transfer process.

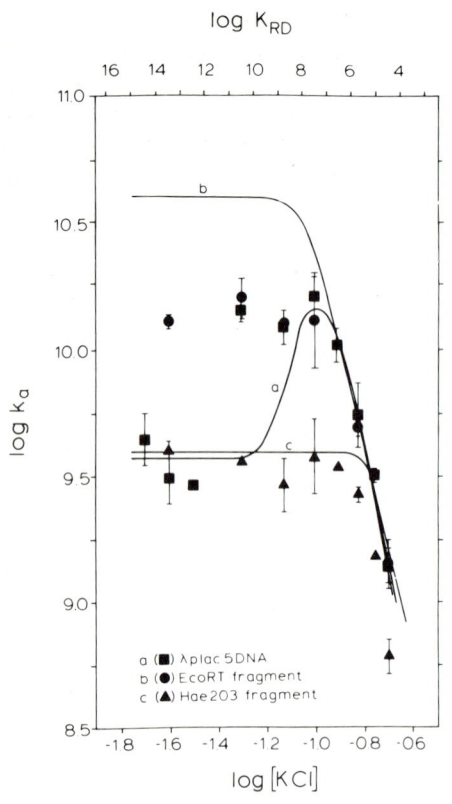

Fig. 2. Plot of log k_a versus log [KCl] for: (a) λplac5 DNA (~ 50,000 base pairs); (b) EcoRI lac-operator-containing DNA fragment (~ 6700 base pairs); and (c) Hae III lac-operator-containing DNA fragment (203 base pairs). Association rates determined by filter binding at ~ 20°C. (Data from Winter et al.[5]. Theoretical curves from Berg et al.[4]).

Careful kinetic and equilibrium measurements of the RO interaction have been made by this technique[5,7,12]. We focus here on the association process.

In Fig. 2 we show some typical values of k_a (plotted as log k_a), as a function of KCl concentration (plotted as log [KCl] or log K_{RD}), for three different sizes of lac operator-containing DNA fragments. The resulting values of k_a depend in a unique manner on DNA length and salt concentration and permit an unambiguous interpretation. In dilute solution, where the DNA 'domains' (see Fig. 1) occupy only a small fraction of the total volume of the solution, the RO interaction clearly proceeds by a two-step process (see eqn. 2). Theoretical expressions[4] for sliding and for intersegment transfer give characteristically different dependences on DNA length and salt concentration.

Repressor slides to the operator target

Fig. 2 shows that the experimental data can be fitted theoretically (solid lines) using sliding as the sole mechanism for the facilitated transfer of repressor, and employing a single value of the sliding rate constant D_1 (i.e., the one-dimensional diffusion coefficient), which is taken as independent of salt concentration. The best-fit value of $D_1 \simeq 9 \times 10^{-10}$ cm^2/s at ~ 20°C*.

Qualitatively, the curves of Fig. 2 can be explained as follows. Starting at the right-hand side of the Fig., the overall forward rate constant can be described as the diffusion-controlled rate of binding to a target with an effective size given by the 'sliding length'; i.e. the distance the protein can slide along the DNA without dissociation into solution. This implies that a protein binding non-specifically within this distance from the operator will bind to the target site without intervening dissociation.

The increase in k_a with decreasing salt concentration (Fig. 2) reflects the increased lifetime of each RD complex, and the concomitant increase in sliding length. When this effective target length approaches and exceeds the dimensions of the DNA molecules in solution, the first step in eq. (2) becomes rate limiting and the specific association rate is given by the diffusion-controlled rate of binding of R to the entire DNA molecule. This accounts for the low-salt plateaus seen for the two shorter DNA fragments (curves b and c of Fig. 2).

For larger DNA molecules (e.g. λplac5;

*Somewhat better theoretical fits to some of the low salt data of Fig. 2 can be obtained by using slightly different values of K_{RD} and/or D_1 at low and high salt concentrations in these calculations (see Winter et al., 1981). Here we show only the theoretical curves obtained for all three DNA fragments using the best single values of D_1 and K_{RD}.

curve a of Fig. 2), at the concentrations of R,O and D used, the first step of eq. (2) is never rate limiting. Instead k_a goes through a maximum with increasing sliding length (decreasing salt concentration); subsequently (at lower salt concentrations), the rate of sliding itself becomes limiting, and k_a decreases with increasing sliding length. At still lower salt concentrations a plateau is also reached for the larger DNA molecules; beyond this point the whole length of the DNA chain is scanned in a single sliding event.

Molecular interpretation of repressor sliding

In the preceding section we have summarized evidence for the existence of a facilitated transfer mechanism for repressor that depends on DNA length because it operates while the protein is (non-specifically) bound to non-operator DNA. This mechanism has been termed 'sliding'. The one-dimensional diffusion coefficient (D_1) calculated for repressor transversing non-specific DNA is approximately 10^{-9} cm²/s, which corresponds to an instantaneous random walk (sliding) rate ($\Gamma_1 = D_1/l^2$) of $\sim 10^6$ 'jumps' between neighboring binding sites (base pairs) per second (ℓ is the length of one base pair; i.e., 3.4×10^{-8} cm). The length 'scanned' by such a random walk process is $(D_1t/l^2)^{\frac{1}{2}}$, which corresponds to the scanning of $\sim 10^3$ base pairs *in one second* (and $\sim 3.2 \times 10^2$ base pairs in 0.1 s., etc.).

How might such a sliding process be visualized in terms of molecules? This question may be broken down further, as follows:

(i) *Is the above sliding rate reasonable for a molecule of the size of repressor, if the rate is limited only by hydrodynamic considerations?* Clearly D_1 ($\simeq 10^{-9}$ cm²/s) is much smaller than the ordinary three-dimensional diffusion coefficient estimated for a protein of this size ($\sim$ 150,000 mol. wt; $D_3 \simeq 5 \times 10^{-7}$ cm²/s). Thus, if the repressor slides linearly along the DNA 'cylinder' and the frictional resistance to this process is comparable to that in bulk solvent, the measured value of D_1 is well below the limit imposed by hydrodynamic considerations. This approach has been carried further by calculating the hydrodynamic limit for D_1 if repressor slides along the DNA by maintaining a *fixed orientation* relative to the sugar-phosphate double-helical backbone (i.e., if it slides by rotating *around* the DNA molecule once for every 34 Å moved along the DNA axis). It has been estimated that D_1 can be no larger than $\sim 5 \times 10^{-9}$ cm²/s in this model[13]; thus, even here the observed value of D_1 is below the hydrodynamic limit.

(ii) *What forces hold repressor to the DNA during the sliding process?* The binding of repressor to non-specific DNA is purely electrostatic, and involves ~ 11 charge–charge interactions per repressor tetramer[14,15]. In terms of the analysis of the interaction of charged ligands with DNA by Record and co-workers[3], which is formulated in terms of the dependence of the interaction on salt concentration, this means that 9–10 monovalent counterions are displaced from the DNA double-helix by the (locally) polycationic repressor. Fig. 3 illustrates this schematically, and also shows that sliding of the repressor along the DNA results in no *net* change in ion displacement; the monovalent cation displaced from in front of the repressor as it moves is simply replaced by one binding to the DNA behind the repressor. Because no net ion displacement (or replacement) is involved, the repressor can be considered to be sliding over the DNA on an isopotential surface. Thus, there is no thermodynamic barrier to repressor sliding along the DNA, since the relaxation of the ion atmosphere is fast relative to the repressor sliding rate. In contrast, dissociation of the repressor from the DNA, either directly or by sliding off the ends of the DNA fragment, requires net counterion replacement and is thermodynamically unfavorable. There should be little or no

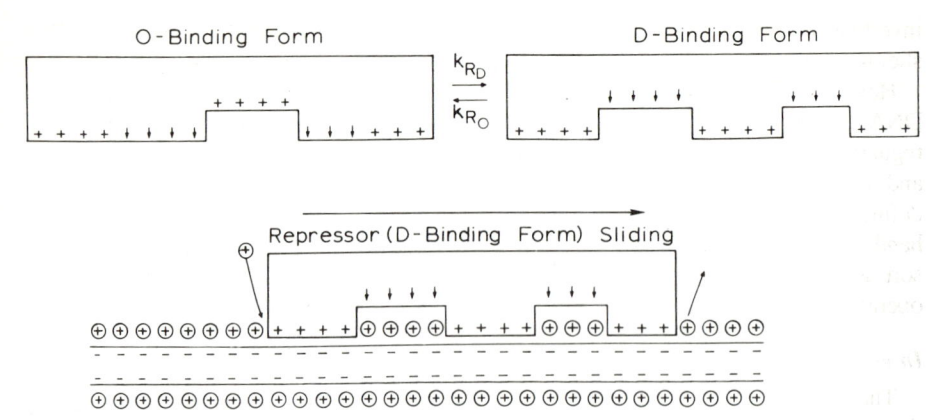

Fig. 3. Schematic models of the operator-binding and non-specific DNA-binding conformations of the lac repressor. The o-binding form (top left) interacts with operator via ~ 7 charge–charge interactions (+), and a number of specific hydrogen-bonding functional groups (→). The D-binding form (top right) interacts with non-specific DNA via ~ 11 charge–charge interactions; the specific hydrogen-bonding groups are shown in the 'withdrawn' state. These two conformations must interconvert with first-order rate constants (kRO and kRD) > 10^6 sec^{-1}. The lower model shows the non-specific DNA-binding form sliding 'through' and displacing the DNA-bound counter-ion layer (see text; this Fig. from Winter et al.[5]).

activation energy barrier to sliding if the positive charges in the protein binding site are effectively 'delocalized' relative to the negative charges of the DNA phosphate groups. This seems likely, since the potential energy (E) of charge–charge interactions is not a strong function of distance ($E \propto 1/r$), and there are ~ 11 charge–charge interactions (between DNA and protein) located within ~ 60–80 Å along the DNA. Furthermore, monovalent cations have been shown to bind to DNA in a delocalized manner[16]. This suggests that the overall sliding rate should approach the hydrodynamically-limited value, as appears to be the case.

(iii) *How does the repressor recognize (and bind to) the operator site when it slides 'over' it?* As Fig. 3 shows, repressor binds to DNA in two binding modes. One, which we call the *non-specific* binding mode, is totally electrostatic and involves ~ 11 charge–charge interactions. The other is the (much tighter) *specific* binding conformation corresponding to RO complex formation; this mode involves only ~ 7 charge–charge interactions, and the binding free energy is only ~ 30% electrostatic at moderate salt concentrations[12]. In the latter the conformation repressor recognizes and articulates with the specific matrix of hydrogen bond donors and acceptors that identifies the operator. For repressor to recognize operator as it slides over it, the rate of interconversion of the two repressor conformations (R$_D$⟷R$_O$; see Fig. 3) must be at least 10^6 per second. Rapid reaction measurements suggest that repressor conformations can indeed interconvert at rates of this order (see Ref. 1).

Do other genome regulatory proteins also locate their target sites by sliding?

In vitro experiments show that RNA polymerase also locates promoters at rates which exceed diffusion controlled limits for one-step processes (M. Chamberlin, H. Bujard; private communications). *Eco*RI endonuclease shows dissociation kinetics from restriction fragments (under conditions in which specific binding, but not cleavage, is possible) which can best be explained by the sliding of this protein in the non-specific binding mode (P. Modrich, private communication). Other proteins exhibit kinetic characteristics which can most easily be explained in the same way, though most have not been investigated enough to determine whether sliding occurs.

However, it *is* clear that non-specific DNA binding of a number of genome-regulatory proteins is largely electrostatic, and thus the conditions for sliding (as defined in this review and in Fig. 3) are at hand. Experiments on other systems, of the sort described here with *lac* repressor and operator, are eagerly awaited.

In vivo implications

The experiments outlined above show that proteins *can* translocate along DNA by sliding, and that this facilitated transport mechanism is used by proteins in locating specific sites under *in vitro* conditions on otherwise 'naked' DNA. To what extent can these ideas be applied *in vivo*? In the *E. coli* cell, the chromosome exists as a compact and supercoiled entity at physiological salt concentrations[17]. Under these conditions stretches of 'naked' DNA involved neither in higher order structure nor encumbered by other proteins may be relatively short. On the other hand, at these salt concentrations the duration of a non-specific binding event (and thus of a 'slide') is also short. Because the DNA is partially 'collapsed' the distances between adjacent DNA segments are reduced, and facilitated translocation by intersegment transfer may become more effective. It seems reasonable to speculate that *both* facilitated transfer mechanisms described here may play a significant role in DNA target location by genome regulatory proteins in the cell (see Discussion in Ref. 5).

Acknowledgements

The work from our laboratory which is described here, and the preparation of this review, were supported in part by USPHS Research Grants GM-15792 and GM-29158 (to PHvH). OGB also gratefully acknowledges partial support from Swedish National Science Research Council. RBW is pleased to acknowledge support as a predoctoral trainee on USPHS Training Grants GM-00715 and GM-07759. We are grateful to Dr R. L. Baldwin for helpful discussions of the manuscript.

References

1 von Hippel, P. H. (1979) *Biological Regulation and Development* (Goldberger, R. F. ed.) Vol. I, pp. 297–347, Plenum Press, New York
2 Miller, J. H. and Reznikoff, W. S. (eds) (1978) *The Operon*, Cold Spring Harbor Laboratory, New York
3 Record, M. T., Jr., Lohman, T. M. and deHaseth, P. L. (1976) *J. Mol. Biol.* 107, 145–158
4 Berg, O. G., Winter, R. B. and von Hippel, P. H. (1981) *Biochemistry* 20, 6926–6948
5 Winter, R. B., Berg, O. G. and von Hippel, P. H. (1981) *Biochemistry* 20, 6961–6977
6 Riggs, A. D., Bourgeouis, S. and Cohn, M. (1970) *J. Mol. Biol.* 53, 401–417
7 Barkley, M. D. (1981) *Biochemistry* 20, 3833–3842
8 Richter, P. H. and Eigen, M. (1974) *Biophys. Chem.* 2, 255–263
9 Berg, O. G. and Blomberg, C. (1976) *Biophys. Chem.* 4, 367–381
10 von Hippel, P. H., Revzin, A., Gross, C. A. and Wang, A. C. (1975) *Protein–Ligand Interactions* (Sund, H. and Blauer, G., eds), pp. 278–288, W. de Gruyter, Berlin
11 Riggs, A. D., Suzuki, H. and Bourgeois, S. (1970) *J. Mol. Biol.* 48, 67–83
12 Winter, R. B. and von Hippel, P. H. (1981) *Biochemistry* 20, 6948–6960
13 Schurr, J. M. (1979) *Biophys. Chem.* 9, 413–414
14 deHaseth, P. L., Lohman, T. M. and Record, M. T., Jr. (1977) *Biochemistry* 16, 4783–4790
15 Revzin, A. and von Hippel, P. H. (1977) *Biochemistry* 16, 4769–4776
16 Anderson, C. F., Record, M. T., Jr. and Hart, P. A. (1978) *Biophys. Chem.* 7, 301–316
17 Pettijohn, D. E. (1976) *CRC Critical Rev. Biochem.* 4, 175–202

Otto Berg is at the Department of Theoretical Physics, Royal Institute of Technology, S–100, 44 Stockholm, Sweden, Robert Winter is at the Department of Molecular, Cellular and Developmental Biology, University of Colorado, Boulder, CO 80302, U.S.A. and Peter von Hippel is at the Institute of Molecular Biology and Department of Chemistry, University of Oregon, Eugene, OR 97403, U.S.A.

The interaction of *E. coli* RNA polymerase with promoters

Hermann Bujard

Among protein–nucleic acid interactions promoter recognition by RNA polymerases is of special interest since it represents an important step in the controlled flux of genetic information common to all biological systems.

During gene expression information is transferred from its storage site – usually a double-stranded DNA – to its site of action, usually a protein. Since only a fraction of the vast biochemical potential of a cell is used at any given time this flux of information must be controlled at various levels. An important level for such control mechanisms is transcription, which can be divided into:

(a) the enzymatic processes by which a predetermined sequence of monomeric units is transferred with high fidelity from the DNA template to the RNA that will be used as a template for protein synthesis;

(b) the recognitory processes by which the transcriptional machinery recognizes start and stop signals encoded in the DNA template.

Selectivity of transcription is due to the recognitory events flanking actual RNA synthesis. In the simplest case a transcriptional unit consists of a start signal – a promoter – the transcribed region, which may comprise the information for one or more polypeptide chains and a stop signal – the terminator (Fig. 1). The extent to which the transcriptional unit is expressed depends primarily upon the efficiency with which the promoter is utilized by the DNA-dependent RNA polymerase. In *E. coli* we have examples of such simple control mechanisms (for instance the *I* gene of the *lac* operon producing constant levels of repressor molecules). In many cases, however, the signals encoded around the transcriptional unit are targets for additional regulatory elements. Thus, negatively or positively acting elements (e.g. repressor or catabolite gene activator protein (CAP) molecules) can diminish or enhance promoter activity, while termination, and anti-termination, factors can influence the efficiency of transcriptional stop signals. These superimposed regulatory mechanisms permit the cell to respond to environmental changes (Fig. 1). Here I shall discuss primarily the interaction of *E. coli* RNA polymerase with promoter sites, paying less attention to modulation of this interaction by additional effectors.

Before RNA synthesis is initiated, the enzyme must recognize a promoter, form a complex and find the information for the precise positioning of the first nucleotide in the nascent RNA chain. The basic questions about this recognitory process therefore are: how does RNA polymerase find a promoter site within the abundance of unspecific sites, and what are the pertinent structural features of the two macromolecules that interact in this process?

The enzyme

DNA-dependent RNA polymerase of *E. coli* has a molecular weight of 500,000. This so-called holoenzyme consists of five subunits β, β', α_2 and σ. It can be dissociated into the core enzyme, β, β', α_2, which is able to perform the basic enzymatic processes of RNA synthesis but does not recognize promoters. The σ-subunit is impor-

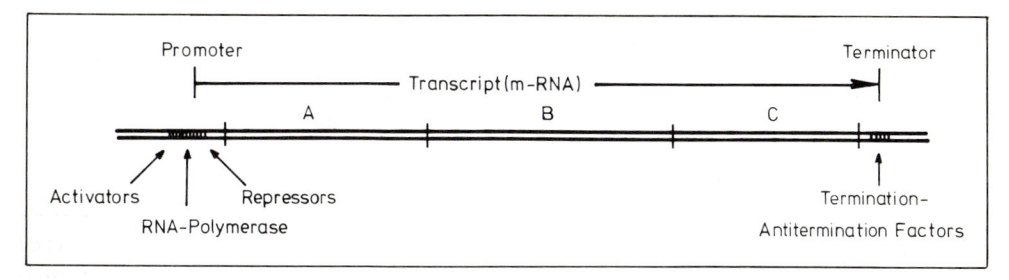

Fig. 1. The transcriptional unit. A stretch of DNA giving rise to a contiguous piece of RNA, the transcript, which may code for one or more polypeptide chains (A, B, C). Such a unit is delimited by a start and a stop signal, promoter and terminator, which are recognized by the DNA-dependent RNA polymerase. Both promoter and terminator are prime targets for additional regulatory factors which may enhance or diminish the interaction between the enzyme and the signal encoded in the DNA.

tant for specific recognition[1], although it does not alone bind to DNA[2]. Little is known about the function of the other subunits except for β which is intimately involved in the catalytic processes of RNA chain initiation and elongation, since, in connection with the α subunit, it binds rifampicin and streptolydigin, drugs known to inhibit these reactions respectively[2]. Furthermore, mutations rendering cells resistant to rifampicin have been mapped within the gene of the β protein[2].

The enzyme–DNA complex

E. coli RNA polymerase binds to a large variety of promoters *in vitro*, forming a stable complex in which the DNA in the promoter region is unwound about one helical turn[3]. The stability of many of these complexes has facilitated the study not only of promoter structure but also of the interaction between the enzyme and DNA. Thus, in the so-called filter binding technique[4] DNA fragments bound to polymerase are retained by nitrocellulose whereas non-complexed double-stranded DNA is not. Using the filter binding technique half-lives of polymerase–DNA complexes of between several minutes and many hours (at 37°C and intermediate ionic strength) have been found[5,6].

Such stable complexes can be utilized for probing biochemical and chemical properties. Thus, DNA fragments protected by RNA polymerase against DNAase I digestion were isolated and their nucleotide sequences determined[7]. The size of the so-called 'p-DNA' (protected DNA) is around 42 nucleotides and covers the promoter region from about +20 to −20 (Fig. 2, sites within the nucleotide sequences of a promoter will be identified henceforth by negative or positive numbers depending on whether their position is upstream or downstream from the initiating nucleotide of the transcript). Although the analysis of p-DNA has led to important insights about promoter structure some properties remain puzzling: the p-DNA–enzyme complex can initiate RNA synthesis in the presence of nucleotide triphosphates and produce a short 'run off' RNA. When the enzyme is dissociated from its short template, however, it is not able to rebind to p-DNA, indicating that regions important for complex formation were not protected against DNAase I. By developing the so-called 'foot printing' method Schmitz and Galas[8] have shown that *E. coli* RNA polymerase actually covers more or less tightly a region of 70 to 80 base pairs (bp) from about position +20 to −60 (Fig. 2). Interestingly, one of the DNA strands is protected less efficiently by the enzyme than the other, indicating a unique orientation of the template-bound protein.

Promoter sequences

p-DNA fragments have provided the first defined probes for the sequencing of promoters, and homologies within such sequences were identified first by Schaller et al.[7] and Pribnow[9]: they pointed out a region of relative homology centred around the −10 position, quite frequently referred to as a 'Pribnow box' (Fig. 2). About 60 promoters have now been sequenced (for review and compilation of 46 sequences see Ref. 10) and although we have little information about the signal strength of the various sequences and their possible dependency upon additional effectors, a number of common structural features can be found. The most conserved sequence is a hexamer in the −10 region. The prototypic sequence is TATAAT, in which the last position, occupied by a thymine, is identical in all promoters investigated so far. Also very stringently conserved are the first two positions (TATAAT) in contrast to the remainder of the hexamer, where a somewhat larger variation is observed. A second region of homology has been identified around position −35[10,11]. Here a highly conserved trinucleotide TTG can be derived from sequence compilation, provided that the distance between the Pribnow box and the −35 region is allowed to vary by two base pairs (Figs 2 and 3). The TTG sequence is followed downstream by three less stringently conserved nucleotides and the resulting hexamer TTGACA is part of a stretch of 12 nucleotides in which additional homologies can be found. Interestingly some promoters, known to require additional factors for their function, show little or no homology to the prototypic sequence of the −35 region, although this finding cannot be generalized.

Common features might also be expected in the sequence surrounding the site of chain initiation, since this reaction is highly selective with respect to the initiating nucleotide: thus RNA chains start predominantly with a purine, A occurring more often than G. In some instances, however, a C or U is the initiating nucleotide although the template would provide an A- or G-start in neighbouring positions. Despite this specificity, the starting nucleotide is located within a rather poorly conserved sequence (CAT) 6 to 9 bp downstream of the −10 hexamer. The highly selective but variable location of this starting nucleotide in close proximity to the most conserved region of a promoter may appear paradoxical. It indicates, however, a sequence-dependent fine adjustment of the initiation region with respect to the reaction centre of the enzyme (there are a few cases where RNA chains are initiated at multiple sites; in most of them, however, the starting sequence consists of several identical nucleotides like GGG or CCC, (see Ref. 10).

Promoter mutation

Picking at random one of the published promoter sequences, the 'unprejudiced' observer might have difficulties in following some of the above conclusions (Fig. 3). Caution is indeed indicated as long as our knowledge of the functional implications of primary structure is as poor as it is today. In the case of *E. coli* promoters, however, the respective conclusions are not only based on an impressive number of well-documented sequences, there is also firm genetic evidence: of 34 promoter point mutations investigated, 29 fall within the defined regions at −10 and −35. In addition 14 of the 16 mutations identified within the −10 hexamer cluster in the three most conserved positions emphasizing the importance of these sites.

The most striking results were obtained by analysing some of the 'promoter-up mutations' which increase the efficiency of the *lac* promoter (Fig. 2). This promoter has in its wild type form the following −10 hexamer: TATGTT. A change of the G/C pair in the fourth position of the hexamer increases the promoter efficiency ten-fold. Since polymerase binding unwinds a promoter about one helical turn and induces a local melting near the −10 position (see below), a change from GC to AT pairs, which reduces hydrogen-bonding, would be expected to have a positive effect on the signal strength of a promoter se-

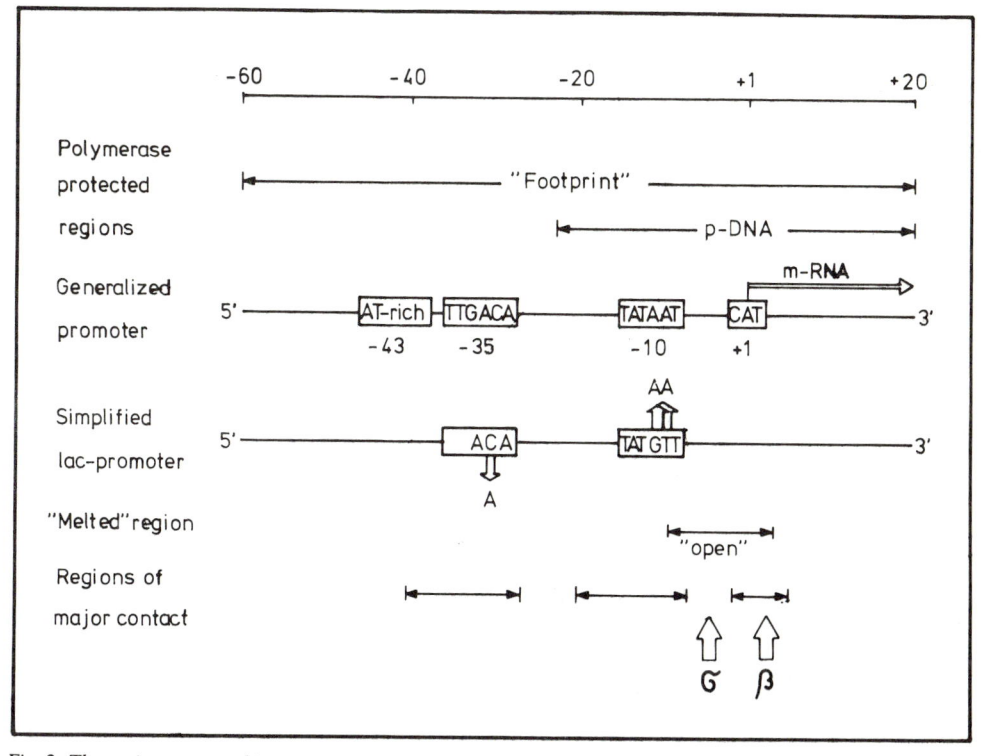

Fig. 2. The major topographic properties of an E. coli *promoter–RNA polymerase complex. The top part of the figure shows the DNA fragment protected by RNA polymerase against limiting digestion with DNAase I ('footprint') as well as the region inaccessible even under conditions of vigorous attack by this enzyme ('p-DNA').*

In the middle part of the figure a simplified version of a generalized promoter sequence is depicted. By convention, only one DNA strand – the non-coding strand – is shown. Transcription is therefore initiated from left to right at position +1, the starting nucleotide of the nascent mRNA chain. This generalized promoter scheme emphasizes the regions where the main sequence homologies are found (boxed-in areas): the so-called 'Pribnow box' at −10, the −35 region, as well as the CAT sequence at the RNA start, and the AT-rich stretch around position −43. The latter homology is predominantly found in promoters of high signal strength (e.g. promoter of coliphage T5, Ref. 10).

The line below shows a small part of the lac-*promoter sequence to demonstrate the effect of mutations in highly conserved regions. The transition in the −10 region from the wild type sequence TATGTT to the one of the p^s (TATATT) and UV5 (TATAAT) mutation results in a 10- and 25-fold increase of* lac *expression*[11] *respectively as the ideal '−10 sequence' is approached (▲ 'promoter-up' mutations). The fact that an exchange of a GC for an AT base pair does not necessarily cause an increase in promoter strength, despite the high AT-content of strong promoters, is demonstrated by the 'promoter-down' mutation (▼) in the −35 region, where the transition from the favoured ACA sequence to AAA reduces* lac *operon expression. (W. Reznikoff, personal communication).*

The bottom part of the figure depicts where the double helix is drastically distorted upon binding of the RNA polymerase ('open' region) and where the major contacts are made between the enzyme and the DNA. It also shows where the lac *promoter and the subunits σ and β of RNA polymerase can be crosslinked indicating the positions of the respective subunits within the promoter–enzyme complex.*

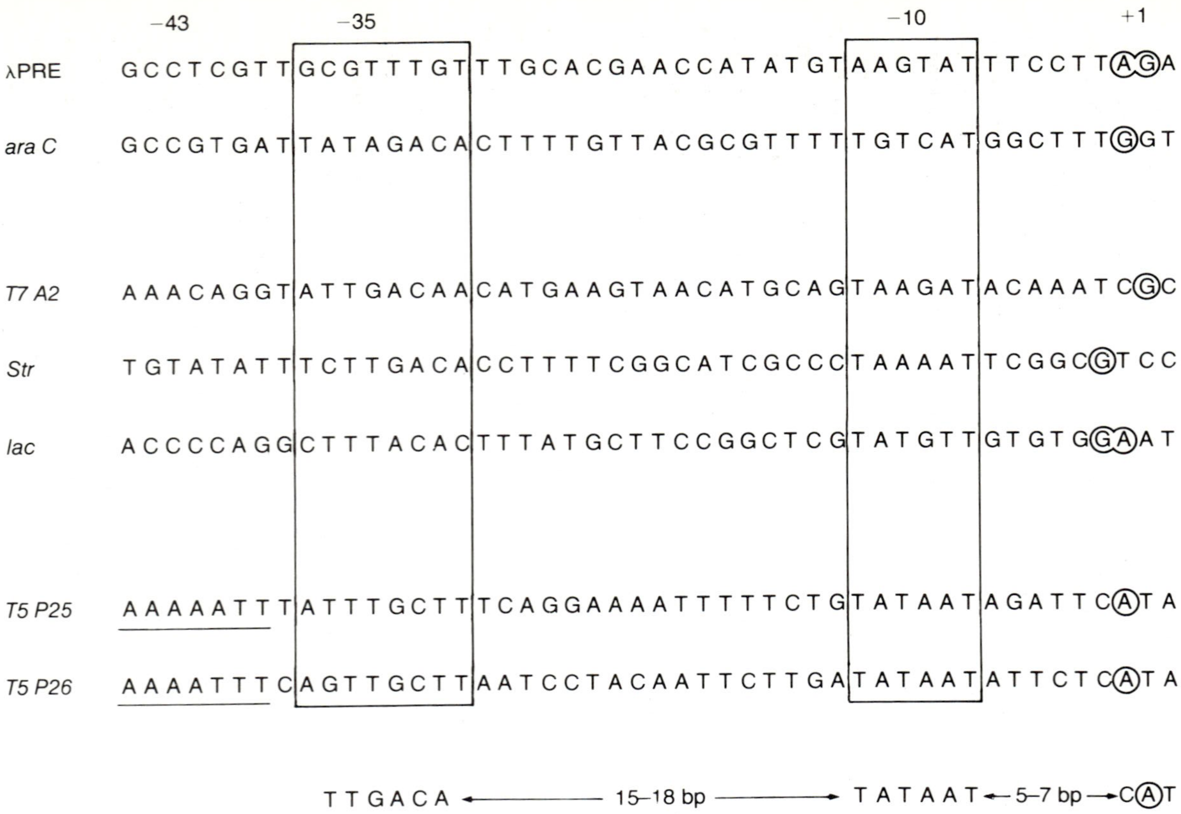

```
           -43              -35                                                  -10         +1
λPRE    GCCTCGTT|GCGTTTGT|TTGCACGAACCATATGT|AAGTATTTCCTT(A)GA
araC    GCCGTGAT|TATAGACAC|TTTTGTTACGCGTTTT|TGTCATGGCTTTT(G)GT
T7 A2   AAACAGGT|ATTGACAA|CATGAAGTAACATGCAGT|AAGATACAAAT C(G)C
Str     TGTATATT|TCTTGACAC|CTTTTCGGCATCGCCC|TAAAATTCGGC(G)TCC
lac     ACCCCAGG|CTTTACACT|TTATGCTTCCGGCTCG|TATGTTGTGTG G(A)AT
T5 P25  AAAAATTT|ATTTGCTT|TCAGGAAAATTTTCTG|TATAATAGATTC(A)TA
T5 P26  AAAATTTC|AGTTGCTT|TAATCCTACAATTCTTGA|TAATAATTCT C(A)TA

         TTGACA  ───── 15-18 bp ─────▶ TATAAT ◀─5-7 bp─▶ C(A)T
```

Fig. 3. The nucleotide sequence of selected promoters. The sequences of the ara D promoter[23] and the PRE promoters of phage λ[24] are examples where to predict a function from sequence information alone appears impossible. In contrast, significant homologies within the conserved regions (around -10 and -35) are found within the promoters depicted in the middle part: T7 A2 from coliphage T7[25], Str and lac from E. coli[26,27]. The two promoters of coliphage T5 (See Ref. 10) exhibit sequences of high signal strength. They show a high overall AT content, an 'ideal' -10 region (TATAAT) and a striking AT-rich region around -43. The generalized sequence indicated at the bottom is derived from about 40 promoter sequences.

quence. However, in the *lac UV 5* promoter mutation the original *lac* −10 hexamer is converted into the prototypic sequence TATAAT increasing the promoter efficiency by another factor of 2.5. Here a change from TA (in the fifth position of the hexamer) to AT, which is not believed to influence the helix stability, still has a profound effect, indicating the importance of specific sites for nucleotide–protein interaction[11]. Another interesting 'up-mutation' is in the −35 region of the *lac I* gene promoter, where change of GCG to GTG in the region of the TTG prototypic sequence increases promoter activity tenfold[12]. Most of the promoter mutations ('down-mutations') diminish promoter activity and although the effects of the various base pair changes are difficult to interpret in all cases, it is obvious that in some positions there is a clear selection for a particular base and any change has adverse effects. In other positions, however, it appears that there is a strong selection against a particular base, and any of the remaining possibilities do not adversely affect promoter activity.

Contacts between RNA polymerase and promoters

The stable association between RNA polymerase and promoters *in vitro* permits studies of the complex with highly specific alkylation reactions: methylation of purines with dimethylsulphate[13] and ethylation of the backbone phosphates with ethylnitrosourea[14].

Two types of question can be answered with such experiments:
(a) Which positions within a promoter sequence are essential for the formation of the complex or its stability?
(b) Which are the positions within a promoter–polymerase complex where the protein is close enough to the DNA to interfere with the chemical attack, either diminishing or enhancing alkylation?

DNA is therefore either alkylated first and its polymerase-binding properties are recorded, or it is alkylated after formation of the enzyme–DNA complex. Alkylation patterns obtained in such experiments carried out mainly with *lac* and phage T7 promoters[15–18] revealed three regions of extensive contact between RNA polymerase and promoters: the 'Pribnow box', the −35 region and the sequence around the RNA initiation sites (Fig. 2). This is in excellent agreement with the picture derived from sequence homologies and analyses of promoter mutants. Looking at the alkylation pattern in a three-dimensional model it appears furthermore that the enzyme recognizes the promoter from just one side of the DNA helix. In a very elegant modification of the methylation protection experiment, Sibenlist succeeded in demonstrating methylation of adenine bases in positions that are accessible to dimethylsulphate only if DNA is melted and in the single-stranded conformation[17]. The region opened upon binding of RNA polymerase, and identified in this way, stretches over about 11 base pairs from the middle of the Pribnow box to just past the RNA start sites (−9 to +2, Fig. 2).

Which subunits of the enzyme are involved in this process? Photochemical probes linking RNA polymerase subunits to bases in defined positions within the promoter sequence[18] revealed that the β-subunit, which is known to be essential in the catalytic process of RNA synthesis, is cross-linked to the nucleotide in the +3 position of the *lac UV 5* promoter, i.e. within the RNA initiation site. The σ-subunit on the other hand was found associated to the −3 position of the same promoter, i.e. in close proximity to the Pribnow box (Fig. 2). No information was obtained about the positions of any of the other enzyme subunits and the complete molecular topology of the RNA polymerase bound to promoter awaits further investigation.

Kinetics of promoter–polymerase interaction

The patterns of RNA synthesis *in vitro* or *in vivo* from phage genomes show that different regions are often transcribed with different intensities; the promoters vary in strength. What makes a promoter strong?

A priori there are at least three features which may be related to promoter strength: promoter recognition by the enzyme, i.e. the rate of complex formation; the stability of the complex; and the rate of RNA chain initiation. With many promoters the rate of initiation of RNA synthesis is quite rapid compared to complex formation[19]. On the other hand complex stability, as well as the rate of complex formation, varies considerably between promoters[5,6,20] and might contribute to the signal strength. When sets of phage promoters differing in their rate of complex formation with RNA polymerase, as well as in the half-lives of the resulting complexes, were analysed with respect to their ability to direct RNA synthesis *in vivo* and *in vitro*, a close correlation between the rate of complex formation and the promoter strength was found[6]. Furthermore the relative rates of complex formation for various *E. coli* phage and plasmid promoters can differ by factors of 10^2 to 10^3[20]. The promoters that react most rapidly with *E. coli* RNA polymerase *in vitro* are located in the 'early' region of the genome of *E. coli* phage T5[6]. Two of these promoters, which outcompete all others studied so far in polymerase binding and RNA synthesis *in vitro*, were sequenced (Fig. 3). Both show a prototypic −10 region and contain the highly conserved TTG sequence with the −35 region. In contrast to most other promoter sequences, however, they are extremely A/T-rich (82%) with clusters of A/T and T/A base pairs, a feature also observed in another promoter which efficiently starts RNA synthesis *in vivo*[21]. Whether the high content of AT pairs, and especially the AT block at −43 region, are general indicators of a strong promoter remains to be seen.

Since the rate of complex formation with RNA polymerase apparently reflects the signal strength, the absolute values of such rate constants are of interest. In combining some of the published absolute rates with the proper relative rate determinations[20] values are obtained (e.g. for some phage T5 promoters) which suggest that the reaction is faster than a diffusion-controlled process would allow. Although a direct proof for such fast reaction rates is still pending it is interesting to speculate about mechanisms such as linear diffusion of RNA polymerase along the DNA template, or a direct displacement of the enzyme from unspecific sites.

Concluding remarks

Among the various proteins that interact specifically with DNA there are those that form specific complexes only when certain sequences are provided. The precision involved in the recognition of DNA-encoded signals is astonishing. For example, the expression of the *lac* operon is turned off by a repressor protein which is able to recognize its signal – the 'lac-operator', a 27 bp stretch of DNA – among 6×10^6 unspecific sites and most strikingly, 10 repressor molecules per *E. coli* cell can reduce 1000-fold the expression of that transcriptional unit (for details see Ref. 22).

The interaction between bacterial promoters and their respective RNA polymerases is clearly a sequence-specific recognition process: on average RNA polymerase has to identify a promoter among 10^3 to 10^4 unspecific sites. However, unlike the interaction between operator and repressor or between restriction nucleases and their targets, RNA polymerase recognizes a wide range of promoter sequences with differing efficiencies. This in turn modulates the intensity with which various transcriptional units are expressed. It is therefore the individual promoter sequence which defines the signal strength, although we should keep in mind that often the ultimate promoter strength is, in addition, determined by the superimposed action of positive control elements, e.g. the CAP-factor.

An 'ideal' or 'generalized' promoter sequence cannot therefore be derived in a straightforward way. One might reduce the problem to 'what does a strong promoter look like?', but even this might turn out to be rather difficult to answer since at least two parameters appear to be involved in the formation of the stable promoter–RNA polymerase complex necessary for RNA chain initiation: (a) to be recognized initially by the enzyme a promoter needs a set of precisely positioned bases and there might be several possible arrangements of 'equivalent sets', resulting in sequence variations; (b) during transition from the first contacts to a stable complex at least a part of the promoter undergoes conformational changes. Consequently, base compositions affecting helix stability will also influence this process. We might therefore envisage promoter families of identical or very similar signal strength but quite different nucleotide sequences.

Promoter recognition is certainly a complex biochemical phenomenon and its elucidation still remains a formidable task. As the first step in the controlled flux of genetic information, however, it represents a protein–nucleic acid interaction central to all biological processes and appears, therefore, to be worth a more detailed analysis.

References

1 Burgess, R., Travers, A., Dunn, J. J. and Bautz, E. K. F. (1969) *Nature (London)* 221, 43–46
2 Zilig, W., Palm, P. and Heil, A. (1976) in *RNA Polymerase* (Losick, R. and Chamberlin, M., eds), p. 100, Cold Spring Harbor Laboratory
3 Saucier, J.-M. and Wang, J. (1972) *Nature (London) New Biol.* 239, 167
4 Jones, O. and Berg, P. (1966) *J. Mol. Biol.* 22, 199
5 Seeburg, P. and Schaller, H. (1975) *J. Mol. Biol.* 92, 261
6 Gabain, v. A. and Bujard, H. (1977) *Mol. Genet.* 157, 301
7 Schaller, H., Gray, C. and Hermann, R. (1975) *Proc. Natl. Acad. Sci. U.S.A.* 72, 737
8 Schmitz, A. and Galas, D. (1979) *Nucleic. Acid Res.* 6, 111
9 Pribnow, D. (1975) *Proc. Natl. Acad. Sci. U.S.A.* 72, 784
10 Rosenberg, M. and Court, D. (1979) *Annu. Rev. Genet.* (in press)
11 Gilbert, W. (1976) in *RNA Polymerase* (Losick, R. and Chamberlin, M., eds), p. 193, Cold Spring Harbor Laboratory
12 Calos, M. (1978) *Nature (London)* 274, 762
13 Gilbert, W., Maxam, A. and Mirzabekow, A. (1976) in *Control of Ribosome Synthesis* (Kijalgaard, N. and Maløe, O., eds), p. 139, Munksgaard, Kopenhagen
14 Sun, L. and Singer (1975) *Biochemistry* 14, 1795
15 Johnsrud, L. (1978) *Proc. Natl. Acad. Sci. U.S.A.* 75, 5314
16 Siebenlist, U. and Gilbert, W. (1980) *Proc. Natl. Acad. Sci. U.S.A.* 77, 122
17 Siebenlist, U. (1979) *Nature (London)* 279, 651
18 Simpson, R. B. (1979) *Cell* 18, 277
19 Chamberlin, M. J. (1976) in *RNA Polymerase* (Losick, R. and Chamberlin, M., eds), p. 17, Cold Spring Harbor Laboratory
20 Gabain, V. A. and Bujard, H. (1979) *Proc. Natl. Acad. Sci. U.S.A.* 76, 189
21 Nakamura, Y. and Inoye, R. (1979) *Cell* 18, 1109
22 Hippel, P. V. (1979) in *Biological Regulation and Development* (Goldberger, R. F., ed.), p. 279, Plenum Press, New York
23 Smith, B. R. and Schleif, R. (1978) *J. Biol. Chem.* 253, 693
24 Rosenberg, M., Court, D., Shimatake, H., Brady, C. and Wulff, D. L. (1978a) *Nature (London)* 272, 414
25 Pribnow, D. (1975) *J. Mol. Biol.* 99, 419
26 Post, L., Arfsten, A., Reusser, F. and Nomura, M. (1978) *Cell* 15, 215
27 Dickson, R., Abelson, J., Barnes, W. and Reznikoff, W. S. (1975) *Science* 182, 27

Hermann Bujard is at the Molekulare Genetik, Universität Heidelberg, Im Neuenheimer Feld 230, D-69 Heidelberg, F.R.G.

Attenuation: translational control of transcription termination

M. D. Watson

The evidence leading to the concept of attenuation as a major control system for operons concerned with amino acid biosynthesis is reviewed.

The repressor/operator model for the control of gene expression in bacteria, first described by Jacob and Monod for the lactose operon, was thought to be basically applicable to most other bacterial operons. The identification of genes specifying repressor proteins for the tryptophan (*trp*), arginine (*arg*) and tyrosine (*tyr*) operons helped to re-inforce this view. This model was also held to be true for the histidine (*his*) and isoleucine–valine (*ilv*) operons, but despite intensive efforts no repressor proteins for these operons have been identified. The concept of termination of transcription as an alternative control system was first proposed for the *his* operon, but it was work on the *trp* operon that revealed the remarkable method by which this control is exerted. It is now clear that operons specifying enzymes for the biosynthesis of several amino acids are regulated by termination of transcription.

The tryptophan operon

The tryptophan operon is regulated by a repressor protein which, when complexed with tryptophan, binds to the operator (*trpO*) and prevents RNA polymerase from transcribing the structural genes (*trpE,D,C,B,A*) (Fig. 1). The repressor–tryptophan complex effectively competes with RNA polymerase for binding at the promoter/operator site although, RNA polymerase already bound to the promoter is not influenced by repressor protein. Thus, under conditions of tryptophan excess competition between repressor and RNA polymerase prevents transcription of the operon. The repressor/operator control of gene expression, and feedback inhibition of enzyme activity were thought to explain adequately the regulation of tryptophan biosynthesis for several years before the phenomenon of attenuation was discovered.

Attenuation

Analysis of a series of strains with small deletions of the *trp* operon, near to, but clearly distinct from, the operator–promotor region, revealed somewhat surprisingly, that some of them caused increased expression of the remaining structural genes[1]. The effective deletions removed a region between the operator site *trpO*, and *trpE* the first structural gene of the operon (see Fig. 1). However, any deletion that was wholly contained within the structural genes did not give rise to increased expression. As all of these mutant strains were capable of normal repressor/operator function, the enhanced expression of structural genes in the deletion mutants implied that normally there was some block to maximum gene expression located between *trpO* and *trpE*, which could be part of another independent regulatory system and had been removed by the deletions. Further deletion analysis showed that this new regulatory site was within 30 bases of the start of the *trpE* structural gene[2]. Analysis of *trp* mRNA showed that the

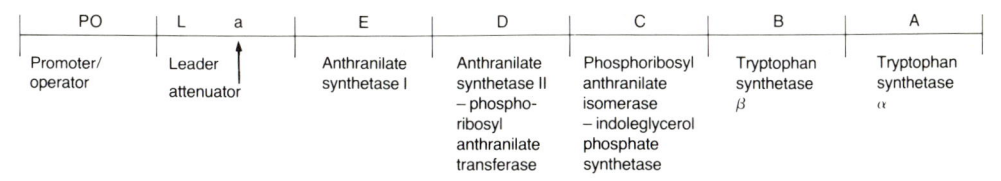

PO	L a	E	D	C	B	A
Promoter/ operator	Leader attenuator	Anthranilate synthetase I	Anthranilate synthetase II – phosphoribosyl anthranilate transferase	Phosphoribosyl anthranilate isomerase – indoleglycerol phosphate synthetase	Tryptophan synthetase β	Tryptophan synthetase α

Fig. 1. *The tryptophan operon.*

transcript started about 160 bases before trpE (Ref. 1). This long transcript before trpE, termed the 'leader' sequence and given the genetic designation *trpL*, must therefore cover the putative regulatory site.

To determine at what level of expression the regulatory site acts, the frequency of transcription of the *trpL* region before the site, was compared with that of a distal region, *trpBA* (see Fig. 1). Transcription of *trpBA* was only 0.12 times as frequent as that of *trpL* (Ref. 2). It appeared that 85% of the events that transcribe *trpL* do not continue onwards across the operon to transcribe *trpBA*. It has long been known that *trpE* and *trpA* are transcribed at the same frequency, so most of the transcripts must terminate before *trpE*. Therefore, the block to maximum gene expression is caused by transcription termination before the structural genes are reached. The site at which termination occurs was called the *attenuator* after a similar site postulated to exist in the *his* operon[3].

During the course of these studies it became apparent that the frequency of termination of transcription (attènuation) at the attenuator varied depending on the growth conditions, with the end product of the pathway, tryptophan, regulating it in some manner[4] possibly by varying the amount of charged tRNA^{trp}(Ref. 5).

How do these molecules interact with the attenuator region and regulate transcription termination? At first it was suggested that tryptophan, or more likely tryptophanyl-tRNA^{trp}, bound to some additional regulatory protein that acted at the attenuator. However, RNA and DNA sequence analyses began to reveal some interesting features about the structure, and possible functions of leader mRNA. First, it has all the information to code for a polypeptide of 14 amino acids containing two tandem tryptophan residues. There is a ribosome-binding site near the beginning of the leader message within which is a translation initiation codon, AUG, at positions 27–29; two tryptophan codons occur at positions 54–59, followed by a stop codon, UGA, at positions 69–71 (position 1 being the first nucleotide of the message). Secondly, the mRNA is capable of forming extensive secondary structures immediately following the stop codon in the leader sequence. Two regions of dyad symmetry, caused by inverted, repeated sequences in the DNA, are capable of forming stem and loop structures[6,7]. The two regions of symmetry partially overlap and are thus mutually exclusive, both structures being unable to form at the same time. The second stem and loop structure, furthest from the stop codon and closest to the first structural gene *trpE*, is G–C rich and immediately followed by eight uridine residues. This combination of a G–C rich stem and loop fol-

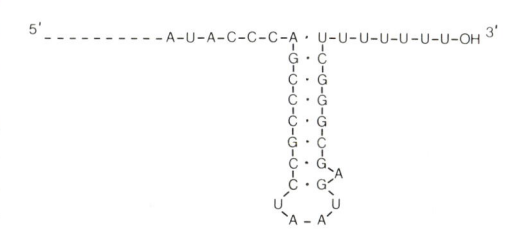

Fig. 2. . *The terminator. The terminator stem and loop structure that occurs at the 3' end of the leader mRNA.*

lowed by poly-U has been found near other transcription termination sites.

The significance of these observations was quickly realized after comparison with other systems and analysis of new regulatory mutations. The sequences of the attenuator regions of the *his* and *phe* operons revealed remarkably similar results, the most important being the presence in both of regions coding for a small translatable polypeptide containing tandem residues of the amino acid which was the end product of that operon[8-10]. Other common features were extensive regions potentially able to form stem and loop structures, the last of which was always G–C rich followed by poly-U.

Proof that the small *trp* leader peptide was translated *in vivo* was obtained by fusing the tryptophan leader sequence, *trpL*, to either the *lacI* or *trpE* gene so that the respective products of the structural genes were made using the translational initiation region of *trpL*. The purified proteins from these fusion strains, either the *lac* repressor or anthranilate synthetase I respectively, had amino-termini corresponding to that predicted for the *trpL* leader peptide[11].

Mutations that alter attenuation were studied and helped to confirm the importance of secondary structure in transcription termination. The normal amount of transcriptional 'read through' of the attenuator *in vitro* under conditions of maximum termination is about 5%, whereas mutations that caused increased expression of the operon were found to allow about 46% 'read through' under identical conditions[12]; DNA sequence analysis revealed that the mutations were located within the G–C rich stem and loop causing it to be less stable. Other mutations were isolated that decreased operon expression by increasing the usual frequency of termination *in vivo* from 85% to 95% – the frequency found *in vitro*. This second class of mutations was located either in the AUG translation initiation codon, preventing translation of the leader peptide, or in another stem and loop structure mutually exclusive with the G–C rich stem and loop. These results indicate that translation of the peptide does occur and is important for the regulation of attenuation, – one stem and loop decreases termination, whereas the G–C rich stem and loop enhances it.

The model

The model for attenuation originally proposed by Yanofsky[13] has now been extended and modified by Keller and Calvo[14] to account for most of the observations. It is presented here in simplified form using the *trp* operon as an example. The attenuator itself consists of the G–C rich stem and loop structure followed by the poly-U stretch. The formation of such a structure, called a terminator (Fig. 2), at the 3′ end of a message will cause RNA polymerase to cease transcription in the

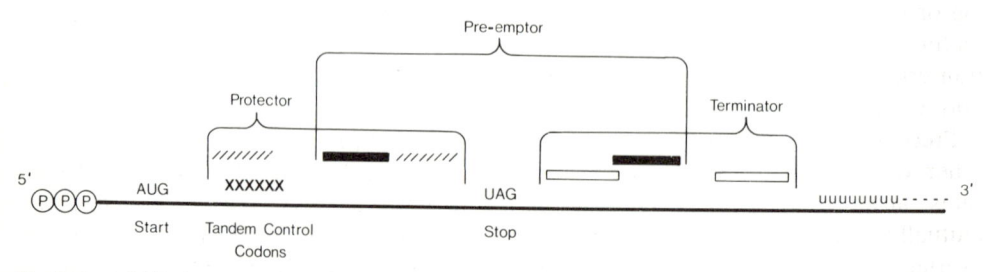

Fig. 3. A model leader region showing regions of symmetry. The important features of leader mRNA are shown. Regions involved in stem and loop formation are indicated thus; /////// *protector stem; pre-emptor stem and* ▭ *terminator stem.* xxxxxx *Indicates the position of the controlling codons e.g. UGGUGG for -trp-trp- in the tryptophan operon.*

through the operon. Overlapping the pre-emptor is yet another stem and loop called the protector, which, if formed, prevents formation of the pre-emptor but allows formation of the terminator. Such structures will form in the order they are transcribed, thus, the protector will always form first, precluding pre-emptor formation and allowing the formation of the terminator, unless it is interfered with by the ribosomes. However, it is easy to see that a large ribosome travelling along the message will interfere with the formation by hydrogen bonding of stem and loop structures.

Under conditions of excess tryptophan there will be large amounts of tryptophanyl-tRNA^{trp} within the cell and the leader peptide will be rapidly synthesized. The ribosome will follow closely behind the RNA polymerase translating the message as soon as it is formed. When the ribosome reaches the stop codon of the leader mRNA all of the protector stem and loop can form (Fig. 4a) whereas the pre-emptor cannot, either because it has not been completely transcribed or because the ribosome is interfering with the formation of its secondary structure. As the RNA polymerase continues to transcribe the leader region the ribosome dissociates from the message leaving the preformed protector. As the remaining message is transcribed it folds into its secondary structure forming the terminator – the pre-emptor being precluded by the protector (Fig. 4b). Because of the presence of the terminator, transcription will cease in the poly-U stretch.

When tryptophan is limiting there will be insufficient charged tRNA^{trp} to completely translate the leader peptide. The ribosome will thus stall at the tandem tryptophan codons (Fig. 4c). As the message continues to be transcribed the first secondary structure formed will be the pre-emptor, the protector being blocked by the stalled ribo-

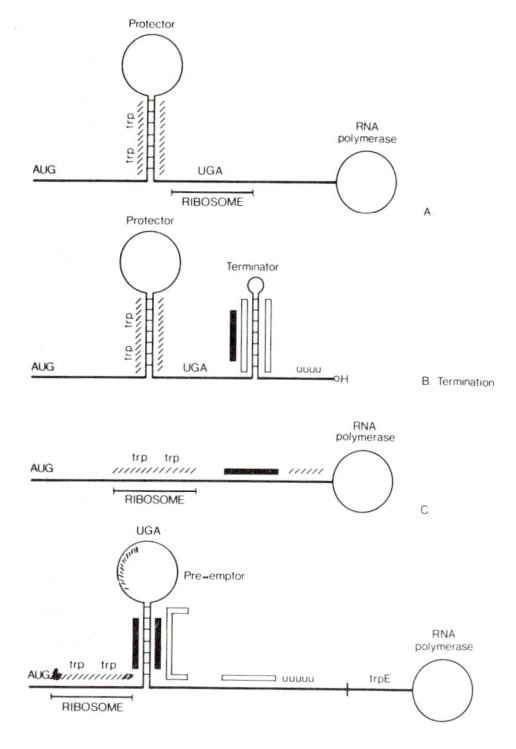

Fig. 4. Secondary structure of the tryptophan leader region under various growth conditions. AUG start codon; trp trp tandem tryptophan codons; UGA stop codon; ⊢———⊣ message protected by the ribosome; ////// protector stem; ▬▬▬ pre-emptor stem; ▭▭▭ terminator stem. The leader peptide and the DNA have been omitted for clarity. The structures are explained in the text; a and b conditions of tryptophan excess, c and d conditions of limiting tryptophan.

poly-U stretch, and dissociate from the DNA releasing the message. The regulation of termination is concerned with allowing or preventing the terminator to form. Failure of the terminator to form will allow transcription to continue through the poly-U stretch into the structural genes.

Preceding the terminator are several other overlapping stem and loop structures; any two adjacent structures being mutually exclusive (Fig. 3). The structure overlapping the terminator is called the pre-emptor, and if it is allowed to form it will prevent the formation of the terminator, allowing transcription to continue some (Fig. 4d). Formation of the pre-emptor prevents the subsequent formation of the terminator once it is completely transcribed. In the absence of the terminator the RNA polymerase will transcribe through the attenuator into the structural genes. If one of the other amino acids in the leader peptide is limiting then the ribosomes will either stall before the protector allowing it and the terminator to form, or stall after the tryptophan codons and obstruct formation of the pre-emptor. Both of these conditions allow the terminator to form. Thus, only a limitation of tryptophan will inhibit termination of transcription and allow maximum expression of the *trp* structural genes.

Prevalance of attenuation

The control regions of biosynthetic operons of several other amino acids have now been sequenced and it appears that attenuation is the major, and probably in some cases (e.g. the *ilv* operon) the only, mode of control. The occurrence of a translatable leader peptide has been shown in all cases where attenuation is found. Repetition of codons for the relevant amino acid in the leader portion of the operon is often more exaggerated than in the *trp* operon. The *phe* operon has seven phenylalanine codons[8], the *his* operon seven histidine codons[9,10] and the *leu* operon four leucine codons[14] in the respective leader regions. But perhaps the most remarkable examples are in those operons under multivalent control. The threonine operon is regulated both by threonine and isoleucine and the isoleucine–valine operon by isoleucine, valine and leucine. The *thr* leader peptide has eight threonine and four isoleucine residues[15], whereas the *ilv* leader peptide has four leucine, five isoleucine and six valine residues[16]. Attenuation is, therefore, a control mechanism for finely tuning the expression of several bacterial operons, including those also subject to the better known repressor/operator control.

References

1 Jackson, E. N. and Yanofsky, C. (1973) *J. Mol. Biol.* 76, 89–101
2 Bertrand, K., Squires, C. and Yanofsky, C. (1976) *J. Mol. Biol.* 103, 319–337
3 Kasai, T. (1974) *Nature (London)* 249, 523–527
4 Bertrand, K. and Yanofsky, C. (1976) *J. Mol. Biol.* 103, 339–349
5 Morse, D. E. and Morse, A. N. C. (1976) *J. Mol. Biol.* 103, 209–226
6 Squires, C., Lee, F., Bertrand, K., Squires, C. L., Bronson, M. J. and Yanofsky, C. (1976) *J. Mol. Biol.* 103, 351–381
7 Lee, F., Bertrand, K., Bennett, G. and Yanofsky, C. (1978) *J. Mol. Biol.* 121, 193–217
8 Zurawski, G., Brown, K., Killingly, D. and Yanofsky, C. (1978) *Proc. Natl Acad. Sci. U.S.A.* 75, 4271–4275
9 Barnes, W. M. (1978) *Proc. Natl Acad. Sci. U.S.A.* 75, 4281–4285
10 Di Nocera, P. P., Blasi, F., Di Lauro, R., Franzio, R., and Bruni, C. B. (1975) *Proc. Natl Acad. Sci. U.S.A.* (1978) 75, 4276–4280
11 Miozzari, G. F. and Yanofsky, C. (1978) *J. Bacteriol* 133, 1457–1466
12 Stauffer, G. V., Zurawski, G. and Yanofsky, C. (1978) *Proc. Natl Acad. Sci. U.S.A.* 75, 4833–4837
13 Zurawski, G., Elseviers, D., Stauffer, G. V. and Yanofsky, C. (1978) *Proc. Natl Acad. Sci. U.S.A.* 75, 5988–5992
14 Keller, E. B. and Calvo, J. M. (1979) *Proc. Natl. Acad. Sci. U.S.A.* 76, 6186–6190
15 Gardner, J. F. (1979) *Proc. Natl Acad. Sci. U.S.A.* 76, 1706–1710
16 Nargang, F. E., Subrahmanyam, C. S. and Umbarger, H. E. (1980) *Proc. Natl Acad. Sci. U.S.A.* 77, 1823–1827

M. D. Watson is at the Department of Botany, The University of Durham, Durham DH1 3LE, U.K.

Initiation of DNA replication in eukaryotic chromosomes

Richard Harland

The large amount of DNA in eukaryotic chromosomes is duplicated exactly for distribution to daughter cells at division. Unlike prokaryotes, the eukaryote must co-ordinate multiple initiations of replication on a single molecule so as to replicate the DNA once, and only once, in a cell cycle.

The chromosome of a prokaryotic cell acts as a single unit of replication, that is, two copies of the chromosome are produced by a single initiation of replication on the circular chromosome followed by propagation of the replication forks and separation of the daughters. Further initiation can take place on the chromosome before the previous cycle is completed, so that multiple replication forks progress round one molecule simultaneously. When the products of one cycle of replication separate, they can be distributed to daughter cells while the next cycle of replication is progressing[1].

In contrast, the much larger DNA content of eukaryotes is organized in separate chromosomes. If each chromosome acted as a single replicon, complete replication of the genome would take many days, even at the fastest observed rates of replication fork movement. The problem of completing replication in a time compatible with rapid growth is solved by initiating replication at many points on the chromosome. Unlike prokaryotes, eukaryotes complete one round of replication before starting another, and initiation must be regulated so that all regions of the chromosome are only replicated once.

In eukaryotic cells, the chromosomes are replicated in a discrete phase of the cell cycle, the DNA synthetic (S) phase. Once in S phase, the rate of replication is controlled by the frequency of initiation rather than by the rate of propagation of a replication fork, or termination of synthesis. Since evidence for this is comprehensively reviewed elsewhere[2,3], I shall concentrate on two aspects of initiation; firstly, the question of whether specific DNA sequences are used to initiate replication and secondly, how re-initiation is prevented on a length of chromosome which has already replicated in one S phase.

Observations on chromosomes

The eukaryotic chromosome initiates replication at multiple sites along its length. This was first deduced by Taylor[4] who showed that different sections of a chromosome replicate simultaneously in S phase. The conclusion from his experiments, which used autoradiography of whole chromosomes labelled with radioactive DNA precursors during S phase, were extended by autoradiography of isolated DNA fibres. Cairns[5] and Huberman and Riggs[6] confirmed that replication was initiated at many sites within a section of the chromosome and that the replication forks progressed bi-directionally until they fused[6]. Evidence that the frequency of initiation of replication was the primary determinant of the length of S phase came from studies in which different cell types of a given organism were compared[7,8]. The rate of elongation in different cell types varies over only a small range but the distance between initiation points varies enormously. Blumenthal, Kriegstein and

Hogness[8] compared the replication of DNA in *Drosophila* embryos during cleavage with that of somatic cells in culture. The rate of replication in the embryo is spectacular. All the DNA is replicated in an interphase of 3–4 min, and the replicating regions are so frequent that their spacing and size can be seen using electron microscopy. In contrast, replicating regions in somatic cells, with an S phase of 10 h, are infrequent but can be picked out from the background of non-replicating DNA by autoradiography of incorporated precursors. Yet in both embryos and somatic cells, the rate of fork progression is about 2600 bases per minute. The embryo evidently speeds up replication by initiating replication at closer intervals along the chromosome.

Such experiments show that the programme of initiation of replication is flexible, which accounts for differences in the length of S phase in different cells. Two different mechanisms whereby the different patterns of initiation could be controlled were proposed[7,8]. Firstly there could be different classes of sequences used as sites for initiation of replication in the different cells, or secondly, there could be a single class of sequences used as sites for initiation which could be masked by binding of chromosomal proteins or coiling of the chromosome to form a compact region where the sites were inaccessible. Both mechanisms involve the idea of a DNA sequence which specifies the initiation of replication, an 'origin of replication'. This idea arises by analogy with prokaryotes, where there is a single origin of replication in the chromosome. Although various lines of evidence do point to the existence of sequence specified origins of replication in eukaryotic chromosomes, the evidence is by no means conclusive.

Is the site of initiation of replication directed by DNA sequence?

The precedent for sequence-specified origins of replication is exemplified by the replication of extrachromosomal molecules such as viral and mitochondrial DNA. DNA replication is always initiated at origins of replication which, in these relatively simple molecules, can be mapped with respect to restriction enzyme cleavage sites by biochemical[9] or electron microscopic[10] analysis of replication intermediates. For the chromosomes the evidence for sequence-specified origins of replication is circumstantial. Sequences in chromosomal DNA which show some homology to viral origins of replication have been isolated[11] but have not been shown to initiate replication in any functional test.

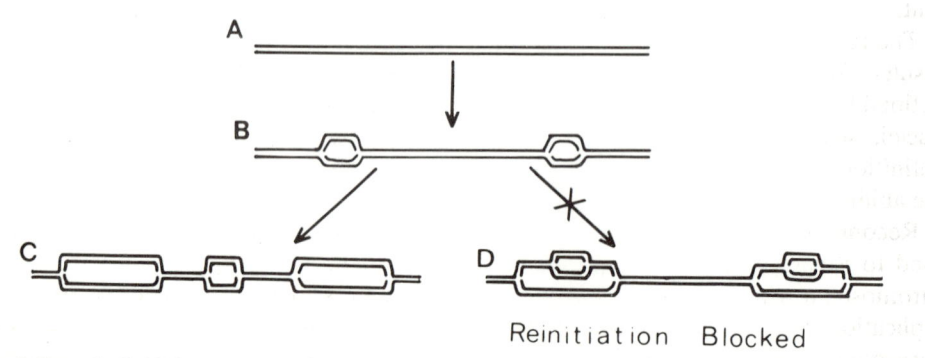

Fig. 1. Control of initiation to prevent chromosomal abnormalities. Initiation on unreplicated DNA (A) gives rise to growing replication 'eyes' (B). Further initiation must be restricted to unreplicated DNA as in C. The possibility in D must be prevented.

Present techniques are unable to detect the initiation of replication from specific sequences in the chromosome. However, some observations on chromosomes show that the initiation of replication is remarkably non-random. The distribution of active origins in *Drosophila* embryo cleavage nuclei, although at a mean distance of 7900 bases apart, revealed a striking periodicity in distribution at multiples of 3400 bases. Such distributions are not unique to embryonic cells; initiation is also non-random in somatic cells in culture[2,3]. The question is: what is the origin of this periodicity? It could be controlled by the presence of origin sequences, but could equally well arise from a periodic structure of the chromosome, so that limited regions of DNA are exposed to initiation enzymes.

Other evidence that sites for the initiation of replication are localized with respect to specific base sequences in the chromosomes has come from experiments based on the observation that transcriptional 'starts' and the origin of replication in the papovavirus SV 40 are in the same region. Seidman, Levine and Weintraub[12] have argued that the same may be true of chromosomes from cultured cells so that the replication fork progresses in the same direction as transcription. They suggest that the fork which replicates a transcription unit must be initiated upstream of that unit.

The remainder of this section deals with results from experiments using small, defined DNA molecules rather than whole nuclei, since they allow a more precise definition of the sequences which confer the ability to replicate.

Recombinant DNA methods have been used to isolate sequences from the yeast chromosome which may act as origins of replication. Such 'autonomously replicating sequences' will allow a plasmid containing a selectable marker to propagate in yeast cells[13,14]. Only a limited subset of sequences allow the plasmid to replicate and thus, it is likely that these sequences may act as origins of replication. These sequences have also been identified in higher eukaryotes[14] but, it has not yet been shown by biochemical methods or electron microscopy that they are sites of replication initiation in the chromosome or recombinant plasmid.

The question of whether specific DNA sequences are required for initiation is emphasized by two examples in which specialized sequences are not required to initiate replication. The first, paradoxically, comes from work with SV 40[15]. When a temperature-sensitive mutant in the viral A gene (required for replication in lytic growth), is grown at the restrictive temperature, replication is much reduced but not eliminated. This residual replication initiates not at the origin, but randomly around the molecule. Thus, the viral gene product is required for specific initiation at the origin, and in its absence initiation is random with respect to nucleotide sequence. It remains to be established whether the residual initiation observed is effected by the altered gene A product or is under cellular control. Nevertheless, this example suggests that initiation need not be sequence specific.

The second line of evidence comes from experiments in which simple, defined DNA molecules are injected into eggs of *Xenopus laevis* where they initiate rounds of semi-conservative replication under strict cell cycle regulation[16]. This occurs with prokaryotic as well as eukaryotic DNA molecules, and in the particular case of SV 40 and polyoma DNA does not require the viral origin of replication used in lytic infection[16].

In theory, the use of recombinant DNA in a reconstructed replication system should provide unequivocal answers to the questions of sequence specificity in initiation. However, progress in this area has

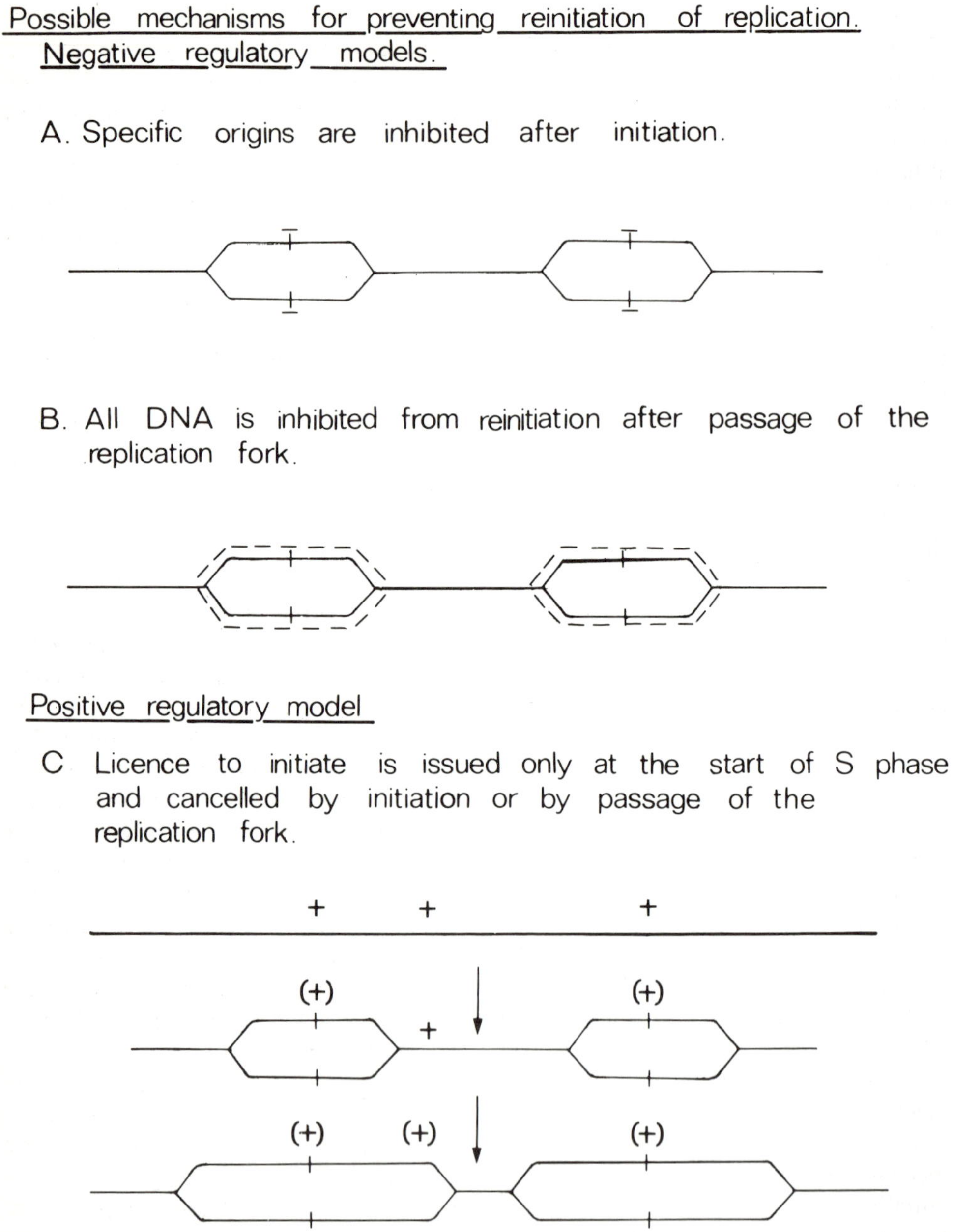

Possible mechanisms for preventing reinitiation of replication. Negative regulatory models.

A. Specific origins are inhibited after initiation.

B. All DNA is inhibited from reinitiation after passage of the replication fork.

Positive regulatory model

C Licence to initiate is issued only at the start of S phase and cancelled by initiation or by passage of the replication fork.

Fig. 2. Reproduced with permission from Laskey, Harland, Earnshaw and Dingwall, in International Cell Biology 1980–81 *(Schweiger, ed.), pp. 162–167, Springer-Verlag.*

been hampered by the lack of systems which efficiently initiate replication on an added DNA template *in vitro*. *In vivo* systems also have their problems; in the case of yeast, only those plasmids which replicate efficiently are selected as autonomous replicating sequences. It is difficult to rule out less efficient initiation on other sequences. There is evidence that in the frog egg different sequences may initiate with slightly different efficiencies[16,17]. The basis of these differences may well be interesting, but so far the differences are small and no plasmid has been shown unequivocally not to initiate replication. In summary, although replication initiates non-randomly in chromosomes, it has not been shown to be due to initiation at specific nucleotide sequences[2,3]. In yeast, however, specific sequences have been shown to confer the ability to replicate on a plasmid[13,14]. In contrast, there are two examples where initiation has been shown not to require a specialized nucleotide sequence[15,16].

The block to re-initiation of replication

Apart from the problem of controlling the sites and spacing of initiation of replication the eukaryote has an additional problem which is not encountered in the single replicon of the prokaryote. Only some of the many possible sites for initiation are actually used and they are not activated synchronously[2,3,6,8]. Therefore, there is a risk that replication will be initiated at a site after the replication fork has passed, leading to over-replication and loss of organization of a particular chromosomal segment (Fig. 1). Such over-replication was not observed by Blumenthal, Kriegstein and Hogness[7] who examined over 1000 replicating regions of DNA. Over-replication does happen in certain terminally differentiating cells, notably cells with polytene chromosomes, but the control mechanisms are probably different.

Thus, in continuously growing cells a mechanism must exist to prevent re-initiation of replication on a sequence that has already replicated in an S phase. Rao and Johnson[18] found that G1 cells are induced to begin DNA synthesis when fused to S phase cells. However, a block to re-initiation of replication persisted in G2 cells: G2 cells were not induced to synthesize DNA by fusion with S phase cells. A similar block to re-initiation is observed during replication of DNA injected into *Xenopus* eggs. When assayed by the classical density labelling approach of Meselson and Stahl[19] replication initiated only once in a cell cycle to produce hybrid heavy–light molecules. Re-initiation to produce a heavy–heavy daughter molecule was only detected when multiple cell cycles occurred[16]. The ability of the egg to discriminate between molecules which have or have not replicated, so that in a pool of molecules only unreplicated ones are templates for initiation, does not require a specialized DNA sequence; even prokaryotic DNA molecules injected into the egg are subject to the block to re-initiation[16].

The mechanism which prevents re-initiation could be negative or positive. In a negative control mechanism possible sites for initiation are inactivated by an inhibitory signal such as bound protein or DNA modification when the replication fork passes. The simplest version of this model (Fig. 2A) inhibits only specific origin sequences but this version is ruled out in the particular case of DNA injected into the frog egg where a specialized sequence to block re-initiation is apparently unnecessary. The block to re-initiation cannot be restricted to specialized sequences but must be pervasive to prevent re-initiation over the entire length of newly replicated DNA (Fig. 2B). In a positive mechanism, a 'licence' to replicate need not be so pervasive and could be distributed to specific sequences, or, more generally, without regard to sequence. The positive signal would then be cancelled by the passage of the replication fork, preventing re-initiation until the next cell cycle.

The observation that controlled initiation in the frog egg does not require a specialized sequence offers an alternative view as to why viruses require an origin of replication for lytic infection. Viral origins of replication, and the viral gene products with which they interact, may be an adaptation to liberate viral replication from the host mechanism which prevents re-initiation in a single cell cycle. If the virus were constrained in the normal way, it could not outgrow its host. The viral origin of replication may, therefore, be an inappropriate model for the sites in the chromosomal DNA where replication is initiated.

To what extent is initiation controlled in chromosomes?

Given that replication can initiate on a wide variety of DNA molecules injected into eggs, and in such a way that DNA is only replicated once in a cell cycle[16], it is not clear what other levels of control are necessary. The precise spatial periodicity of initiation in the chromosomes of embryos and somatic cells[2,3,8] needs to be explained. Either control at the level of base sequence can be invoked, or there is some higher order feature of the chromatin. In this latter connection it is interesting to note that Pardoll, Vogelstein and Coffey[20] have found that newly replicated DNA is attached to a structure called the nuclear matrix, which can be isolated from interphase nuclei. By extrapolation, initiation sites might be those parts of the chromosome, perhaps a subset of sequences, which are attached to the matrix before S phase. The repeat units in *Drosophila* embryos[8] and other cells[2] may, therefore, be a result of structural constraints in the attachment of DNA to the matrix.

Apart from the control of initiation to allow variable length of S phase, it is possible that the sites of initiation are important in determining differentiation of the cell. A model on these lines has been proposed by Weintraub *et al.*[21], based on the observation that the histone octamers of the nucleosome segregate to one side of the repli-

Drawn for TIBS by TAB

WHERE DO YOU START??! THIS AIN'T NO YEAST CELL, BUDDY! START WHERE THE HECK YOU LIKE!

cation fork. Thus, a particular replicated gene may inherit all, or none of the parental histones and, by extension, any proteins which determine gene activity. Depending on the siting of origins of replication relative to a set of genes, particular programmes of gene expression may be inherited by daughter cells.

Clearly, the question of the role of specific DNA sequences in initiating replication is by no means resolved. In yeast, such sequences are required in a plasmid to allow replication[13,14]. In contrast replication of DNA injected into frog eggs does not require a specialized sequence, yet is regulated by the cell cycle[16]. It is possible that the embryo allows the initiation of replication on a wide variety of DNA sequences so as to complete replication rapidly, whereas, initiation is directed in a much more specific way in the chromosomes of adult cells, because of structural constraints, or to permit ordered gene expression.

References

1. Kornberg, A. (1980) *DNA Replication*, W. H. Freeman and Co., San Francisco
2. Hand, R. (1978) *Cell* 15, 317–325
3. Edenberg, H. J. and Huberman, J. A. (1975) *Ann. Rev. Genet.* 9, 245–284
4. Taylor, J. H. (1960) *J. Biophys. Biochem. Cytol.* 7, 455–464
5. Cairns, J. (1966) *J. Mol. Biol.* 15, 372–373
6. Huberman, J. A. and Riggs, A. D. (1968) *J. Mol. Biol.* 32, 327–348
7. Callan, H. G. (1972) *Proc. Roy. Soc. Ser. B.* 181, 19–41
8. Blumenthal, A. B., Kriegstein, H. J. and Hogness, D. S. (1974) *Cold Spring Harbor Symp. Quant. Biol.* 38, 205–223
9. Nathans, D. and Danna, K. (1972) *Nature N. Biol.* 236, 200–202
10. Fareed, G. C., Garon, C. F. and Salzman, N. P. (1972) *J. Virol.* 10, 484–491
11. Jelinek, W. R., Toomey, T. P., Leinwand, L., Duncan, C. H., Biro, P. A., Choudary, P. V., Weissman, S. M., Rubin, C. M., Houck, C. M., Deininger, P. L. and Schmid, C. W. (1980) *Proc. Natl. Acad. Sci. U.S.A.* 77, 1398–1402
12. Seidman, M. M., Levine, A. J. and Weintraub, H. (1979) *Cell* 18, 439–449
13. Struhl, K., Stinchcomb, D. T., Scherer, S. and Davis, R. W. (1979) *Proc. Natl. Acad. Sci. U.S.A.* 76, 1035–1039
14. Stinchcomb, D. T., Thomas, M., Kelly, J., Selker, E. and Davis, R. W. (1980) *Proc. Natl. Acad. Sci. U.S.A.* 77, 4559–4563
15. Martin, R. G. and Setlow, V. P. (1980) *Cell* 20, 381–391
16. Harland, R. M. and Laskey, R. A. (1980) *Cell* 21, 761–771
17. Watanabe, S. and Taylor, J. H. (1980) *Proc. Natl. Acad. Sci. U.S.A.* 77, 5292–5296
18. Rao, P. N. and Johnson, R. T. (1970) *Nature (London)* 225, 159–164
19. Meselson, M. and Stahl, F. W. (1958) *Proc. Natl. Acad. Sci. U.S.A.* 44, 671–682
20. Pardoll, D. M., Vogelstein, B. and Coffey, D. S. (1980) *Cell* 19, 527–536
21. Weintraub, H., Flint, S. J., Leffak, I. M., Groudine, M. and Grainger, R. M. (1978) *Cold Spring Harbor Symp. Quant. Biol.* 42, 401–407

Richard Harland is at the MRC Laboratory of Molecular Biology, Hills Road, Cambridge, U.K.

Eukaryotic DNA polymerases

Anna I. Scovassi, Paolo Plevani and Umberto Bertazzoni

The biological properties, classification and phylogeny of eukaryotic DNA polymerases are reviewed.

There are several animal DNA polymerases; α- and β-nuclear polymerases and γ-type polymerases in both the nucleus and mitochondria[1], which differ both structurally and functionally. This nomenclature used not to be so clear-cut but in the last three or four years a large number of studies have made it easier to discriminate between the different eukaryotic DNA polymerases and have led to a better understanding of their biological roles.

We do not intend to review here the biochemical characteristics of these eukaryotic DNA polymerases (for recent reviews see Refs 2–4), but rather to focus our attention on their biological properties, classification and phylogeny.

Major progress has been made in the following directions: an understanding of α and β functions in DNA replication and repair, respectively; the identification of the mitochrondrial (mt) DNA polymerase as a γ type; the finding of a new type of activity called δ with properties similar to the α polymerase, but having associated exonuclease activity[5].

Function of α-, β- and γ-polymerases and effect of inhibitors

The lack of mutants with altered DNA polymerase genes in animal cells has made it difficult to assign a definite role to each of these enzymes. However, a wealth of physiological experiments has indicated that α-polymerase responds to variations in the rate of the DNA synthesis whereas the level of β-polymerase does not change significantly[2–4]. Recently, two different approaches have proved useful in understanding the functions of polymerase: first SV40 and polyoma DNAs have been used as model systems for host cell DNA replication and second, specific inhibitors of polymerase activity have been found. Only DNA polymerase α seems to be associated with the viral chromosome, and its replication is hampered by the addition of α inhibitors[2–4].

In Table I are listed the substances that exert a major effect on α-, β- and γ-polymerases. It is evident that the behaviour of α- and β-polymerases with respect to inhibition by NEM, ddTTP and aphidicolin is strictly antithetic, thus allowing a clear distinction between these two enzymes *in vitro* and permitting the development of meaningful physiological experiments. The γ activity can be distinguished from α by its aphidicolin resistance, and from β by its sensitivity to NEM. By using these different inhibitors it has recently been possible to accumulate strong evidence for a direct involvement of α-polymerase in the replication of nuclear DNA, of β-polymerase in DNA repair and of γ-polymerase in the replication of mitochondrial DNA. In all the physiological systems examined, replicative DNA synthesis is completely blocked by low concentrations of aphidicolin[4,6], whereas unscheduled DNA synthesis[13] and the replication of mitochondrial DNA (S. Spadari, personal communication) appear to be insensitive to aphidicolin. The latter two processes are affected by appropriate concentrations of ddNTPs[4].

The possibility that α-polymerase is involved in the DNA repair process has not

TABLE I
DNA polymerase inhibitors.

Inhibitor	% of inhibition			Ref.
	α	β	γ	
NEM (N-Ethylmaleimide) (1 mM)	Sensitive (100%)	Resistant (0%)	Sensitive (85%)	4
ddTTP (ddTTP/dTTP = 10)	Resistant (10%)	Sensitive (90%)	Sensitive (90%)	4
Aphidicolin (2 µg/ml)	Sensitive (90%)	Resistant (0%)	Resistant (0%)	6
araCTP (125 µM)	Sensitive (85%)	Resistant (15%)		7
Butylanilinouracil (100 mM)	Sensitive (60%)	Resistant (0%)	Resistant (0%)	8
β-Lapachone (83 µM)	Sensitive (60%)	Resistant (10%)		9
Pyridoxal-5-phosphate (0.4 mM)	Sensitive (96%)	Resistant (55%)	Sensitive (92%)	10
Heat inactivation (10', 45°C)	Resistant (0%)	Sensitive (90%)		4
Heparin (0.1 µg/ml)	Sensitive (70%)	Resistant (10%)	Resistant (10%)	11
Ethidium bromide (20 µM)	Resistant (10%)	Resistant (10%)	Sensitive (60%)	12

yet been clarified since unscheduled DNA synthesis in some systems is not affected by aphidicolin[13], while other systems exhibit a reduced rate of DNA repair[14].

The possible function of the nuclear γ enzyme also needs to be clarified. Replication of adenovirus DNA requires the participation of γ-polymerase[15], but no evidence has been obtained so far of a definite role for this enzyme in the replication of the nuclear DNA.

Although still the subject of debate, there appears to be a DNA polymerase (δ) which is distinct from the α, β and γ forms[5]. This enzyme has many of the features of DNA polymerase α, in that it is sensitive to NEM and aphidicolin but it can be distinguished from the α-polymerase by its chromatographic behavior and by peptide mapping[16].

The importance of factors which can interact with eukaryotic DNA polymerases

TABLE II
DNA polymerases from eukaryotes.

Organisms	Type	Function	Mass (kdal)	Associated exo	NEM sensitivity
Animals	α	DNA replication	130–280	No	Yes
	β	DNA repair	30–50	No	No
	γ	Adenovirus DNA replication	150–300	No	Yes
	δ	Unknown	190	Yes	Yes
	mt-γ	mt-DNA replication	150–300	No	Yes
Euglena gracilis[4]	A	Unknown	190	No	Yes
	B	Unknown	240	Yes	Yes
Tetrahymena pyriformis[22]	I	DNA replication	80–130	Yes	Yes
	II	Unknown	70		No
Trypanosoma brucei[23]	Major	Unknown	100		Yes
	Minor	Unknown	50		No
Yeast[4]	I; A	DNA replication	150	No	Yes
	II; B	Unknown	150	Yes	Yes
Ustilago maydis[4]	Major	DNA replication	180	Yes	Yes
Neurospora crassa[24]	A	Unknown	150	No	
	B	Unknown	110	No	
Dictiostelium discoideum[23]	Single	Unknown	130	No	Yes
Physarum polycephalum[25]	Major	Unknown	120–200		Yes

has recently been pointed out. Of particular interest are those which stimulate specifically α-polymerase (such as Ap4A[17], DNA helicases, Helix Destabilizing Protein) and β-polymerase[4]. The finding that α-polymerase often co-purifies or interacts with proteins involved in the replication machinery, further supports its role in the process of DNA replication.

Intracellular localization of DNA polymerases

The localization of the different DNA polymerases could be inferred from their roles inside the cell. DNA polymerases α and β should be found only in the nuclear fraction, while the mitochondrial γ-polymerase should be confined within this organelle. However, α-polymerase, though it should reside in the nucleus, cannot be extracted from nuclei using aqueous buffers. Many approaches have been used in attempts to locate this enzyme in the nucleus, but a definitive answer has not been obtained. Using immunofluorescence techniques it has recently been shown that the enzyme is located mainly in the perinuclear region[18]. The β-polymerase fractionates with chromatin, but a certain amount can also be found in the cytoplasmic fraction. The mitochrondria contain a unique DNA polymerase, which has properties typical of the γ enzyme[3,4]; however, a certain amount of the γ-polymerase can also be detected in the nuclear fraction[4,15]. One should therefore consider the possibility that a mechanism exists whereby the γ enzyme, synthesized in the cytoplasm, can migrate into the mitochondria or into the nucleus. Possible explanations could involve differential mRNA splicing or proteolytic processing of the precursor molecules.

Type of DNA synthesis

The type of DNA synthesis catalysed *in vitro* by the different DNA polymerases gives biochemical information and clues about the ways these enzymes function *in vivo*. Several different types of DNA polymerization can be visualized: (i) replicative nuclear DNA synthesis, which is characterized by being semiconservative, symmetrical, bidirectional and with short RNA-primed intermediates in the lagging strand; (ii) repair DNA synthesis, occurring on both strands and restricted to short DNA gaps; (iii) mitochondrial and adenovirus DNA synthesis, which is known to be continuous and proceeds in an

TABLE III
Phylogeny of eukaryotic DNA polymerases

Phyla	Classes	α-like	β-like	γ-like
Vertebrates	Mammals	+	+	+
	Birds	+	+	+
	Reptiles	+	+	+
	Amphibians	+	+	+
	Fishes	+	+	+
Arthropods	Insects (embryos)	+	−	+
Echinoderms	Echinoids	+	+	+
Mollusks	Cephalopods	+	+	
Coelenterates	Hydrozoas	+	+	
Porifera	Sponges	+	+	+
Protozoa	Ciliates	+	−	
	Flagellates (parasitic)	+	+	
Thallophytes	Ascomycetes	+	−	+
Spermaphytes	Monocotyledones	+	−	+

asymmetrical way.

The polymerization promoted by DNA polymerase α shows that this enzyme is particularly well suited to carry out replication of nuclear DNA. This enzyme is capable of making relatively short products (3-4 S) on ribo-primers in a quasi-processive way and may also be able to synthesize much longer DNA chains with the aid of accessory proteins, which might improve its affinity for the DNA template[19] as has been shown in prokaryotic systems. By contrast, DNA polymerase β works in a distributive way and can add single nucleotides to activated DNA, properties consistent with its role in a repair type of DNA synthesis[4]. The DNA polymerase γ synthesizes DNA *in vitro* in a highly processive fashion[20], suggesting that it could replicate mitochondrial and adenovirus DNAs, as well as synthesize the long, single-stranded DNA regions visualized in human cells, which might represent displaced strands of parental DNA being replicated asymmetrically[21].

Classification

We propose the following definitions for animal DNA polymerases:

(1) DNA polymerase α: high molecular weight; sensitive to NEM; sensitive to aphidicolin; insensitive to ddNTPs; quasi-processive type of synthesis; using of RNA-primed DNA.

(2) DNA polymerase β: low molecular weight; resistant to NEM; resistant to aphidicolin; sensitive to ddNTPs; distributive type of synthesis; elongation of deoxy-primers on ribo-templates in the absence of phosphate.

(3) γ and mt DNA polymerases: high molecular weight; sensitive to NEM; resistant to aphidicolin; sensitive to ddNTPs; highly processive type of DNA synthesis; using of deoxy-primers on ribo-templates in the presence of phosphate.

(4) DNA polymerase δ: high molecular weight; sensitive to NEM and aphidicolin; associated exonuclease activities.

DNA polymerases from lower eukaryotes

In Table II we outline some properties of DNA polymerases from vertebrates down to the lower eukaryotes. At present it is not possible to draw up a unified classification for the polymerases of the lower eukaryotes. However, some common features can be seen: multiple extramitochondrial DNA polymerases exist, which are both of high molecular weight and are sensitive to NEM; the presence in some of the polymerases of associated exonucleases indicates a possible relationship to prokaryotes.

Many questions remain about the biological function, the localization and the type of synthesis carried out *in vivo*; the number of DNA polymerases found within the cell might also be influenced by different factors, such as the growth conditions and the phase of the cell cycle. However, at present, it might be of interest for future studies to take the available information and define at least two types of extramitochondrial DNA polymerases in lower eukaryotes. We propose to extend to these classes the nomenclature used for the yeast enzymes: DNA polymerase I is the major enzyme, which is very probably responsible for DNA replication, is capable of using a ribo-primer and lacks associated exonucleases; DNA polymerase II is the minor polymerase with an associated exonuclease and is antigenically distinct from DNA polymerase I.

Phylogeny

A promising way to classify eukaryotic DNA polymerases is to study their occurrence in a variety of organisms that cover a wide evolutionary scale. A phylogenetic survey has been carried out for the α- and β-polymerases[23] and to a lesser extent for the γ-polymerase[26]. In Table III we have outlined the main results obtained so far. The polymerases have been assigned as α-, β- or γ-like according to the criteria discussed above. An α-like enzyme has been found in all of the different phyla examined, from vertebrates to protozoa, fungi and plants, while a β-like polymerase is present in all multicellular organisms but not in fungi or in plants. One interesting early observation was the absence of β-like polymerase in *Drosophyla* embryos, whereas a low-molecular-weight, NEM-insensitive enzyme has been demonstrated in parasitic protozoa (*Trypanosoma*) which are known to spend part of their life in the insects[23]. The γ-polymerase is widespread in the vertebrate classes where it is the only polymerase activity found within the mitochondria[26]. In the three invertebrate phyla listed in Table III, in fungi and in plants, a γ-like polymerase activity is certainly present but the mitochondrial polymerase seems to differ from that of vertebrates since it responds not only to the γ assay but also to the α assay[27].

Acknowledgements

We thank S. Spadari, P. van der Vliet and P. Borst for helpful discussions, and R. Porreca for editorial assistance.

This paper is contribution No. 1709 of the Radiation Protection Programme of Eur. Commun. Commission.

References

1 Weissbach, A., Baltimore, D., Bollum, F. J., Gallo, R. C. and Korn, D. (1975) *Science* 190, 401–402
2 Falaschi, A. and Spadari, S. (1978) in *DNA synthesis: present and future* (Molineux, I. and Kohiyama, M., eds), pp. 487–515, Plenum Press, New York
3 DePamphilis, M. L. and Wassarman, P. M. (1980) *Ann. Rev. Biochem.* 49, 627–666
4 Kornberg, A. (1980) *DNA replication*, Freeman and Co., San Francisco
5 Byrnes, J. J. and Black, V. L. (1978) *Biochemistry* 17, 4226–4231
6 Ikegami, S., Taguchi, T., Ohashi, M., Oguro, H., Nagano, H. and Mano, Y. (1978) *Nature (London)* 275, 458–460
7 Wist, E. (1979) *Biochem. Biophys. Acta* 562, 62–69
8 Wright, G. E., Baril, E. F. and Brown, N. C. (1980) *Nucleic Acids Res.* 8, 99–110
9 Schuerch, A. R. and Wehrli, W. (1978) *Eur. J. Biochem.* 84, 197–205
10 Oguro, M., Nagano, H. and Mano, Y. (1979) *Nucleic Acids Res.* 7, 727–734
11 Di Cioccio, R. A. and Srivastava, B. I. S. (1978) *Cancer Res.* 38, 2401–2407
12 Tarrago-Litvak, L., Viratelle, O., Darriet, D., Dalibart, R., Graves, P. V. and Litvak, S. (1978) *Nucleic Acids Res.* 5, 2197–2210
13 Pedrali-Noy, G. and Spadari, S. (1980) *Mutation Res.* 70, 389–394
14 Ciarrocchi, G., Jose, J. G. and Linn, S. (1979) *Nucleic Acids Res.* 7, 1205–1219
15 van der Vliet, P. C. and Kwant, M. M. (1978) *Nature (London)* 276, 532–534
16 Chang-Chen, Y., Bohn, E. W., Planck, S. R. and Wilson, S. H. (1979) *J. Biol. Chem.* 254, 11678–11687
17 Grummt, F., Walte, G., Jantzen, H. M., Hamprecht, K., Hübscher, U. and Kuenzle, C. C. (1979) *Proc. Natl. Acad. Sci. U.S.A.* 76, 6081–6085
18 Bollum, F. J. (1979) in *Antiviral mechanisms in the control of neoplasia* (Chandra, P., ed.), pp. 587–601, Plenum Publishing Co.
19 Fichot, O., Pascal, M., Mechali, M. and deRecondo, A. M. (1979) *Biochem. Biophys. Acta* 561, 29–41
20 Yamaguchi, M., Matsukage, A. and Takahashi, T. (1980) *Nature (London)* 285, 45–47
21 Bjursell, G., Gussander, E. and Lindahl, T. (1979) *Nature (London)* 280, 420–423
22 Furukawa, Y., Yamada, R. and Kohno, M. (1979) *Nucleic Acids Res.* 7, 2387–2398
23 Chang, L. M. S., Cheriathundam, E., Mahoney, E. M. and Cerami, A. (1980) *Science* 207, 510–511
24 Joester, W., Joester, K. E., van Dorp, B. and Hofschneider, P. H. (1978) *Nucleic Acids Res.* 5, 3043–3055
25 Baer, A. and Schiebel, W. (1978) *Eur. J. Biochem.* 86, 77–84
26 Scovassi, A. I., Wicker, R. and Bertazzoni, U. (1979) *Eur. J. Biochem.* 100, 491–496
27 Bertazzoni, U. and Scovassi, A. I. in *Organisation and expression of the mitochondria genome* (Kroon, A. M. and Saccone, C., eds), Elsevier/North-Holland Biomedical Press (in press)

Anna I. Scovassi and Umberto Bertazzoni are at the Istituto di Genetica, Biochimica ed Evoluzionistica del Consiglio Nazionale delle Ricerche, Via S. Epifanio, 14-27100 Pavia, Italy, and Paolo Plevani is at the Istituto di Biologia Generale, Ente Universitario Lombardia Orientale, Via Valsabbina, 19-25100 Brescia, Italy.

Addendum – A. I. Scovassi et al.

The evidence accumulated earlier for the involvement of DNA polymerases α, β and γ in the replication of chromosomal DNA, repair DNA synthesis and replication of mitochondrial DNA, respectively, has been confirmed. More recently, specific inhibitors have been used to investigate the role of eukaryotic DNA polymerases in other physiological processes. Thus, DNA polymerase α has been suggested to participate in the amplification of ribosomal DNA[28], DNA polymerase β in DNA endoreduplication in giant trophoblast cells[29] and DNA polymerase γ in the replication of parvovirus H-1[30].

The study of the progress of DNA polymerase α on natural DNA templates indicates that the enzyme tends to pause at certain DNA sites, characterised by specific DNA sequences and obstacles of secondary structure[31,32]. These arrest sites may serve as signals for attenuation or promotion of DNA replication.

The recent purification of chick embryo DNA polymerase α and γ to near homogeneity has given new information about the structure of these enzymes: DNA polymerase α is composed of two clusters of peptides, one of Mr 13,000–15,500 and the other of Mr 51,000–59,000, each resulting from the microheterogeneity of a single polypeptide[33]; DNA polymerase γ consists of four identical subunits of Mr 47,000 (Ref. 34).

The study of the evolutionary conservation of DNA polymerases has been greatly promoted by procedures which renature catalytic activities *in situ* after NaDodSO$_4$ polyacrylamide gel electrophoresis[35]. A high Mr polypeptide ($>$125,000), responsible for chromosomal DNA replication and closely related to DNA polymerase α, is maintained in higher and lower eukaryotes[36]. A recent survey has shown that the structure and immunoreactants of DNA polymerase β are conserved from mammals to parasitic protozoa as a single polypeptide chain of Mr 39,000–47,000 (Ref. 37).

We have recently shown that the 'activity gel' technique is also useful for detecting and discriminating between the active DNA polymerase peptides in crude extracts of a variety of eukaryotic organisms. Several high Mr DNA polymerase fragments were found in extracts, from mammals to plants, having the same migration pattern. Among these, the Mr 70,000 fragment is particularly evident and appears to be conserved from prokaryotes to eukaryotes. The low Mr active peptide corresponding to DNA polymerase β was found only in animal cells, suggesting that it appeared in evolution at the same time as the development of the metazoans[38].

These technologies will certainly aid in the resolution of the protein components of replication systems and in the study of DNA polymerases from organelles and animal viruses.

References
28 Zimmermann, W. and Weissbach, A. (1981) *Molec. and Cell. Biol.* 1, 680–686
29 Siegel, R. L. and Kalf, G. F. (1982) *J. Biol. Chem.* 257, 1785–1790
30 Kollek, R. and Goulian, M. (1981) *Proc. Natl Acad. Sci. U.S.A.* 78, 6206–6210
31 Weaver, D. T. and De Pamphilis, M. L. (1982) *J. Biol. Chem.* 257, 2075–2086
32 Kagumi, L. S. and Clayton, D. A. (1982) *Proc. Natl Acad. Sci. U.S.A.* 79, 983–987

33 Yamaguchi, M., Tanabe, K., Takahashi, T. and Matsukage, A. (1982) *J. Biol. Chem.* 257, 4484–4489
34 Yamaguchi, M., Matsukage, A. and Takahashi, T. (1980) *J. Biol. Chem.* 255, 7002–7009
35 Spanos, A., Sedgwick, S. G., Yarranton, G. T., Hübsher, U. and Banks, G. R. (1981) *Nucleic Acids Res.* 9, 1825–1839
36 Hübsher, U., Spanos, A., Albert, W., Grummt, F. and Banks, G. R. (1981) *Proc. Natl Acad. Sci. U.S.A.* 78, 6771–6775
37 Chang, L. M. S., Plevani, P. and Bollum, F. J. (1982) *Proc. Natl Acad. Sci. U.S.A.* 79, 758–761
38 Bertazzoni, U., Scovassi, A. I., Torsello, S., Badaracco, G. and Plevani, P. (1982) in: *Cell Function and Differentiation* (Evangelopoulos, A., ed.), A. R. Liss Inc., New York (in press)

Superhelical DNA

James C. Wang

Since the finding of the late J. Vinograd and his co-workers 15 years ago that the DNA of the animal virus polyoma is superhelical, DNAs of this class have emerged as the most abundant form of genetic material. Some unique properties of such DNAs and the discovery of enzymes that can relax or supercoil DNA are summarized in this review.

In the well-known Watson–Crick structure of DNA, the two antiparallel strands are coiled around each other. A direct consequence of this intertwining is that if a double-stranded DNA molecule is in the form of a ring with no discontinuity in the backbone bonds of either strand, the complementary single-stranded rings are linked. The parameter that describes quantitatively the linking of the pair of single-stranded rings in such a DNA is the linking number α or Lk. Roughly speaking, α or Lk is the number of times one strand goes around the other in the duplex ring; it is an integer. A more rigorous definition can be found in the review by Crick[1].

The appearance and hydrodynamic properties of a covalently closed DNA is strongly affected by its linking number

For a circular DNA duplex containing one or more single-chain scissions, there is no topological constraint on the linking number; α is not a topological invariant. When all scissions are sealed by DNA ligase, the DNA becomes 'covalently closed' and α is an invariant so long as none of the backbone bonds is broken either transiently or permanently. If there are n base pairs (bp) per DNA molecule and $h°$ bp make a complete helical turn in the absence of any constraint, then it seems intuitively reasonable that the linking number $\alpha°$ of the DNA closed by ligase is $n/h°$*. This DNA will be referred to as the relaxed DNA. It should be noted that $h°$ is dependent on temperature, counter-ion concentrations, hydration, the binding of intercalating agents etc. Therefore, a DNA relaxed under one set of conditions is usually not relaxed under a different set of conditions.

The properties of a covalently closed DNA are much affected by its linking number. The difference between its linking number α and the linking number $\alpha°$ of the same DNA when it is relaxed, $\Delta\alpha = \alpha - \alpha°$, has been referred to in the literature as the number of superhelical turns†. Frequently, the deviations of the properties of a supercoiled DNA from those of a relaxed DNA are best correlated with $\Delta\alpha/\alpha°$ rather than $\Delta\alpha$ itself. The quantity $\Delta\alpha/\alpha°$ will be referred to as the superhelical density or the specific linking difference‡.

As $\Delta\alpha$ deviates from zero, the appearance of the DNA molecules changes. More crossovers are seen and the molecules appear twisted when viewed with an electron microscope (Fig. 1). The term supercoiled, superhelical, or supertwisted has been used to describe such molecules; depending on whether $\Delta\alpha$ is positive or negative, the DNA is referred to as positively or negatively supercoiled.

---

* The relation $\alpha° = n/h°$ results if the average writhing number of the relaxed DNA is zero. See the review by Crick, Ref. 1.

† Vinograd *et al.* originally defined the number of superhelical turns as $\alpha - \beta$, where β is the number of 'helical turns'. The identification of β as $\alpha°$ provides a more rigorous definition of β.

‡ The superhelical density was originally defined as $(\alpha - \beta)$ per 10 bp.

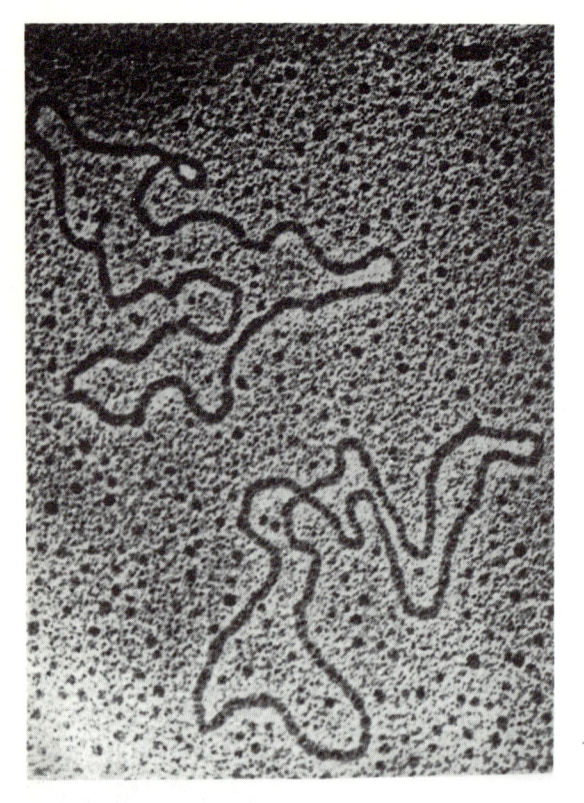

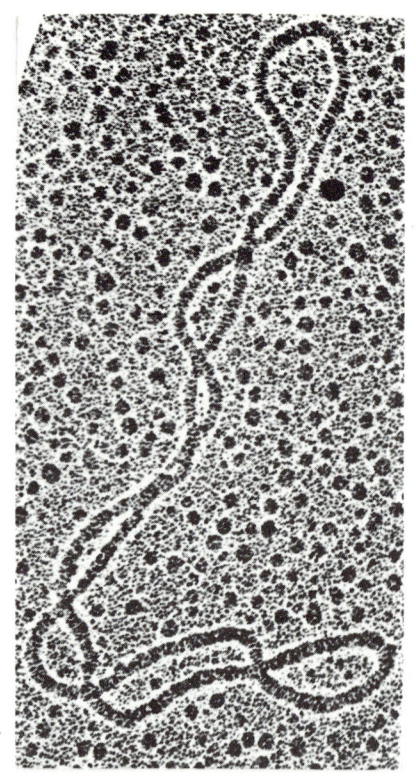

Fig. 1. Electron micrographs showing two relaxed (left) and one supercoiled (right) phage PM2 DNA molecules.

The configuration changes of a DNA as $\Delta\alpha$ deviates from zero are also reflected in changes in its hydrodynamic properties. Although sedimentation and viscosity measurements have been used extensively to monitor such changes, in recent years gel electrophoresis has become the method of choice for DNAs under 10^7 daltons. Fig. 2 depicts the electrophoretic patterns of a family of four phage PM2 DNA samples with different $\Delta\alpha$ values. The sample on the far left is highly negatively supercoiled, with a specific linking difference of about -0.11. The DNA migrates much faster than the nicked form, a trace of which is present in this sample and shows as a faint slower migrating band. The sample run in the second lane is less negatively supercoiled. Here the striking feature is that a series of discrete bands is seen. It has been shown that each of these is a DNA of a given linking number[2]. Within a certain range of $|\Delta\alpha|/\alpha°$, two topological isomers (topoisomers) which differ by 1 in their linking numbers are well-resolved. When $|\Delta\alpha|/\alpha°$ is large, this fine resolution is lost and the topoisomers run as a single band, as is the case for the sample on the far left. When $|\Delta\alpha|/\alpha°$ approaches zero, the resolution is again much reduced and the mobilities of the topoisomers approach that of the nicked species (Fig. 2, lanes 3 and 4).

The standard free energy $\Delta G_s°$ of supercoiling

Supercoiling of a DNA affects the stability of the DNA with respect to strand separation and cruciform formation, as well as its interactions with certain small and large molecules. The energetics of supercoiling provides an important basis for understanding these effects[3-6].

The ability to resolve individual topoisomers by gel electrophoresis provides a direct method of examining free energy differences, ΔG_T° (see Refs 7 and 8). If a nicked DNA is converted into the covalent closed form by a DNA ligase, one would expect a distribution of topoisomers in the product if the difference in ΔG° between two topoisomers that differ by 1 in α is of the same order of magnitude as the thermal energy RT. This turns out to be the case. The free energy differences can be deduced from the experimentally measured Boltzmann distribution, and it is found that ΔG_T°, the difference in ΔG° between a particular topoisomer with a linking number α and the relaxed DNA, is equal to $K(\Delta\alpha)^2$, where K is a constant roughly inversely proportional to α°. It should be noted that because of the distribution of topoisomers, the linking number of the relaxed DNA should be taken as the population average $<\alpha^\circ>$; it need not be an integer. Although the calculation of ΔG_T° from the Boltzmann distribution involves species with $\Delta\alpha$ close to zero, the dependence of ΔG_T° on $(\Delta\alpha)^2$ has been shown to be correct for $|\Delta\alpha|/\alpha^\circ$ at least as high as 0.1.

To illustrate how ΔG_T° affects the properties of a DNA, consider the binding of a bacterial RNA polymerase to a DNA of a linking difference $\alpha - \alpha^\circ$. When a polymerase binds to a DNA, it unwinds the double helix by about one turn, depending on the DNA and the counter-ion concentrations. It is therefore expected that α° will be reduced by about 1. The linking number α, being a topological invariant, is unaffected. If the DNA is negatively supercoiled, a reduction in α° lowers $(\Delta\alpha)^2$ and therefore lowers ΔG_T°. In other words, the binding of the polymerase molecule is favored by a lowering of ΔG_T° of the negatively supercoiled DNA in addition to the favorable intrinsic binding free energy. For a moderately supercoiled DNA with a $\Delta\alpha/\alpha^\circ$ of -0.05, the lowering of ΔG_T° contributes about 6 kcal if the binding of an RNA polymerase lowers α° by 1. This corresponds to a factor of 1.7×10^4 in the relative binding constants to the negatively supercoiled and relaxed DNA. In general, processes that lower $|\Delta\alpha|$ are more favored for a supercoiled DNA substrate. For a negatively supercoiled DNA, these include unpairing of bases or unwinding of the double helix. Reasonable estimates can be made on the magnitudes of the ΔG_T° contributions to the various processes, but these will not be discussed here.

Measuring the helical repeat of DNA in solution

Studies on covalently closed DNAs led to the confirmation of the double-helix

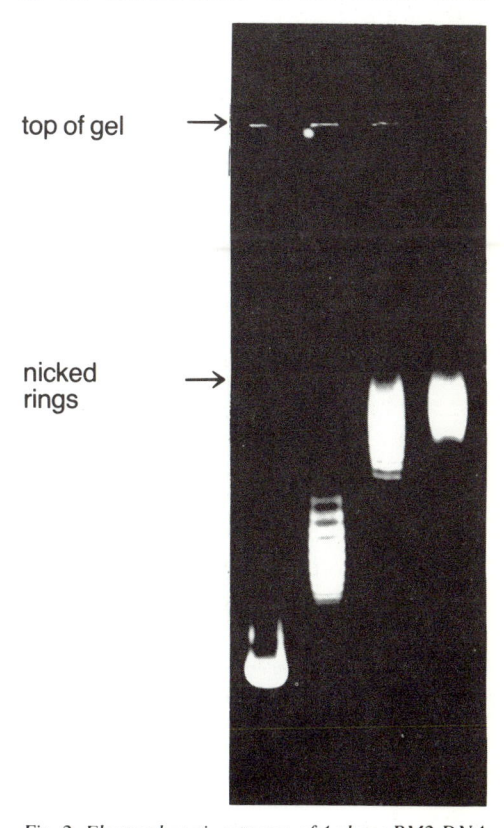

Fig. 2. Electrophoretic patterns of 4 phage PM2 DNA samples in 0.7% agarose gel. See text for explanations.

structure of DNA[2] and the development of two methods that can determine accurately the helical repeat of the number of base pairs per helical turn of a DNA segment of a given nucleotide sequence in solution[9,10]. The experimentally more simple method of the two, the band shift method, is described below.

Suppose two nicked circular DNAs, one containing n bp/molecule and the other the same n bp plus an inserted sequence of x bp, are treated with ligase. The case $x = 6$ is taken as an example. As discussed in the section above, a distribution in α results for each DNA. Because of the integral nature of α, if the set of linking numbers of the ligase closed DNA with n bp/molecule is $\ldots i - 1, i, i + 1 \ldots$, the same set of integers must represent the linking numbers of the topoisomers in the DNA sample with $n + 6$ bp/molecule. The linking differences or the numbers of superhelical turns of the two samples during gel electrophoresis are not the same, however, since the population average linking numbers of the DNAs relaxed under the electrophoresis conditions are clearly different. For the DNA with n bp/molecule, the distribution in $\Delta\alpha$ is $\ldots i - 1 - n/h°, i - n/h°, i + 1 - n/h° \ldots$; for the DNA with $n + 6$ bp/molecule, the distribution in $\Delta\alpha$ is $\ldots i - 1 - (n + 6)/h°, i - (n + 6)/h°, i + 1 - (n + 6)/h° \ldots$. Clearly, for a topoisomer with a given linking difference k in the first set, there is a corresponding topoisomer with a linking difference $k - 6/h°$ in the second set. Since the spacing between two adjacent topoisomer bands of either sample represents a difference of 1 in $\Delta\alpha$, it follows that all topoisomer bands of the DNA with $n + 6$ bp/molecule are shifted from the corresponding topoisomer bands of the DNA with n bp/molecule by a fraction $6/h°$ of the spacing between bands. This is indeed observed as indicated in Fig. 3. By extending this line of reasoning, it can be generalized that if $x = $ integer $\cdot h° + y$, where $y < h°$ is the non-integral residue, all topoisomer bands in the DNA with $n + x$ bp/molecule are shifted by a fraction $y/h°$ of the spacing between bands. This dependence on the non-integral residue is reminiscent of the interference phenomenon in optics. By using cloned inserts of larger and larger x, a high accuracy in the measured $h°$ can be achieved. The conclusion above is arrived at by assuming that the only effect on mobility by the addition of x bp is due to the dependence of the electrophoretic mobility on $\Delta\alpha$. The mobility is actually slightly affected also by the lengthening of the DNA. This length effect can be corrected, however[10]. For inserts of several different nucleotide sequences, the measurements so far show that at 37°C, in a dilute aqueous buffer containing a few mM MgII, 10.5 ± 0.1 bp makes a helical turn. The sequence $dA_n : dT_n$ provides an exception: the measured $h°$ is 9.9 ± 0.1 bp[11].

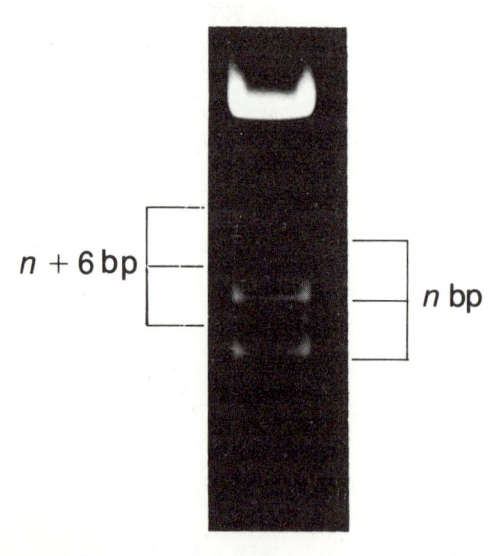

Fig. 3. Electrophoretic pattern of a mixture of two DNA samples that differ by 6 bp in their sizes. n is about 4300 bp. The bright band on top contains the nicked species; the length difference between the two samples is too small to be resolved in this band. The two groups of the covalently closed topoisomers are well-resolved. Three from each group are indicated in the figure. See Ref. 10 for more details.

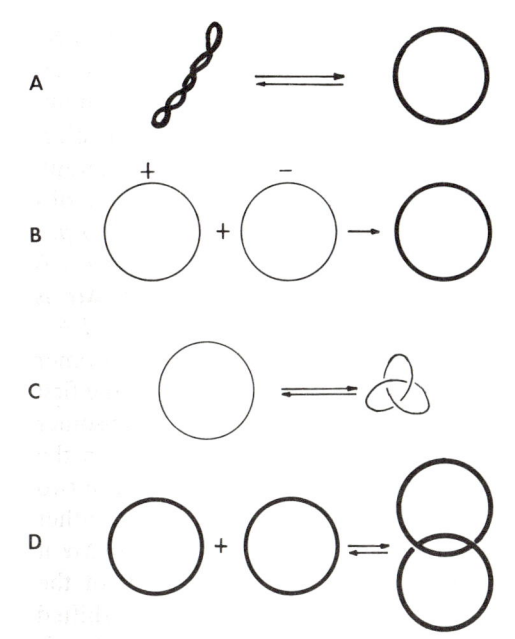

Fig. 4. Topoisomerization reactions that are catalysed by DNA topoisomerases. The thicker line represents double-stranded DNA and the thinner line single-stranded DNA. Reactions B and C as drawn are known to be catalysed by the type I enzymes; the double-stranded equivalent of Reaction C is catalysed by the type II enzymes. The linking and unlinking of double-stranded DNA rings with or without gross sequence homology are catalysed by both types of topoisomerases. The type I enzymes, however, require the presence of a pre-existing single-chain scission in one of the participating rings.

There are enzymes, the DNA topoisomerases, that can relax or supercoil covalently closed DNAs

An interesting outcome of studies on covalently closed DNAs is the discovery of enzymes that can relax or supercoil DNA[12,13]. These enzymes can be subdivided into two categories[14,15]. The type I DNA topoisomerases require no energy cofactor such as ATP; they can reduce but cannot increase $|\Delta\alpha|$, and α is changed in steps of ± 1 by these enzymes. The type II DNA topoisomerases usually require ATP. Some of the type II enzymes, the DNA gyrases, can catalyse the negative supercoiling of DNA, and $|\Delta\alpha|$ is increased. Other type II enzymes, such as phage T4 DNA topoisomerase or the ATP-dependent drosophila DNA topoisomerase, can only reduce $|\Delta\alpha|$ at least under the assay conditions devised so far. The values of α are changed in steps of ± 2 by the type II enzymes[15-17].

The DNA topoisomerases not only catalyse the reduction or increase in $|\Delta\alpha|$, but also interconversion between other types of topoisomers. Some of these reactions are depicted in Fig. 4.

It is increasingly clear that the DNA topoisomerases play important roles in many vital processes involving DNA[18]. The blocking of DNA gyrase *in vivo* leads to cessation of replicative DNA synthesis, inhibition of transcription of certain operons and certain repair processes. This multitude of effects might all result from the role of gyrase in maintaining the DNA in a negatively supercoiled state *in vivo*. The ATP-dependent T4 topoisomerase is required in replication. The phage λ integrase, which is responsible for the integration of the viral DNA into the host genome, has a topoisomerase activity[19]. The *in vivo* roles of many other DNA topoisomerases are at present unknown, and their studies are likely to offer further insights into how nature deals with the topological problems of DNA.

References

1 Crick, F. H. C. (1976) *Proc. Natl. Acad. Sci. U.S.A.* 73, 2639–2643
2 Crick, F. H. C., Wang, J. C. and Bauer, W. R. (1979) *J. Mol. Biol.* 129, 449–461
3 Bauer, W. R. and Vinograd, J. (1970) *J. Mol. Biol.* 47, 419–435
4 Hsieh, T.-S. and Wang, J. C. (1975) *Biochemistry* 14, 527–535
5 Davidson, N. (1972) *J. Mol. Biol.* 66, 307
6 Vologodskii, A. V., Lukashin, A. V., Anshelevich, V. V. and Frank-Kameneskii, M. D. (1979) *Nucleic Acids Res.* 6, 967–982
7 Depew, R. E. and Wang, J. C. (1975) *Proc. Natl. Acad. Sci. U.S.A.* 72, 4275–4279
8 Pulleyblank, D. E., Shure, M., Tang, D.,

Vinograd, J. and Vosberg, H. P. (1975) *Proc. Natl. Acad. Sci, U.S.A.* 72, 4280–4284
9 Wang, J. C. (1979) *Cold Spring Harbor Symp. Quant. Biol.* 43, 29–33
10 Wang, J. C. (1979) *Proc. Natl. Acad. Sci, U.S.A.* 76, 200–203
11 Peck, L. and Wang, J. C. (to be published)
12 Reviewed in Wang, J. C. and Liu, L. F. (1979) in *Molecular Genetics* (Taylor, J. H., ed.), Part 3, pp. 65–88, Academic Press, New York
13 Gellert, M., Mizuuchi, K., O'Dea, M. H. and Nash, H. A. (1976) *Proc. Natl. Acad. Sci. U.S.A.* 73, 3872–3876
14 Liu, L. F., Liu, C.-C. and Alberts, B. M. (1979) *Nature (London)* 281, 456–461
15 Liu, L. F., Liu, C.-C. and Alberts, B. M. (1980) *Cell.* (in press)
16 Brown, P. O. and Cozzarelli, N R. (1979) *Science* 206, 1081–1083
17 Mizuuchi, K., Fisher, M., O'Dea, M. H. and Gellert, M. (1980) *Proc. Natl. Acad. Sci. U.S.A.* 77, 1847–1851
18 Reviewed in Cozzarelli, N. R. (1980) *Science* 207, 953–960
19 Kikuchi, Y. and Nash, H. A. (1979) *Proc. Natl. Acad. Sci, U.S.A.* 76, 3760–3764

James C. Wang is at the Department of Biochemistry and Molecular Biology, Harvard University, Cambridge, MA 02138, U.S.A.

Addendum – J. C. Wang

Supercoiling-induced cruciform formation has been demonstrated conclusively by the elegant experiments of Gellert et al. and Mizuuchi et al.[20,21]. The discovery of the left-handed 12-fold Z helical structure of alternating CG[22] has prompted several studies in which negative super-coiling induced flipping of such a sequence from the right-handed B structure to the left-handed Z structure has been demonstrated under physiological conditions[23,24]. Two publications on the use of the band-shift method to determine the dependence of the helical periodicity of DNA in solution have appeared[25,26]. In a completely different approach, the helical repeats have also been evaluated from the spacings of nuclease cleavage sites on DNA absorbed to flat surfaces[27,28]. The results obtained by these different methods are in excellent agreement. The gene *top*A encoding for *Escherichia coli* DNA topoisomerase I has been identified[29,30], and the complete nucleotide sequence of the gene has been determined[31]. It is now certain that *top*A is identical to a gene *Sup*X that was initially identified in *Salmonella typhimurium* as a mutation that reduces the level of transcription of the leucine operon[32]. A new type I topoisomerase from *E. coli* has also been reported[33].

References

20 Gellert, M., Mizuuchi, K., O'Dea, M. H., Ohmori, H. and Tomizawa, J. (1978) *Cold Spring Harbor Symp. Quan. Bio.* 43, 35–40
21 Mizuuchi, K., Mizuuchi, M. and Gellert, M. (1982) *J. Mol. Biol.* 156, 229–243
22 Wang, A. H.-J., Quigley, G. J., Kolpak, F. J., Crawford, J. L., Van Boom, J. H., van der Marel, G. and Rich, A. (1979) *Nature* 282, 680–686
23 Peck, L. J., Nordheim, A., Rich, A. and Wang, J. C. *Proc. Natl Acad. Sci. U.S.A.* (in press)
24 Wells, R. D., Hart, P. A., Kilpatrick, M., Klysik, J., Larson, J. E., Miglietta, J. J., Singleton, C. K., Stirdivant, S. M., Wartell, R. M. and Zacharias, W. (1982) *Cold Spring Harbor Symp. Quan. Biol.* 47 (in press)
25 Peck, L. J. and Wang, J. C. (1981) *Nature* 292, 375–377
26 Strauss, F., Gaillard, C. and Purnell, A. (1981) *Eur. J. Biochem.* 118, 215–222
27 Rhoades, D. and Klug, A. (1980) *Nature* 286, 573–578
28 Rhoades, D. and Klug, A. (1981) *Nature* 292, 378–380
29 Sternglanz, R., DiNardo, S., Voelkel, K. A., Nishimura, Y., Hirota, Y., Becherer, K., Zumstein, L. and Wang, J. C. (1981) *Proc. Natl Acad. Sci., U.S.A.* 78, 2747–2751
30 Trucksis, M. and Depew, R. E. (1981) *Proc. Natl Acad. Sci., U.S.A.* 78, 2164–2168
31 Tse, Y.-C. and Wang, J. C. (unpublished results)
32 Margolin, P., Overbye, K. M. and Basu, S. (1982) *Cold Spring Harbor Symp. Quan. Biol.* 47 (in press)
33 Krasnow, M., Dean, F., Otter, R., Pastorcic, M., Matzuk, M. M. and Cozzarelli, N. R. (1982) *Cold Spring Harbor Symp. Quan. Biol.* 47 (in press)

DNA unwinding enzymes

M. Abdel-Monem and H. Hoffmann-Berling

DNA unwinding enzymes use the energy released by ATP hydrolysis for separating the strands of a DNA. They are representatives of a large group of chemomechanically active DNA enzymes.

The idea that replicative unwinding of DNA is driven by reactions other than the dephosphorylation of the deoxynucleoside-triphosphates [1] has recently been strengthened by the identification of two new types of enzymes. The DNA unwinding enzymes separate the strands of a linear duplex [2,3] while DNA gyrase, in a reaction involving transient interruption in the helix, generates negative superhelical turns in a circular DNA [4]. The activity of both types of enzymes is coupled to the hydrolysis of the $\beta\gamma$-phosphate bond of ATP. A DNA unwinding enzyme and DNA gyrase together are theoretically capable of dissociating the strands of very long DNA molecules, including circular ones.

DNA unwinding ATPases have been found in bacteria, phage-infected bacteria, plant and animal cells. Two groups of these ATPases can be distinguished, DNA polymerase accessory enzymes and DNA helicases. Classified in the first group are the products of genes 44/62 and 45 of phage T4 [5] and the multifunctional gene 4 protein of phage T7 [6,7]. These enzymes cannot unwind the DNA except in the presence of the replicating viral DNA polymerase. Since only polymerase activity and not that of the unwinding ATPase is measured the actual mechanism of action of these ATPases has not been well documented. The same is true for two eukaryotic DNA unwinding enzymes [8,9]. They open only a limited length of a double strand.

DNA helicases are independent of the activity of other enzymes. Furthermore, they are able to dissociate the strands of a duplex completely. Four such enzymes have been isolated from *Escherichia coli* (Table I). DNA helicase I is a large fibrous peptide noted for its tendency to form aggregates [2,10]. DNA helicase II [11], *rep* protein, the product of the *rep* gene [3,12] and DNA helicase III (recently discovered) [13], are globular proteins existing either as monomers or dimers. A helicase dependent on gene *dda* of phage T4 is similar to DNA helicase II [14]. It is not identical to any of the T4 polymerase accessory proteins mentioned above. DNA helicase II is probably responsible for the major part of the DNA-dependent ATPase activity in *E. coli*.

The action of the DNA helicases can be measured in several ways. The unwinding of DNA is detected conveniently by the degradation of the denatured fraction of the DNA molecules with a single-stranded specific endonuclease followed by the determination of the amount of double-stranded DNA remaining.

Common aspects of DNA helicase action

Despite differences in the action of the helicases (see below) they share several common properties [12,13,15,16]. They exploit the hydrolysis of ATP for separating the DNA strands, not for recycling the enzyme (as appears to be the case with

TABLE I. DNA Helicases

Organism	Protein	Gene	Mol. wt	Molecular shape	Copies/cell
E. coli	DNA helicase I		180,000	fibrous	500–700
	DNA helicase II		75,000	globular	5000–8000
	rep protein	rep	65,000	globular	50
	DNA helicase III		2 × 20,000	globular	
T4	dda helicase	dda	56,000	globular	

DNA gyrase). They require a region of unpaired DNA to initiate DNA unwinding; DNA helicase I uses approximately 200 nucleotides, DNA helicase II and T4 DNA helicase use 12 nucleotides, and rep protein uses a nick. Binding to single-stranded DNA is independent of ATP. The enzymes unwind DNA unidirectionally, all in the 5' to 3' direction of the DNA strand to which they are bound (opposite to the action of a nucleotide polymerase) except for the rep protein which unwinds in the 3' to 5' direction. DNA helicase I, II, and T4 DNA helicase can unwind a continuous double strand of 119,000 base pairs (ligated T5 phage DNA) while rep protein can unwind a duplex of at least 6400 base pairs (double-stranded fd phage DNA). The rate of DNA unwinding measured for the three former enzymes is in the range of 1000 base pairs/s, comparable to the rate of advancement of a replication fork in E. coli.

Thus far, little is known about DNA helicase III.

Evidence for different mechanisms of DNA helicase action

Single strand binding enzymes can be expected to unwind the DNA in one of two ways: by active translocation of the enzyme on the DNA strand or by saturation of the strand with protein. Differences in the kinetics of the enzyme–DNA complex formation and in the stoichiometry of the enzyme–DNA reaction distinguish these two mechanisms from one another.

Based on these criteria, DNA helicase I unwinds DNA processively [10,15]. A group of 70 to 80 enzyme molecules appears to bind to a single-stranded portion of the DNA and, because of their tendency to aggregate, to form a unit. Following the supply of ATP these molecules are thought to migrate actively along the DNA thereby displacing a complementary strand (Fig. 1A). The high molecular weight and fibrous shape of the enzyme protein together with its tendency to aggregate, its slow dissociation from DNA and the apparently processive mode of action of the ATPase point to a similarity with myosin sliding along an actin filament.

DNA helicase II and T4 DNA helicase appear to unwind DNA non-processively [11,15]. Most enzyme adsorbs to the DNA after the start of DNA unwinding and the amount of enzyme needed is stoichiometric: one helicase II molecule can unwind approximately five base pairs and one T4 helicase molecule approximately two base pairs. To explain these findings it has been suggested that enzyme bound to a single-stranded DNA region promotes further binding of enzyme to the DNA. Complete dissociation of the DNA strands is thought to occur when one of them is saturated (Fig. 1B). The role of ATP hydrolysis in this reaction is not clear. Furthermore, since complexes between the enzymes and DNA dissociate rapidly it has not been possible to determine whether the enzymes bind cooperatively to DNA, as the given hypothesis predicts.

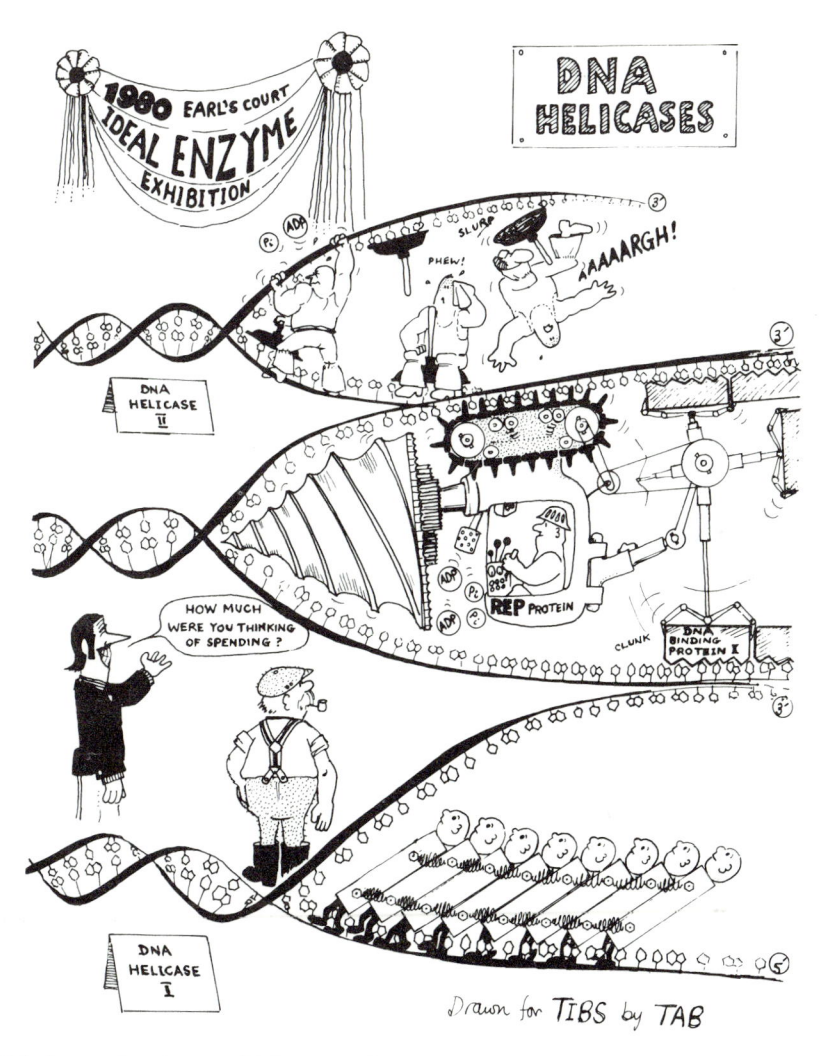

A third mechanism which can be regarded as a combination of the above two is used by *rep* protein [3,12] and possibly also by DNA helicase III [13]. The fact that only a few *rep* ATPase molecules are needed to unwind DNA suggests a processive mode of action of this enzyme. On the other hand, the presence of a stoichiometric amount of DNA binding protein I (the 'helix destabilizing' protein of *E. coli* [17,18]) is also necessary. This protein which is enzymatically non-active and one of several proteins required for single strand replication, binds preferentially to single-stranded DNA. Electron microscopy indicates that it saturates the unwound DNA strands thus keeping them apart (Fig. 1C) (the problem of spontaneous renaturation is not posed in DNA unwinding catalysed by DNA helicase I, II or T4 DNA helicase; the reason is probably the large number of ATPase molecules involved in these reactions). A. Kornberg and collaborators have assayed for the consumption of ATP in *rep*-promoted double strand replication of ϕX174 phage DNA (see below). They have found that two ATP molecules are hydrolysed for each nucleotide copied in the strand leading in DNA synthesis [12].

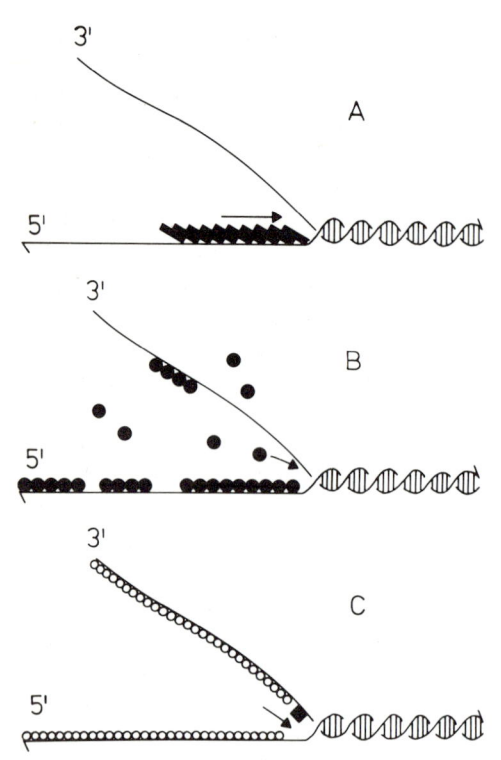

Fig. 1. Proposed mechanisms of DNA helicase action. (A) DNA helicase I (solid rectangles); (B) DNA helicase II or T4 DNA helicase (solid circles); (C) rep protein (solid rectangle), DNA binding protein I (open circles).

The different modes of action of DNA helicase II and *rep* protein are reflected, apparently, by their different concentrations in the cell (Table I).

Biological functions of DNA helicases

As shown in Table I genes are known for the *rep* function and the T4 *dda* helicase but not for DNA helicase I, II and III. *rep* Mutant *E. coli* cells are viable but are unable to propagate several bacteriophages including the small phages φX174 and fd [19]. *In vitro* studies have shown that *rep* protein is required to unwind the duplex DNA replicative forms (RF) of these phages during replication. Furthermore, studies on T4 *dda* mutants have demonstrated that they yield an almost normal burst of progeny phage. T4 *dda* mutants have thus contributed little to an understanding of the biological role of the *dda* enzyme [20].

Interestingly, recent *in vitro* studies [21] have shown that DNA replication in *E. coli*, when initiated in a closed circle, is inhibited by antibody directed against DNA helicase II. Cases studied include the bacterial DNA, phage λ DNA and ColE1 plasmid DNA. The *rep*-dependent replication of fd RF which is initiated in a nicked circle was not inhibited by the antibody. All these processes are known to be carried out largely by bacterial enzymes. None of them was inhibited by antibody against DNA helicase I.

Like the bacterial DNA, λ DNA and ColE1 DNA can replicate in *rep*-deficient cells. It is therefore of interest to ask whether replication in closed circles *in general* requires the activity of DNA helicase II as would replication in open (rolling) circles the *rep* protein.

Although no conclusion can be drawn at this stage there might be some reason for such a specialization. The replication cycle of *E. coli* DNA, λ DNA and ColE1 DNA in its early stage is sensitive to rifampicin suggesting that transcription of the DNA by RNA polymerase is needed [22]. In ColE1 DNA this synthesis of RNA is followed by the synthesis of a DNA leader fragment (catalysed by DNA polymerase I) [23]. Conceivably the early synthesis of RNA (or RNA and DNA) leading to the displacement of a loop provides a single-stranded DNA from which DNA helicase II (which requires a single-stranded region of 12 nucleotides) can initiate unwinding. The 5' to 3' direction of action of this enzyme requires that it attaches to the parental strand lagging in DNA synthesis (Fig. 2A).

rep Protein, on the other hand, can initiate unwinding at a nick (in the presence of the site-specific viral endonucleases which

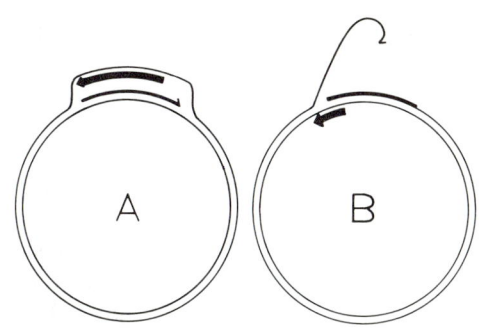

Fig. 2. Schemes of DNA replication. (A) Closed circle replication; (B) open (rolling) circle replication. Hooks represent 5' DNA termini, thick lines the newly synthesized RNA or DNA, and arrows the position and direction of action of DNA helicase II (A) or rep *protein (B). The given scheme (A) is oversimplified in the initiation of λ DNA replication since the transcription of both strands is required here* [29].

activate φX174 RF and fd RF for replication [3,12]). The *rep* protein is therefore preferable to DNA helicase II in promoting the replication of the viral RF molecules. *rep* Protein, acting in the 3' to 5' direction, has to bind to the parental strand leading in DNA synthesis (Fig. 2B).

This interpretation does not exclude the possibility of *rep* protein functioning in the replication of *E. coli* DNA or in other closed circle replication. In fact, *rep* mutant cells replicate their DNA at only half the rate of wild type cells [24]. Furthermore, in a system containing branched λ DNA molecules (to mimic replication forks), *E. coli* DNA replicase (the DNA polymerase III holoenzyme), DNA binding protein I and DNA helicase II, DNA synthesis is enhanced when *rep* protein is also present. In the absence of DNA helicase II, *rep* protein does not promote DNA synthesis (B. Kuhn, unpublished results).

Concluding remarks

DNA unwinding enzymes use the energy released by the hydrolysis of ATP for rotating the DNA strands around each other. The same holds for other ATP-dependent enzymes like the *rec*BC-type nucleases (which unwind the duplex as a step in the degradation of the DNA) [25], DNA gyrase and the recently characterized *E. coli rec*A gene product. *rec*A Protein, essential for general genetic recombination, catalyses the pairing of a DNA single strand with a homologous superhelical DNA – apparently by melting part of the double strand [26,27]. Furthermore, an ATPase capable of active translocation on the DNA is apparently the *dna*B gene product of *E. coli*. This enzyme is thought to provide a mobile promotor signal for the RNA-primed initiation of replication synthesis [28]. DNA unwinding enzymes are therefore representatives of a large class of mechanically active DNA enzymes.

The reason why so many different DNA unwinding enzymes exist is not clear. The identification of further genes coding for DNA unwinding enzymes in *E. coli* – and, accordingly, mutants in these genes – is essential for solving this problem.

References

1 Cairns, J. and Denhardt, D. T. (1968) *J. Mol. Biol.* 36, 335–342
2 Abdel-Monem, M., Dürwald, H. and Hoffmann-Berling, H. (1976) *Eur. J. Biochem.* 65, 441–449
3 Scott, J. F., Eisenberg, S., Bertsch, L. L. and Kornberg, A. (1977) *Proc. Natl. Acad. Sci. U.S.A.* 74, 193–197
4 Gellert, M., Mizuuchi, K., O'Dea, M. H. and Nash, H. A. (1976) *Proc. Natl. Acad. Sci. U.S.A.* 76, 126–130
5 Liu, C. C., Burke, R. L., Hibner, U., Barry, J. and Alberts, B. (1979) *Cold Spring Harbor Symp. Quant. Biol.* 43, 469–487
6 Hillenbrand, G., Morelli, G., Lanka, E. and Scherzinger, E. (1979) *Cold Spring Harbor Symp. Quant. Biol.* 43, 449–459
7 Richardson, C. C., Romano, L. J., Kolodner, R., LeClerc, J. E., Tamanoi, F., Engler, M. J., Dean, F. B. and Richardson, D. S. (1979) *Cold Spring Harbor Symp. Quant. Biol.* 43, 427–440
8 Hotta, Y. and Stern, H. (1978) *Biochemistry* 17, 1872–1880
9 Cobianchi, F., Riva, S., Mastromei, G., Spadari, S., Pedali-Noy, G. and Falaschi, A. (1979) *Cold Spring Harbor Symp. Quant. Biol.* 43, 639–647
10 Abdel-Monem, M., Lauppe, H. F., Kartenbeck, J., Dürwald, H. and Hoffmann-Berling, H. (1977)

J. Mol. Biol. 110, 667–685
11 Abdel-Monem, M., Dürwald, H. and Hoffmann-Berling, H. (1977) Eur. J. Biochem. 79, 39–45
12 Scott, J. F. and Kornberg, A. (1978) J. Biol. Chem. 253, 3292–3297
13 Yarranton, G. T., Das, R. H. and Gefter, M. L. (1980) J. Biol. Chem. (in press)
14 Krell, H., Dürwald, H. and Hoffmann-Berling, H. (1978) Eur. J. Biochem. 93, 387–395
15 Kuhn, B., Abdel-Monem, M., Krell, H. and Hoffmann-Berling, H. (1979) J. Biol. Chem. 254, 11343–11350
16 Yarranton, G. T. and Gefter, M. L. (1979) Proc. Natl. Acad. Sci. U.S.A. 76, 1658–1662
17 Sigal, N., Delius, H., Kornberg, T., Gefter, M. L. and Alberts, B. (1972) Proc. Natl. Acad. Sci. U.S.A. 69, 3537–3541
18 Alberts, B. and Sternglanz, R. (1977) Nature (London) 269, 655–661
19 Denhardt, D. T., Dressler, D. H. and Hathaway, A. (1967) Proc. Natl. Acad. Sci. U.S.A. 57, 813–820
20 Behme, M. T. and Ebisuzaki, K. (1975) J. Virol. 15, 50–54
21 Klinkert, M., Klein, A. and Abdel-Monem, M. (manuscript submitted)
22 Tomizawa, J. and Selzer, G. (1979) Annu. Rev. Biochem. 48, 999–1034
23 Staudenbauer, W. L. (1978) Curr. Top. Microbiol. Immunol. 83, 93–156
24 Lane, H. E. D. and Denhardt, D. T. (1975) J. Mol. Biol. 97, 99–112
25 MacKay, V. and Linn, S. (1976) J. Biol. Chem. 251, 3716–3719
26 McEntee, K., Weinstock, G. M. and Lehman, I. R. (1979) Proc. Natl. Acad. Sci. U.S.A. 76, 2615–2619
27 Shibata, T., DasGupta, C., Cunningham, R. P. and Radding, C. M. (1979) Proc. Natl. Acad. Sci. U.S.A. 76, 1638–1642
28 McMacken, R., Ueda, K. and Kornberg, A. (1977) Proc. Natl. Acad. Sci. U.S.A. 74, 4190–4194
29 Hobom, G., Grosschedl, R., Lusky, M., Scherer, G., Schwarz, E. and Kössel, H. (1979) Cold Spring Harbor Symp. Quant. Biol. 43, 165–178

Mahmoud Abdel-Monem and Hartmut Hoffmann-Berling are at the Max-Planck-Institut für Medizinische Forschung, 6900, Heidelberg, F.R.G.

Addendum – M. Abdel-Monem and H. Hoffmann-Berling

Recent results (M. Abdel-Monem, G. Taucher-Scholz and M. Klinkert, manuscript submitted) suggest a role for helicase I in bacterial conjugation: ATPase and DNA unwinding activity inhibitable by antibody against the enzyme are found in F$^+$ cells (bacteria bearing the E. coli F sex factor), not F$^-$ cells. The enzyme protein is further indistinguishable in size from the product of traI, one of the so-called transfer genes of the F factor. F$^-$ cells harboring this gene cloned in a plasmid vector liberate helicase I activity upon lysis, but they lack this activity following inactivation of the cloned gene by inserting a transposon into the gene (the plasmids [30] were constructed in the laboratory of Dr M. Achtman, Berlin). Helicase I is, therefore, probably the traI gene product of the F factor which would imply that the enzyme is involved for transferring DNA between conjugating cells.

Reference
30 Manning, P. A., Kuseck, B., Morelli, G., Fisseau, C. and Achtman, M. (1982) J. Bacteriol. 150, 76–88

The role of DNA topoisomerase I in transcription and transposition
by Martin D. Watson

Gene expression at the transcriptional level depends not only on RNA polymerase and the concentration of regulatory effectors, but also the structure of the DNA template. The chromosomes of both eucaryotes and procaryotes are supercoiled, i.e. the double helix itself forms a helix or superhelix. Superhelicity in eucaryotes is obtained by folding the DNA around the histone nucleosome core particles. Procaryotes do not have histones but achieve superhelicity by over- or underwinding the covalently closed circular chromosome, forming positive of negative supercoils, respectively. Negative supercoiling generates tension in the DNA duplex such that it tends to unwind. For a covalently closed circular chromosome supercoiling can only be achieved by transiently breaking the phosphodiester backbone and rotating the strands about each other. Enzymes that catalyse such actions are know as DNA topoisomerases[1,2] and can either induce negative supercoiling (DNA gyrase) or relax supercoiling

Many cellular functions involving DNA operate most efficiently when it is negatively supercoiled. It is, therefore, crucial that the degree of supercoiling is maintained. DNA topoisomerase I, also known as omega protein, is biochemically well characterised. It can relax supercoiled plasmid DNA, knot and unknot single stranded DNA, link and unlink rings of duplex DNA. However, its role *in vivo* is poorly understood. Sternglanz *et al.*[3] searched for mutations in the structural gene for DNA topoisomerase I by screening a bank of temperature-sensitive strains of *Escherichia coli*. The enzyme was assayed at the non-permissive temperature by using agarose gels to monitor the conversion of supercoiled phage PM2 DNA to its relaxed form. From 800 isolates, two, designated *top10* and *top250*, were found to have low topoisomerase activity. Mapping of the *top* gene showed that it and the temperature-sensitive mutation were unrelated and that *top* lies closely linked to the well characterized *cysB* gene at 28 min on the linkage map.

The proximity of *top* to *cysB* prompted a search for *top* deletions among strains known to be deleted in *cysB*. Four isolates with no detectable topoisomerase I activity were found, showing *top10* and *top250* mutants produce polypeptides with residual activity. The existence of *top* deletions indicates that DNA topoisomerase I is not essential for viability. The growth rates under various conditions were identical for isogenic *top*⁺ and *top*⁻ strains. The plating efficiencies of phages P1, T4 and T7 were unaffected. These results imply that DNA topoisomerase I is not involved in DNA replication unlike the *rep* protein and DNA helicase II.

top⁻ strains have increased sensitivity to UV light and the mutagen methyl methanesulphonate. Whether or not this increased sensitivity is due to a change in the conformation of the DNA or to the direct involvement of DNA topoisomerase I in a repair pathway is unknown.

The frequency of transposition of many transposons is lowered by the *top* mutations; Tn5, Tn9 and Tn10 are affected, but not Tn3. The plating efficiency of phage Mu is also reduced. This is perhaps not surprising as Mu is thought to replicate by a method analogous to transposition[4]. It is not known if the *top* product is directly involved in transposition, but the intertwining, unknotting and alignment of DNA strands are functions of the enzyme.

top mutations increase expression of

catabolite-sensitive operons. An increase in negative superhelicity, due *in vivo* to the disrupted balance between DNA gyrase and DNA topoisomerase I, causes an increase in transcription. The opening of promoters by increased negative superhelicity may allow initiation of transcription in the absence of the usual positive effector, cAMP/cAMP receptor protein. It is not known if operons under the positive control of regulatory systems other than cAMP/CRP are affected.

Thus a gene for one of the many DNA topoisomerases has been identified and the role of DNA topoisomerase I *in vivo* has been indicated. At present this role is still somewhat hazy, but appears to be one of maintaining the superhelicity of the chromosome at the required level. The existence of *top* deletions will allow the genes for other topoisomerases to be identified and so help explain why deletions in these genes is non-lethal.

References

1 Abdel-Monem, M. and Hoffmann-Berling, H. (1980) *Trends Biochem. Sci.* 5, 128–130
2 Cozzarelli, N. R. (1980) *Cell* 22, 327–328
3 Sternlanz, R., Di Nardo, S., Voelkel, K. A., Nishimura, Y., Hirota, Y., Becherer, L., Zumstein, L. and Wang, J. C. (1981) *Proc. Natl Acad. Sci. U.S.A.* 78, 2747–2751
4 Shapiro, J. A. (1979) *Proc. Natl Acad. Sci. U.S.A.* 76, 1933–1937

MARTIN D. WATSON

Department of Botany, University of Durham, Durham DH1 3LE, U.K.

Nucleosome structure

A. D. Mirzabekov

Nucleosomes are the elementary repeating units of chromatin and much of their structural organization has now been elucidated. A knowledge of nucleosome structure may lead to a better understanding of the organization and function of chromatin.

The discovery of nucleosomes in 1973–1974 by the joint efforts of several laboratories[1] marked a turning point in chromatin research. The nucleosomes are the elementary repeating unit of chromatin structure which underlies higher levels of chromatin organization. There is common agreement that the main role of nucleosomes is a structural one: to pack DNA in the nuclei of eukaryotic cells. However, this rather mundane concept of nucleosomal function has been challenged by recent experiments demonstrating that some regulation of chromatin activity may occur at the level of nucleosome structure[2].

Nucleosomes are prepared by mild digestion of nuclei with staphilococcal nuclease. Nucleosomal repeats in chromatin usually contain DNA about 200 base pairs long, but in different cell types this may vary from 160 to 240 base pairs[1]. In most cases, nuclease digestion of chromatin initially produces nucleosomes (termed also 'chromatosomes')[3], as stable discrete particles containing 160–180 (about 165, 175, and 185[4]) base pairs of DNA complexed with histone H1 and a histone octamer composed of two molecules of each histone H2A, H2B, H3, and H4. Further nuclease digestion converts the nucleosomes into a nucleosome core particle containing DNA 146 base pairs long and the histone octamer but no histone H1[1,2].

Nucleosome conformation

The most detailed data on the three-dimensional structure of core particles have so far been reported by the Cambridge group. Studies of core particle crystals using X-ray diffraction and electron microscopy have shown, with a resolution of about 20 Å, that the particle looks like a wedge-shaped disc with dimensions of about 110×57 Å[5]. The density maps of the protein and DNA moieties of the core particles have been produced by contrasting neutron diffraction from their crystals in 65% and 39% D_2O, respectively[6]. These data confirm the earlier suggestions[2,5] that DNA forms about one and three-quarter turns of a left-handed superhelix on the surface of the protein core. The conformation of the core particle is evidently organized by proteins since the wedge-shaped appearance of the protein core was found to be similar in the core particle and isolated histone octamer[7]. According to the image reconstruction analysis of electron micrographs of the histone octamer crystallized as regular tubes, the octamer is not tightly packed, contains many cavities and, probably, a hollow centre[7]. ^{31}P-NMR spectroscopy has not revealed any essential differences in the configuration of DNA within the core particle as compared to naked DNA. This indicates that nucleosomal DNA remains in the B-family of configurations and is more or less smoothly folded[8]. Nucleosomes and core particles have a similar shape and the nucleosomal DNA is assumed to be folded into about two full superhelical turns[9].

The DNA in nucleosomes is tightly packed and, in order to participate in transcription and replication, one would expect it to undergo some, as yet unknown, conformational transition. Conformational changes and the partial or complete unraveling of the core particle can be induced in many ways; for example, by elevating the temperature, urea and changes in the ionic strength. Another interesting property of the core particle is related to the ability of the histone octamer to slide along DNA or exchange DNA molecules at increased salt concentrations[2].

The arrangement of histones on DNA

Histones H2A, H2B, H3, and H4 of the core particle are rather similar, small and very basic proteins. Histone H1 is larger and differs from the other histones in several ways. For instance, basic amino acids residues are located mainly within the N-terminal third in core histones and in the C-terminal half of the H1 molecule.

DNA–histone interactions primarily involve salt bridges between the positively charged lysine and arginine residues in histones and the negatively charged phosphate groups of DNA. However, only about 50% of DNA phosphates are shielded with histones and are not available therefore for titration with, for example, Mn^{2+} (Ref. 10). In nucleosomes the unshielded regions are likely to be arranged on one, apparently the external, side of the DNA. This conclusion is based mainly on the following evidence: first, the sites of nucleosomal DNA that are accessible to digestion with DNase I and other nucleases[2] are arranged with a periodicity of about 10.4 bases along the DNA and seem to be located on one side of the DNA surface. This periodicity probably corresponds to the regularity of the DNA double helix[11]. Second, the location of the sites accessible to DNase I coincides well with the location of histone-free gaps in the DNA of the core particles as revealed in histone–DNA crosslinking experiments[12].

It is surprising that histones do not essentially block the grooves of the DNA double helix. For example, the minor groove of DNA is fully accessible and the major groove is shielded against dimethyl sulfate (a chemical probe[13,14]) by only about 15–20% in chromatin and nucleosomes. Dimethyl sulfate methylates the N3 of adenine and the N7 of guanine exposed within the minor and major grooves, respectively. The rather low shielding of the major groove by histones reflects the state of the whole DNA groove in the chromatin rather than the strong blocking of a few sites. Thus, no guanines have been found to be effectively protected against methylation in the core DNA[15] in the same kind of experiments that demonstrated the strong blocking of specific bases in the *lac* repressor–operator complex[16].

The comparatively low shielding of the DNA phosphate groups and both of its grooves by histones exposes the core DNA for interaction with other proteins. A remarkable example of such an interaction is found in the specific binding of the *lac*-repressor to operator DNA, covered with histones, within the reconstituted core particle[17].

Primary organization of nucleosomes

The method for sequencing the linear order of histones along the DNA in nucleosomes[12,14] is basically very similar to Maxam and Gilbert's procedure for DNA sequencing which was first proposed for adenine and guanine bases[16]. In both cases, the measured length of 5'-terminal DNA fragments (produced either by crosslinking histones to DNA through their lysine residues or by splitting DNA at specific bases) directly shows the position of crosslinked histone molecules or DNA bases, on the DNA. Both methods make use of the chemistry of DNA methylation with dimethyl sulfate, and the subsequent breaking of DNA by excised methylated purine bases leading to specific fragmentation[13].

The sequential arrangement of histones on DNA has been determined for both the core particle[12] and nucleosomes containing a DNA sequence of about 165–175 base pairs long[4] and is shown in Figs. 1–3. One can see in the generalized model for the core particle (Fig. 1) that the core histones are arranged in the same symmetric way on two DNA strands, which is consistent with the presence of a 2-fold axis of symmetry in the histone octamer[7]. Histones H2B, H3 and H4 evidently should be in a quite extended configuration, being bound to three turns of the DNA helix. It is of interest that the histone tetramer (H3$_2$, H4$_2$), that plays a key role in the core structure[1,2], covers mainly the central region and is also attached by histone H3 to both ends of the core DNA. Thus the tetramer clamps together one and three-quarter turns of the core DNA superhelix (Fig. 2) and may therefore fold the DNA to form a nucleosome-like particle[18]. Two histone dimers (H2A, H2B) flank the central DNA regions, but the first 20 nucleotides from the 5'-termini of each DNA strand are not covered with histones.

The detailed model of a nucleosome in Fig. 3 shows that core histones are bound

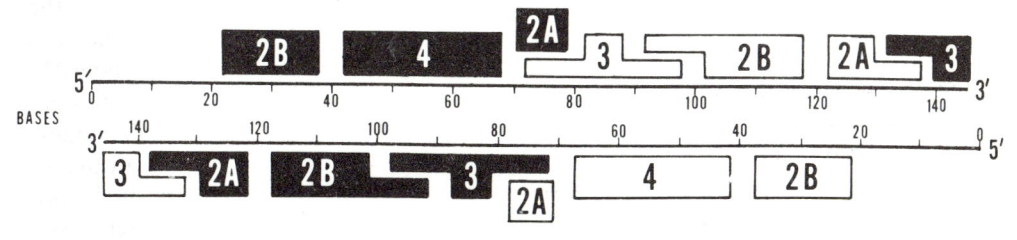

Fig. 1. A generalized model for symmetrical arrangement of eight histone molecules on two DNA strands in a core particle[12]. The position of each of two histone molecules 2A, 2B, 3, and 4 of the histone octamer is designated with black or white boxes. The order number of bases on each DNA strand is indicated with small figures from the 5'- to 3'-termini; the values of these figures have been measured to be indeed about 20, 41, 63, 82, 103, 124 and 142[11].

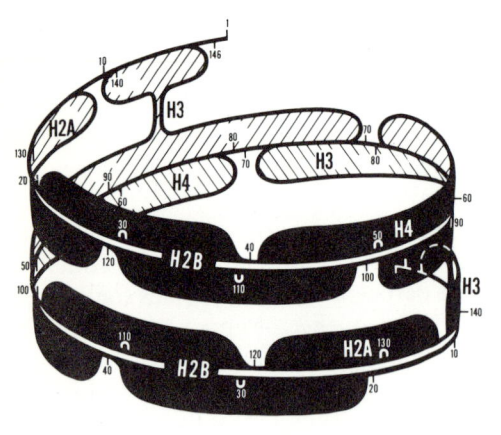

Fig. 2. A model for arrangement of histones on two DNA strands in a folded core particle[12]. DNA forms a left-handed superhelix containing about 80 base pairs per turn[5,6]. For the sake of simplicity, the histone-binding sites H2A(75), H2B(95), and H3(135) are not shown.

and thus stabilizes nucleosomes against unraveling induced, for example, by low ionic strength[9]. Secondary, weaker binding sites for histone H1 (probably for its extended basic C-terminal half) have also been identified along the whole length of DNA, within regions that are free of core histones and located apparently on the external side of nucleosomes[4].

Organization of histones inside nucleosomes

Histone–histone crosslinking in chromatin and nucleosomes yielded all of the ten possible dimers although, in most reports, the homodimers H2B–H2B and H4–H4 were found either in insignificant amounts or not at all[1,2]. As a promising approach to evaluating the internal nucleosomal structure and histone–histone contacts, zero-length crosslinks have been produced in the core particle between proline 26 in H2A and tyrosine 37, 40 or 42 in H2B[20] as well as between cysteins 110 in two H3 molecules[21].

The histone octamer is stable in 2 M NaCl but dissociates into a tetramer (H3$_2$, H4$_2$) and two dimers (H2A, H2B) upon lowering the salt concentration[2]. The polypeptide chains of the octamer either within the core particle or isolated, are composed of about 50% of α-helices, probably located beyond the basic N-terminal tails of histones, but no β-sheets[22].

within several discrete DNA segments, about six nucleotides in length, which are arranged with periodicity of about ten nucleotides and spaced with histone-free gaps[4,12]. The histone octamer is arranged in a similar way on the DNA in the symmetric core particle (Fig. 1), in the symmetric nucleosome containing about 165 base pairs[3,4] and in the asymmetric nucleosome with 175 base pairs (Fig. 3)). In nucleosomes, the histone octamer is stretched over 165–175 base pairs, a lot more than the 146 base pairs of the core DNA.

Histone–DNA crosslinking[4,19] and DNase I digestion[3] studies provide strong evidence that histone H1 simultaneously occupies both ends of nucleosomal DNA

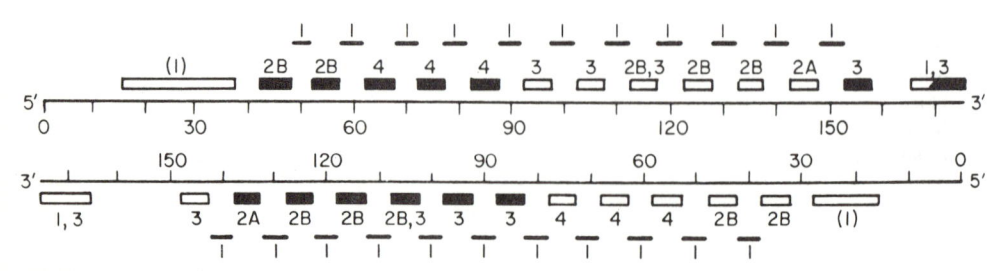

Fig. 3. A detailed model for arrangement of histones on two DNA strands in nucleosome containing DNA 175 base pairs long[4]. The binding sites for histones 2A, 2B, 3, and 4 and the main binding sites for histone 1 on the DNA ends are indicated with boxes; the weaker binding sites for histone 1 on the internal DNA regions within the gaps that are not occupied with core histones are shown with heavy lines.

Nucleosomes in transcribed chromatin

The increased sensitivity of transcribed regions of chromatin towards digestion with DNase I[23] and other nucleases is a striking structural feature of active chromatin. Isolated core particles retain this increased sensitivity which was found to stem from the presence of non-histone chromosomal proteins HMG-14 and HMG-17[24] associated with transcribed DNA sequences[25,26]. The analysis of core particles has revealed that they may reversibly bind two molecules of HMG-14/17 to the terminal DNA regions[27] where one DNA strand is free from histones (Fig. 1).

In conclusion, significant progress in our understanding of nucleosomal structure has been made within the last few years. This provides new and exciting possibilities for the study of the role of nucleosomes at a higher level of chromatin organization and in chromatin function.

Acknowledgement

The author is indebted to V. Kolesnick for composing the cartoon.

References

1. Kornberg, R. D. (1977) *Annu. Rev. Biochem.* 46, 931–954
2. McGhee, J. D. and Felsenfeld, G. (1980) *Annu. Rev. Biochem.* 49, 1115–1156
3. Simpson, R. T. (1978) *Biochemistry*, 17, 5524–5531
4. Belyavsky, A. V., Bavykin, S. G., Goguadze, E. G. and Mirzabekov, A. D. (1980) *J. Mol. Biol.* 139, 519–536
5. Finch, J. T., Lutter, L. C., Rhodes, D., Brown, R. S., Rushton, B., Lebitt, M. and Klug, A. (1977) *Nature (London)*, 269, 29–36
6. Finch, J. T., Lewit-Bentley, A., Bentley, G., Roth, M. and Timmins, P. (1980) *Philos. Trans. R. Soc. London* (in press)
7. Klug, A., Rhodes, D., Smith, J., Finch, J. T. and Thomas, J. O. (1980) *Nature (London)*, 287, 509–516
8. Cotter, R. I. and Lilley, D. M. (1977) *FEBS Lett.* 82, 63–68
9. Thoma, F., Koller, T. and Klug, A. (1979) *J. Cell. Biol.* 83, 403–428
10. Girardet, J. L. and Lawrence, J. J. (1979) *Nucl. Acids Res.* 7, 2419–2429
11. Prunell, A., Kornberg, R. D., Lutter, L., Klug, A., Levitt, M. and Crick, F. H. C. (1979) *Science*, 204, 855–858
12. Shick, V. V., Belyavsky, A. V., Bavykin, S. G. and Mirzabekov, A. D. (1980) *J. Mol. Biol.* 139, 491–517
13. Mirzabekov, A. D., Melnikova, A. F. (1974) *Mol. Biol. Reports* 1, 385–390
14. Mirzabekov, A. D., Shick, V. V., Belyavsky, A. V., Karpov, V. L. and Bavykin, S. G. (1977) *Cold Spring Harbor Symp. Quant. Biol.* 42, 149–155
15. McGhee, J. and Felsenfeld, G. (1979) *Proc. Natl Acad. Sci. U.S.A.* 76, 2133–2137
16. Gilbert, W., Maxam, A. and Mirzabekov, A. D. (1976) in *Control of Ribosome Synthesis. The Alfred Benzon Symp. IX* (Kjeldgaard, N. C. and Maaloe. O. eds), pp. 139–148, Copenhagen; Munksgaard
17. Chao, M. V., Gralla, J. D. and Martinson, H. G. (1980) *Biochemistry*, 19, 3254–3260
18. Thomas, J. O. and Oudet, P. (1979) *Nucl. Acids Res.* 7, 611–623
19. Sperling, J. and Sperling, R. (1978) *Nucl. Acids Res.* 5, 2755–2773
20. DeLange, R. J., Williams, L. C. and Martinson, H. G. (1979) *Biochemistry*, 18, 1942–1946
21. Camerini-Otero, R. D. and Felsenfeld, G. (1977) *Proc. Natl Acad. Sci. U.S.A.* 74, 5519–5523
22. Thomas, G. J., Prescott, B. and Olins, D. E. (1977) *Science*, 197, 385–388
23. Weintraub, H. and Groudine, M. (1976) *Science*, 193, 848–856
24. Weisbrod, S., Groudine, M. and Weintraub, H. (1980) *Cell*, 19, 289–301
25. Levy, B., Connor, W. and Dixon, G. H. (1979) *J. Biol. Chem.* 254, 609–620
26. Bakayev, V. V., Schmatchenko, V. V. and Georgiev, G. P. (1979) *Nucl. Acids Res.* 7, 1525–1540
27. Sandeen, G., Wood, W. I. and Felsenfeld, G. (1980) *Nucl. Acids Res.* 8, 3757–3778

A. D. Mirzabekov is at the Institute of Molecular Biology, Academy of Sciences of the U.S.S.R., Moscow, U.S.S.R.

Models of DNA transposition

A. I. Bukhari

An exciting development in molecular genetics in the 1970s has been the realization that spatial arrangements of genetic material are not inherently permanent. The genetic material may undergo frequent changes and different DNA segments may fuse or dissociate by mechanisms occurring commonly in nature. Transposable elements are the main agents of such DNA–DNA interaction. These elements are specific segments of DNA that are normal constituents of the genome of an organism and are yet movable or transposable to different sites. They exhibit extraordinary recombinational versatility. They are endowed with the ability to join or disjoin different DNA segments.

The history of transposable elements can be traced to the pioneering work of McClintock, who studied high frequency genetic changes in maize[1]. The precise molecular nature of the transposable element has, however, only become clear from studies on genetic rearrangements in bacteria (see Ref. 2) and more recently from studies on movable elements in yeast and Drosophila[3]. Transposable elements in bacteria include the IS elements, transposons and bacteriophage Mu. IS elements are simple units of DNA (many are about 1200–1400 base pairs in length) found in various numbers in bacterial chromosomes; transposons are more complex structures (several thousand base pairs in length) and are made up of one or more genes. Bacteriophage Mu which has many genes in its 37 kilobases of DNA undergoes transposition at a very high frequency in the host cell, and causes all the genetic rearrangements characteristic of a transposable element. It is in effect a giant transposon.

Genetic changes, brought about by transposable elements, can have dramatic biological consequences. Transposons for drug resistance carried by plasmids render bacterial species resistant to drugs. In the yeast *Saccharomyces cerevisiae* the cells can change their mating types by transposition of the mating type genes[4]. In higher organisms rearrangements of antibody genes can lead to differentiated cells that make different antibodies (see Ref. 3). What are the underlying principles of this genetic flexibility?

Central observations on transposable elements

(1) The specificity of transposable elements lies in their ends. What distinguishes one transposable element from another is the base sequences at their ends. These sequences which are faithfully preserved are apparently recognized by proteins during the transposition process. Most transposable elements have been found to have repetitious sequences generally arranged in an inverted order, at the ends.

(2) Transposition generates small duplications of host DNA at sites of insertion. All of the elements studied so far cause a small base pair sequence in the host DNA to be repeated on either side of the inserted element (Fig. 1). The duplications range from 3 to 12 base pairs but each element duplicates a characteristic number of host base pairs (see Refs 5,6). A popular hypothesis is that duplications result from staggered cuts in host DNA during insertion of the element (Fig 1, Ref. 7). Since one transposable element can integrate at many different sites in host DNA, the duplications generated also differ from site to site. Most elements fall in two general

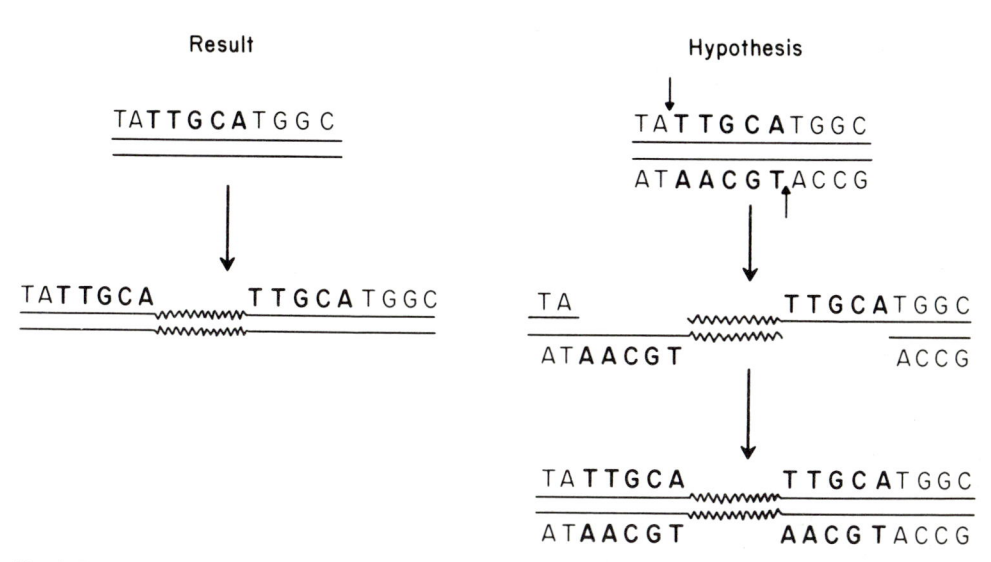

Fig. 1. Duplication of host base pairs at the target site. Duplication of a hypothetical 5 base pairs 'TTGCA' is shown. The transposable element is shown by a wavy line. On the right is depicted the staggered cut hypothesis for generation of such a duplication[7]. The element is ligated to the protruding ends of host DNA and the gaps are filled in to give the duplications.

groups; those that cause duplications of 5 base pairs and those that cause duplications of 9 base pairs of host DNA. The 5 base pair group includes Tn3 (a transposon for ampicillin resistance) γδ sequence, Mu, and Ty1 element of yeast. The 9 base pair group includes IS1 (a prokaryotic insertion sequence), Tn10 (a transposon for tetracycline resistance), and Tn5 (a transposon for kanamycin resistance). Some transposable sequences in *Drosophila melanogaster* also cause small duplications, as does the integration of DNA proviruses of RNA tumor viruses.

(3) Transposable elements cause rearrangements in adjacent host DNA. The elements can cause deletions and inversions of host DNA. A predominant class of deletion is one in which the deletions begin at either end of the inserted element and remove various amounts of adjacent genetic material[5]. The deletions generated by IS1, for example, remove one set of the duplicated 9 base pair host sequences. Thus the ends of the elements are the active structures in the generation of deletions.

(4) Transposable elements cause fusion of replicons generating co-integrates. One of the primary consequences of the activity of the transposable element is the fusion of different DNA molecules. Thus, when a transposable element on a plasmid moves to the chromosomal DNA it takes the plasmid along with it (the transposable element and the plasmid are co-integrated). The transposable element in the co-integrate structure is repeated in a direct manner at the end of the inserted plasmid. To date every transposable element in prokaryotes generates co-integrate structures.

(5) Some transposable elements cause the generation of heterogeneous circles containing host DNA and the element. Formation of heterogeneous circles has been studied mainly with bacteriophage Mu since genetic rearrangements occur at a very high frequency during the lytic growth of the phage (see Ref. 8). During its lytic cycle, as Mu actively transposes itself, covalently closed circles of heterogeneous lengths are generated. These circles con-

tain host DNA of various lengths linked to at least one copy of Mu DNA; the high frequency of these genetic rearrangements contributes to the cell death. Such circles may be formed by all transposable elements but it would not be simple to detect them because of very low frequencies of transposition of other elements as compared to Mu.

Mechanism of transposition

There are two basic models of integration of a genetic element. In the well-known Campbell model for λ integration, a circular form of λ DNA undergoes recombination with the host chromosome; the recombination, brought about by specific proteins, occurs between specific sites on the chromosome and the circular phage DNA resulting in the linear insertion of λ DNA into the chromosome. Under appropriate conditions reversal of this reaction results in excision of λ DNA. There is no net synthesis of DNA during integration or excision[9]. In summary λ + Host (H) = λH.

By contrast Ljungquist and Bukhari, who examined prophage Mu DNA after induction, found that the prophage was not excised but had moved to different places in the host chromosome[10]. They inferred that replicas of the original Mu prophage are inserted at new sites and that replication of Mu is required for transposition. In most, if not all, transposable element (TE) systems the element duplicates itself in the process of transposition[5,6,11,12]. In summary, Host 1. TE + Host 2 = Host 1. TE + Host 2. TE.

Various molecular models have been proposed to account for the duplication of the element during transposition. These

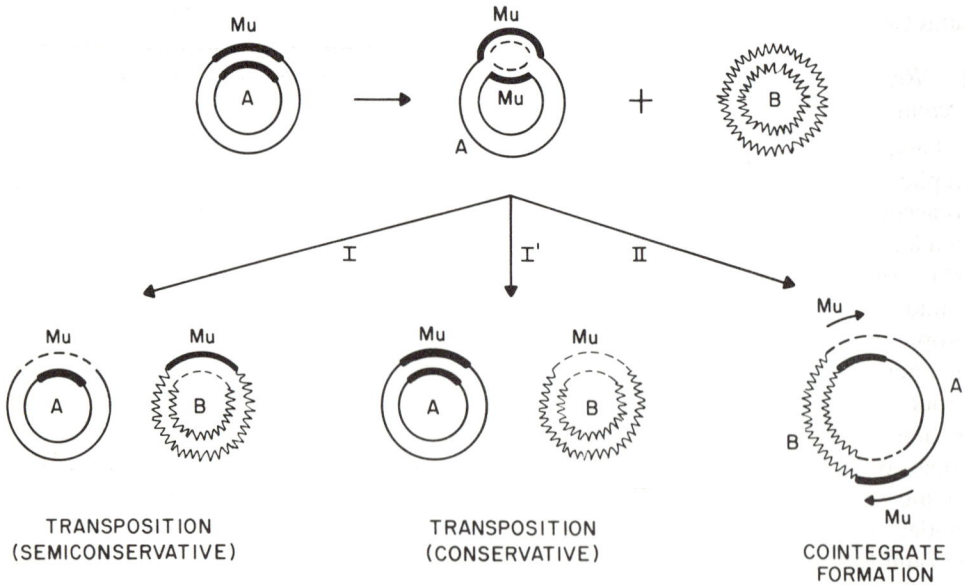

Fig. 2. The prototype 'replication first – recombination second' model. The model shown has been modified from Toussaint et al.[14]. Mu (thick lines) in host DNA 'A' is replicated (new strands are shown by broken lines) and recombines with the recipient DNA 'B'. If the two ends of one copy of Mu recombine, the result is transposition (I). If only the new strands recombine, the result is transposition, with both the parental strands remaining in the donor molecule (I') (Leach and Symonds model). If one end from one copy of Mu and the other end from the second copy of Mu act together to recombine with the recipient DNA the result will be a cointegrate (II). If this type of recombination occurred within the same host DNA, the result will be a deletion (generating a circle with Mu) (see Ref. 14).

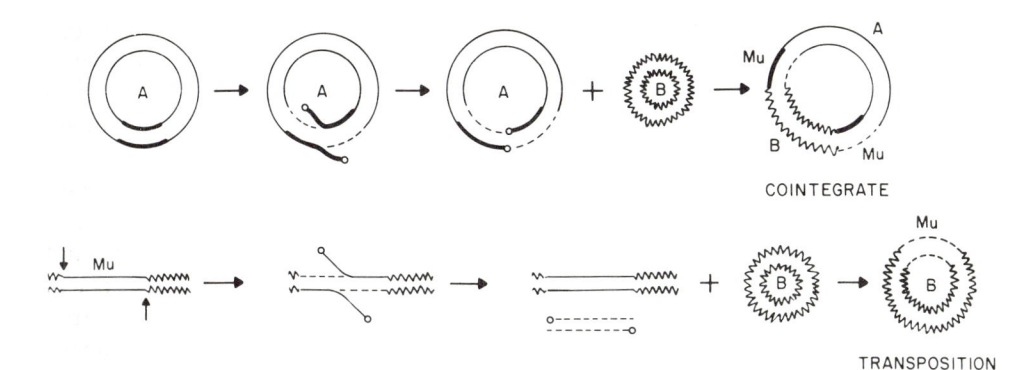

Fig. 3. Displacement synthesis and generation of the integrative intermediate. The diagram is based on the proposals of Kamp and Kahmann; and Mizuuchi. (Upper panel) Cointegrate formation (Mu = thick lines). Displacement synthesis finally results in head to tail tandem copies held by proteins. The ends held by proteins are active in recombination. (Lower panel) Transposition from linear Mu DNA (Mu = straight lines). A copy of Mu DNA consisting of only newly synthesized strands is produced, the ends of which have the proteins bound (the proteins are transferred from the old strands to the new strands).

models can be divided into two groups; those in which the replication of the transposable element occurs before recombination at the target site in host DNA, and those in which the element attaches to the target site first, and then replicates.

(i) *'Replication first – recombination second' models*

Faelen et al.[13] proposed a prototype of 'replication, then recombination' models to account for the Mu-mediated insertion of a λ *gal* transducing phage at the site of a Mu prophage. Upon integration, λ *gal* is found sandwiched between the two Mu prophages in the same orientation (in the manner of a co-integrate structure). They proposed, therefore, that the Mu prophage is first duplicated, the ends of the two copies of Mu recombine with λ giving rise to the sandwich structure (Fig. 2). This model was not originally proposed to account for Mu transposition. However, it can be readily modified to explain the integration of Mu at a new site[14] (Fig. 2). When a prophage is duplicated, one copy can be integrated at a new site, leaving behind one copy. Is it only newly replicated strands (conservative segregation of replicas), or a copy made up of one old and one new strand, that move? The prototype model has recently been favored by Leach and Symonds[3] to explain co-integrate formation, but for simple transposition they proposed insertion of only newly replicated strands at a new site (Fig 2).

An essentially conservative segregation model has also been proposed by Kamp and Kahmann (see Ref. 3). They proposed a strand-displacement mechanism for Mu DNA synthesis with a new copy being released with the integration proteins bound at the ends (Fig. 3). These proteins facilitate the insertion of the replicas at a target site. To account for co-integrate formation, they suggest interaction between the newly replicated copies as shown in the upper panel of the figure. An essentially similar model was also proposed by K. Mizuuchi, at the 1980 Cold Spring Harbor Symposium (personal communication).

Fitts and Taylor[15] have also suggested a conservative segregation mechanism that results in the release of a single strand. The single strand then joins the replication fork and is ligated to the replicating strand of host DNA. This radically different model relies entirely upon the normal host DNA replication to complete the insertion reaction. Unlike other models, it proposes that 3-12 base pair duplications arise from a distortion in the host DNA synthesis when a single strand is ligated to the replicating strand of host DNA. Co-integrate formation also involves insertion of a single strand of the entire donor molecule which must then be replicated to generate a co-integrate structure.

Finally, Jaskunas and co-workers[16] have proposed that an element is first replicated to produce a tandem duplication which can then recombine to give rise to a circular intermediate able to insert at a new site. If there is no recombination between the tandem duplicated copies, and the junction of the elements acts as a recombination site, a co-integrate can result.

(ii) *'Recombination first – replication second' models*

The intimate association of replication and transposition has led to models in which the parental transposable element replicates while attached to the new target site. Grindley and Sherratt[7] formally extended the observations on the Mu replication–integration cycle by proposing that one of the ends of a transposable element is first transferred to one 5' end of the recipient DNA, which has been cut in a staggered manner. The 3' OH free group of the host DNA is then used as a primer to copy the strand of the element ligated to the host DNA, displacing the complementary strand. After the replication of the strand is completed, the other 3' OH extension of the host DNA is used as a primer with the newly synthesized strand as a template. The parental strand transferred originally to the target site returns to its homologue at the donor site. The process amounts to a conservative segregation of the replicas.

Shapiro[17] proposed that both ends of the element get ligated to the target site (Fig. 4). The target site is cut in a staggered manner, both ends of the element are nicked and are then transferred to the target site, replication occurs from both ends and when it is completed it results in the formation of co-integrates. The co-integrates then are resolved by recombination between the two copies of the element. The recombination may be specific, as in Tn*3*, or resolution may occur by generalized recombination between the two elements. (It should be noted that Heffron, Gill and Falkow[18] first pointed out that in Tn*3* co-integrates may be the intermediates in transposition.) Shapiro's model readily explains co-integrates. It also explains the generation of heterogeneous circles if transposition occurs within the same replicon (in much the same manner

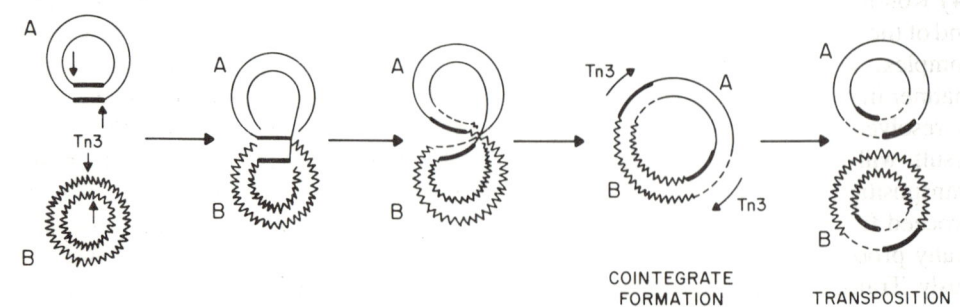

Fig. 4. Shapiro model of cointegrate formation. The donor molecule A (containing Tn3, indicated by thick lines in this example) recombines with the recipient B at both ends in such a manner that an end of a Tn3 strand is ligated to a host DNA strand, the opposite end of the second Tn3 is ligated to the other host DNA strand. The molecules are now fused, the transposable element has not yet replicated and four nicks are present in the structure. Once the replication of the element is completed all nicks are closed. This would give rise to a co-integrate structure, which has to be resolved by recombination within the elements to give transposition.

as the model in Fig. 2 does). The manner in which the ends of element are ligated to the staggered target site within the same replicon (that is, which strand of the element is ligated to which strand of the host DNA) would determine whether a heterogeneous circle with a copy of the element is formed, or whether transposition is accompanied by inversion of the host sequences in between the two copies (see Refs 17 and 20 for detailed explanation). Essentially the same model was also proposed by Arthur and Sherratt[19].

Harshey and Bukhari[20] have recently suggested a mechanism of integrative replication, based on DNA structures observed in the E.M. during Mu DNA transposition, which simplifies the earlier models.

As shown in Fig. 5 they envisage four steps: (1) A protein-mediated *association* is brought about between the donor molecule and the target DNA. (2) One end of the element is nicked and *attached* to a site that undergoes a double-stranded cleavage. (3) Roll-in Replication. While one strand of the target DNA is linked to the nicked strand of the element, the complementary strand of the target DNA is used as a primer for replication into the element, such that the replicating DNA is threaded through the replication complex. (4) Roll-in Termination. When the distal end of the element arrives at the replication complex, replication is terminated. The manner in which the termination structure is resolved, will determine whether the result will be a co-integrate or simple transposition. In this model replication can proceed from either end; it does not normally proceed from both ends simultaneously. Transposition within the same replicon would result either in an inversion of a chromosomal segment or a deletion accompanied by the formation of a heterogeneous circle. Furthermore, the roll-in replication explains a generation of tandem repeats, that is, amplification of circular sequences as shown in Fig. 6.

Klar *et al.*[21] have proposed a 'recombination first – replication second' model for substitution of mating type genes in *Saccharomyces cerevisiae* in which the DNA strands from the silent loci invade the mating type locus, a copy is then synthesized and replaces the old information in the manner of gene conversion.

Assessment of the models

The prototype 'replication first – recombination second' model, of Toussaint, Faelen and Bukhari[14], and both the Shapiro and Harshey and Bukhari models, differ little in terms of the consequences of transposition. These models can explain co-integrate formation, as well as formation of the heterogeneous circles. The 'replication first – recombination second' model uses the two ends of the transposable element at the target site of host DNA simultaneously for recombination (three point recombination). However, the model is vague, both in terms of replication and recombination. Shapiro also uses three point recombination, the only difference being that replication occurs after the recombinational step. Harshey and Bukhari describe in more detail the manner of replication and postulate an initial two point recombination between one end of the element and the target site. Recombination at a third point occurs at the end of the replication cycle. The idea that the ends of a transposable element are both the sites for recombination and for initiation for replication stems from the observations that the middle part of the elements can be removed without affecting their replication –transposition (if the necessary proteins are provided).

'Recombination first – replication second' models were proposed because no replicating forms of Mu DNA free of host DNA can be seen[10,20,22,23]. If replication is

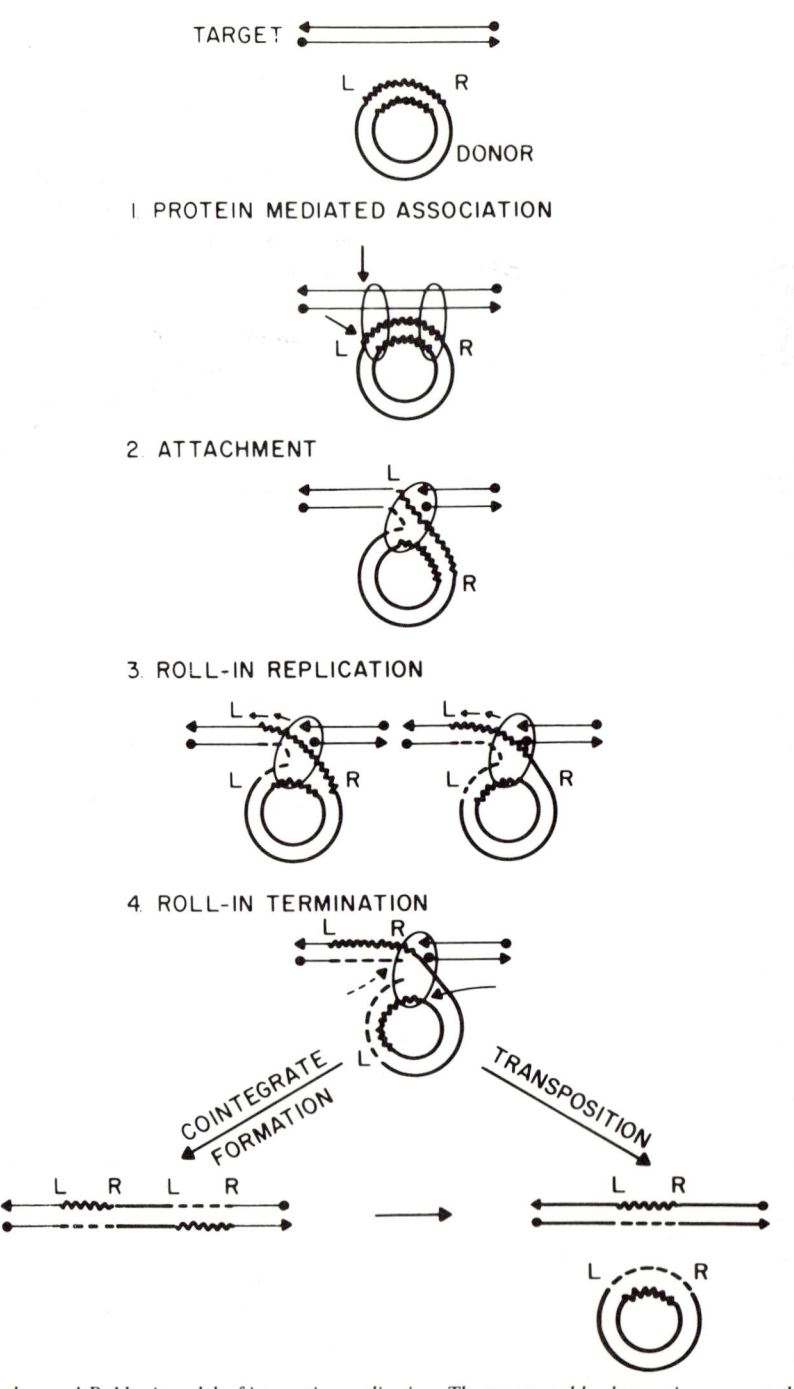

Fig. 5. Harshey and Bukhari model of integrative replication. The transposable element is represented by wavy lines. Cleavage sites are indicated by arrows. L and R are the ends of the element. Proteins (shown by circles) hold the DNA complex in place. DNA is replicated at a fixed complex as it reels through it. At termination, a nick is made on the parental strand opposite to the one (solid arrow at right) on which the initiating nick was made and is ligated to the free end of the target strand — this would generate a co-integrate structure. If, however, the 3' end of the newly synthesized strand (broken arrow at left) is recognized for ligation this would result in direct transposition of the element.

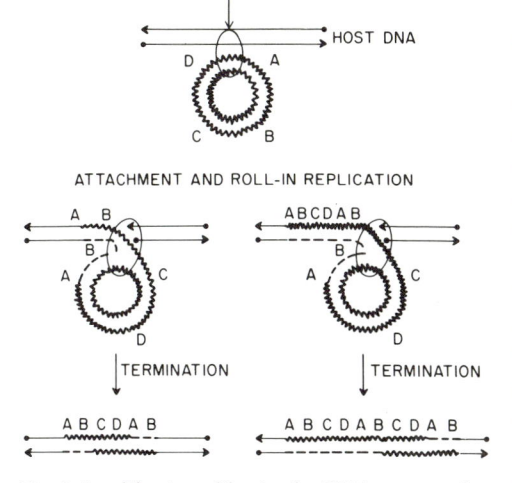

Fig. 6. Amplification of the circular DNA sequences by the roll-in mechanism. The replication process is the same as in Fig. 5, but the replication termination site is not fixed. This results in insertion of tandem copies.

completed first, then we might expect to see these structures before they recombine with the target site. But if the half-life of the integrative intermediate that arises from replication is so short as to preclude the possibility of its isolation it would be a major technical feat to decide between the molecular models. It should be noted, however, that the structures described by Harshey and Bukhari[20] are apparently compatible only with the models in which the element is first attached to the target site by recombination, then follows replication and recombination is completed after the replication.

One of the essential features of Shapiro's model is first the formation of co-integrates and then their resolution by recombination. The Harshey and Bukhari model is essentially a one-step model in which the end result can be either a co-integrate or a simple transposition of the element. If Shapiro's model is modified such that the recombination which resolves the co-integrates occurs at the very ends of the element, then the model can no longer be distinguished from various others which are based on recombinational events at the ends (for example, Faelen et al.; Harshey and Bukhari models).

The element duplication model (the general Mu model) is gaining wide acceptance, although molecular details of the duplicative reaction remain to be worked out. It is generally believed that for most transposable elements excision of the element is not a prerequisite for transposition. We can imagine, however, that some elements can undergo λ type excision and integration, that is, they have evolved mechanisms by which they can be precisely excised at a high frequency and then reinserted at another site. Clear-cut examples of such transposable elements have not been found in prokaryotes so far.

References

1 McClintock, B. (1957) *Cold Spring Harbor Symp. Quant. Biol.* 21, 197–216
2 Bukhari, A. I., Shapiro, J. and Adhya, S. (eds) in *DNA Insertion Elements, Plasmids and Episomes*, Cold Spring Harbor Laboratory
3 *Cold Spring Harbor Symp. Quant. Biol.* Vol XLV. (1980) *Movable Genetic Elements*, Cold Spring Harbor Laboratory (in press)
4 Hicks, J., Strathern, J. and Klar, A. (1979) *Nature (London)* 282, 478–483
5 Calos, M. and Miller, J. H. (1980) *Cell* 20, 579–595
6 Starlinger, P. (1980) *Plasmid* 3, 241–259
7 Grindley, N. D. F. and Sherratt, D. (1979) *Cold Spring Harbor Symp. Quant. Biol.* 43, 1257–1261
8 Bukhari, A. I. (1976) *Ann. Rev. Genet.* 10, 389–412
9 Campbell, A. (1976) *Sci. Am.* 235, 102–113
10 Ljungquist, E. and Bukhari, A. I. (1977) *Proc. Natl. Acad. Sci. U.S.A.* 74, 3143–3147
11 Bukhari, A. I. (1977) *Brookhaven Symp.*, Associated Universities, Inc., pp. 218–232
12 Bukhari, A. I., Ljungquist, E., de Bruijn, F. and Khatoon, H. (1977) in *DNA Insertion Elements, Plasmids and Episomes*, pp. 249–261, Cold Spring Harbor Laboratory
13 Faelen, M., Toussaint, A. and DeLafonteyne, J. (1975) *J. Bact.* 121, 873–882
14 Toussaint, A., Faelen, M. and Bukhari, A. I. (1977) in *DNA Insertion Elements, Plasmids and Episomes*, pp. 275–285, Cold Spring Harbor Laboratory
15 Fitts, R. A. and Taylor, A. (1980) *Proc. Natl.*

Acad. Sci. U.S.A. 77, 2801–2805
16 Holly, R. A., Das Sarma, S. and Jaskunas, R. (1980) *Proc. Natl. Acad. Sci. U.S.A.* 77, 2514–2518
17 Shapiro, J. A. (1979) *Proc. Natl. Acad. Sci. U.S.A.* 76, 1933–1937
18 Gill, R., Heffron, F. and Falkow, S. (1978) *J. Bact.* 136, 742–756
19 Arthur, A. and Sherratt, D. (1979) *Mol. Gen. Genet.* 175, 267–274
20 Harshey, R. and Bukhari, A. I. (1981) *Proc. Natl. Acad. Sci. U.S.A.* (in press)
21 Klar, A., McIndoo, J., Strathern, J. and Hicks, J. (1980) *Cell* 22, 291–298
22 Ljungquist, E. and Bukhari, A. I. (1979) *J. Mol. Biol.* 133, 339–357
23 Chaconas, G., Harshey, R. M. and Bukhari, A. I. (1980) *Proc. Natl. Acad. Sci. U.S.A.* 77, 1778–1782
24 Farabaugh, P. J. and Fink. G. R. (1980) *Nature (London)* 274, 765–769

A. I. Bukhari is at the Cold Spring Harbor Laboratory, P.O. Box 100, Cold Spring Harbor, New York 11724 U.S.A.

Mitochondrial genomes

The petite mutation in yeast

Giorgio Bernardi

The cytoplasmic 'petite' mutation in yeast, which is characterized by the loss of respiratory functions, is caused by large deletions in the mitochondrial genome. The 'petite' mutants arise spontaneously when a segment of the wild-type mitochondrial genome, excised by an illegitimate, site-specific recombination process, is amplified to form a defective mitochondrial genome.

In 1948, the Centre National de la Recherche Scientifique, organized a colloquium entitled 'Unités Biologiques Douées de Continuité Génétique' (Biological Units Endowed with Genetic Continuity). Thirty years later, the proceedings still make most interesting reading. Among the contributions of several outstanding biologists including Lwoff, Delbrück and Monod, there is a paper which can be considered the starting point of extrachromosomal genetics. In this paper, Boris Ephrussi gave the first account of investigations, started three years earlier, on the 'petite colonie' mutation in *Saccharomyces cerevisiae* [1]. It is difficult to describe the initial observation more clearly than in Ephrussi's own words: 'When a culture of baker's yeast, whether diploid or haploid, is plated, each of the cells gives rise in the course of the next few days to a colony. The great majority of these colonies are of very nearly identical size, but one usually finds also a very small number – say 1 or 2% – of distinctly smaller colonies (Fig. 1). These facts suggest that the population of cells which was plated was heterogeneous and that it may be possible to purify it by taking cells from either the big or the small colonies only. The results of such a selection show, however, that cells from the big colonies again and again produce the two types of colonies, while the cells from the small colonies give rise to small colonies only' [2]. Besides describing the mutation and its irreversibility, this paper also reported a number of fundamental observations: (a) that acriflavine treatment increases the number of 'petite' mutants from 1–2 to 100% (Fig. 1); (b) that the mutants grow slowly because they cannot respire, owing to the loss of their ability to synthesize a whole series of respiratory enzymes; and that in anaerobiosis wild-type cells and petite mutants grow at the same slow rate using fermentative pathways; (c) that crosses of wild-type with petite mutants show a non-Mendelian segregation of the mutation, in that they lead either to wild-type progeny exclusively or to both wild-type cells and petite mutants in different proportions; the petite mutants entering the cross were called neutral in the first case and suppressive in the second one [3]. The conclusion drawn by Ephrussi was that wild-type cells and petite mutants differed by 'the presence in the former and the absence in the latter of cytoplasmic units endowed with genetic continuity and required for the synthesis of certain respiratory enzymes' [2].

The petite mutation is due to gross alterations of mitochondrial DNA

The cytoplasmic units postulated by Ephrussi in 1948 were only identified as mitochondrial genes 20 years later. Indeed, the first hard facts about the molecular basis of the petite mutation were published in 1968 [4,5], when mitochondrial DNAs

This paper is dedicated to the memory of Boris Ephrussi.

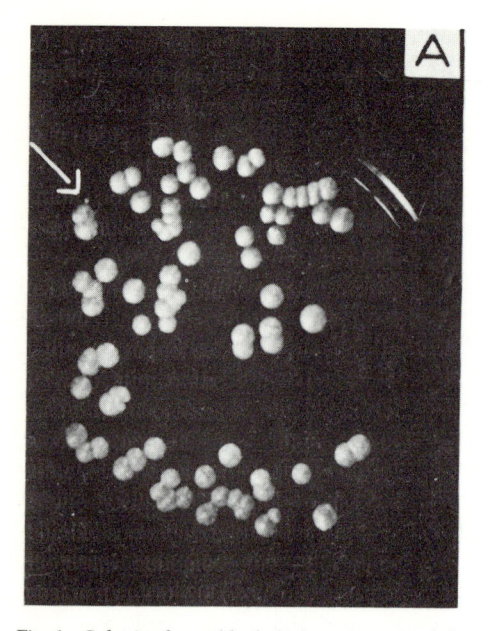

Fig. 1. Colonies formed by baker's yeast on a solid medium. (A) Colonies of a normal yeast, showing one small colony. (B) Colonies formed by the same yeast grown prior to plating in the presence of acriflavine [2].

from two genetically unrelated, acriflavine-induced, petite mutants were shown to have a grossly altered base composition (GC = 4%) compared to DNA from the parent wild-type strain (GC = 18%). These findings unequivocally established that massive alterations in the nucleotide sequences of the mitochondrial genome may accompany the petite mutation and be responsible for it; conclusions that were quickly confirmed by more detailed investigations, which probed the structure of mitochondrial DNA from both wild-type cells and petite mutants [6,7].

The AT spacers and the deletion hypothesis

Mitochondrial DNA from wild-type yeast cells was found to be extremely heterogeneous in base composition, about half of it melting at a very low temperature and being almost exclusively formed by long stretches of short alternating and non-alternating AT:AT and A:T sequences, and the rest melting over an extremely broad temperature range. Interestingly, the existence of two sorts of AT sequences in the AT-rich stretches (later called AT spacers) had been predicted for the first petite genome investigated [4]. Compared with mitochondrial DNA from wild-type cells, DNAs from three spontaneous suppressive petite mutants [6,7] were shown to have lower amounts of GC, to lack a number of components that melt at high temperatures and to renature very rapidly. At that time, I interpreted these results as indicating that petite mutants had defective mitochondrial genomes, in which large segments of the parental wild-type genomes were deleted; I suggested that such deletions arose by a mechanism of the Campbell type [8], involving illegitimate, site-specific recombination events in the AT spacers which I supposed to contain sequence repetitions because of their composition. It was evident that the loss of any known mitochondrial gene products (ribosomal RNAs, tRNAs, the sub-units of enzymes involved in respiration and oxidative phosphorylation) would have a pleiotropic effect and lead to a loss of respiratory functions.

The petite mutation is due to large deletions

Further work [9–12] showed that the AT spacers formed 50% of the wild-type mitochondrial genome, had a GC content lower than 5%, were indeed repetitive in nucleotide sequence and were, therefore, likely to be endowed with sequence homology over stretches long enough to allow illegitimate site-specific recombination. Five years ago, direct evidence was provided for both a deletion mechanism [13] and an accompanying amplification of the excised genome segment [13,14]: only a fraction of the restriction fragments of wild-type mitochondrial DNA were present in petite genomes and these were present in multiple copies per genome unit. These findings disposed of a number of strange *ad hoc* hypotheses put forward to explain the petite mutation [15–17] and led to the scheme shown in Fig. 2 in which the excised segment from the wild-type genome becomes the repeat unit of the petite genome. This may in turn undergo further deletions leading to secondary petite genomes having simpler repeat units. Incidentally, the analysis of restriction patterns of mitochondrial DNA from wild-type cells [13] provided the first unequivocal estimate of the size of the mitochondrial genome unit, about 50×10^6 [13].

The GC clusters

Another important advance in our knowledge of the organization of the mitochondrial genome was the discovery of short segments of mitochondrial DNA extremely rich in GC, the GC clusters [18–20]. Operationally, two sorts of GC clusters can be distinguished, the (CCGG, GGCC) clusters, present in 60–70 copies per genome unit and recognizable because they are degraded by two particular restriction enzymes, Hpa II and Hae III, and the GC-rich clusters which do not contain

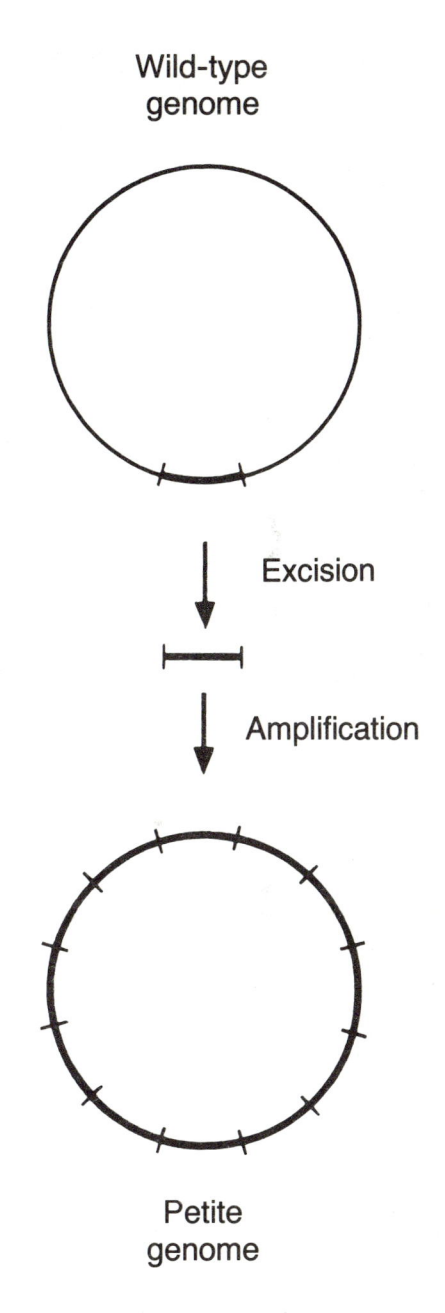

Fig. 2. Scheme of the process leading to the formation of spontaneous petite genomes. A segment of the mitochondrial genome unit from wild-type yeast cells is excised and amplified to yield the mitochondrial genome unit of a petite mutant. The excised segment from the wild-type genome becomes the repeat unit of the petite genome. This may in turn undergo further deletions leading to secondary petite genomes having simpler repeat units [25].

CCGG or GGCC sequences, but are often close to (CCGG, GGCC) clusters and to isolated CCGG sequences. A certain number of these clusters were likely to be endowed with sequence symmetry and to be homologous in sequence. Very recent sequence analyses carried out in several laboratories (see below) indicates that the GC clusters are located in the middle of AT spacers and not, as first suggested [20], at one end of them so that an overall scheme of the organization of the mitochondrial genome of yeast is that given in Fig. 3.

The excision sites

The next step was the precise definition of the sequences involved in the excision process. The basic idea of the deletion model mentioned above was that the instability of the mitochondrial genome of yeast was due to the existence in each genome unit of a number of nucleotide sequences having enough homology to allow illegitimate, site-specific recombination to take place. Clearly the newly discovered GC clusters were at least as good candidates as the AT spacers.

Detailed investigations of a number of spontaneous petite mutants [21] showed that most frequently the ends of the repeat units were formed by (CCGG, GGCC) clusters; less frequently, they appeared to correspond to GC-rich clusters and to AT spacer sequences.

For example, in the petite genomes of Fig. 4, excision of the repeat units occurred at, or very near (CCGG, GGCC) clusters in the top four cases; and at GC-rich cluster or AT spacers in the other two. Furthermore, it was shown that repeat units were organized in a perfect tandem (head-to-tail) fashion. Interestingly, secondary excision of simpler repeat units from the petite genomes originally derived from the parental wild-type genome appears to take place at the same kind of sites used in the primary process, the end product being rather simple and stable petite genomes, such as those presented in Fig. 4. The spontaneous petite mutation should, therefore, be visualized as a cascade of excisions, which is slowed down or stopped only by the fact that sequences appropriate for excision are used up in the process.

Such a simple excision mechanism was not found in petites induced by ethidium bromide, practically the only ones studied in most other laboratories. These contained inverted repetitions, multiple deletions, sequence rearrangements and, above all, were caused by much less specific excisions [22]. This is not surprising if one considers that the tremendous increase in petite formation upon mutagenization is accompanied by extensive genome fragmentation [23] and that petites lacking mitochondrial DNA altogether are frequently formed [23,24].

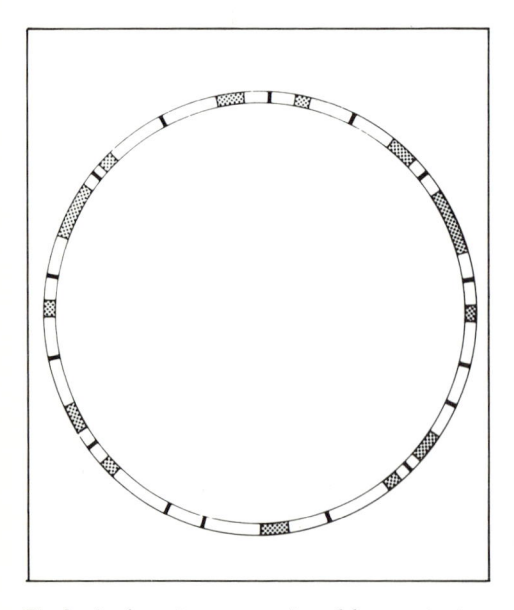

Fig. 3. A schematic representation of the organization of a mitochondrial genome unit of yeast. Grey stretches represent genes, white stretches AT spacers, black bars GC clusters.

The excision mechanism

It is obvious that what is needed now is

detailed knowledge, at the nucleotide level, of the sequences involved in the excision of petite genomes. Such work is in progress and hopefully will lead to a more precise understanding of the excision process. The sequence data already available strongly suggests that excision is due to a crossing-over process. If this is so, the

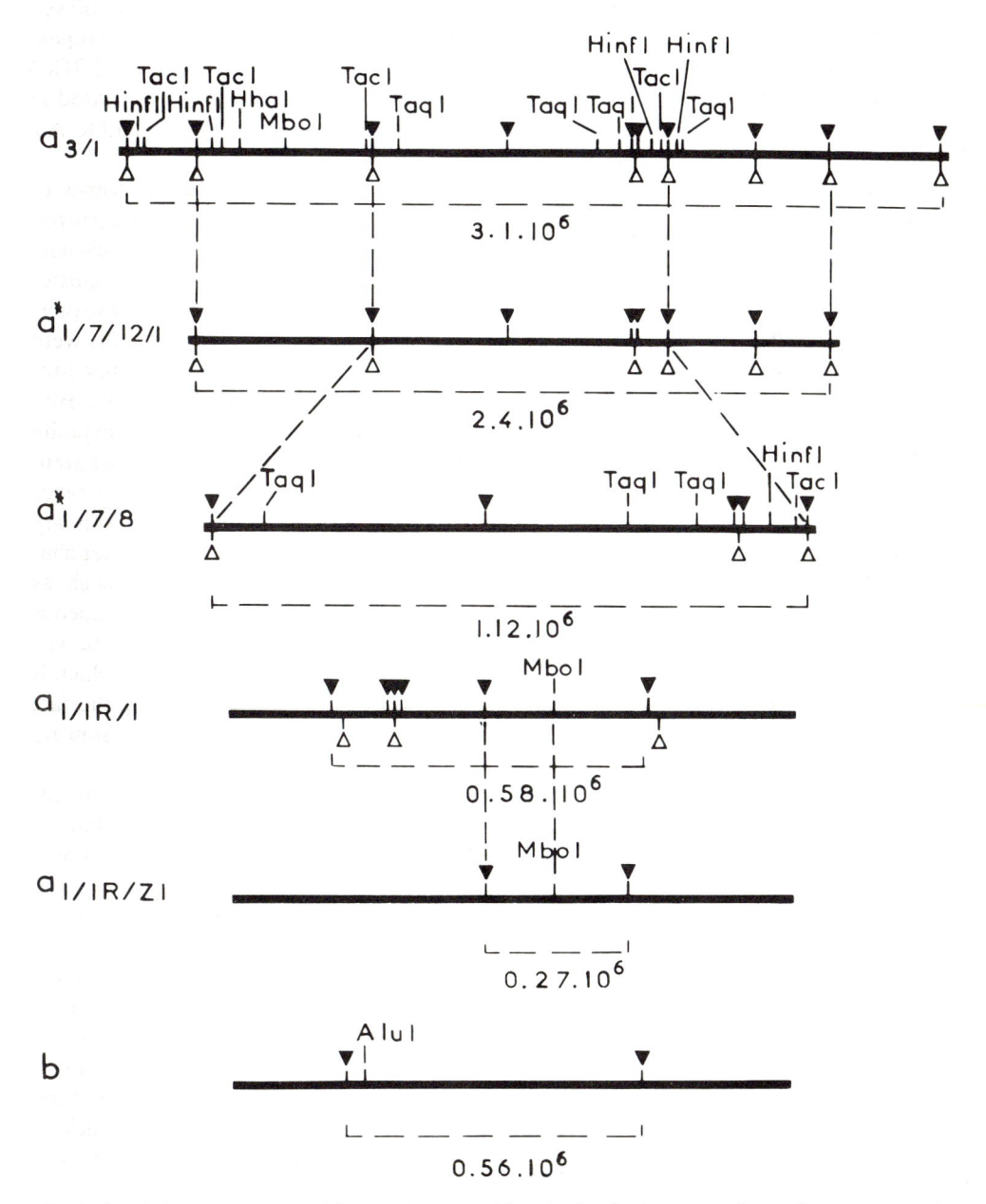

Fig. 4. Restriction enzyme maps of the repeating units of the mitochondrial genomes of several spontaneous petite mutants. The molecular weights of the repeat units are indicated, along with the positions of Hae III ($\triangle$), Hpa II ($\blacktriangledown$) and other restriction sites. In the case of $a_{3b/1}$, five isolated Hpa II sites and a Hinc II site are not shown. The broken lines indicate corresponding restriction sites in different repeat units [21].

primary event in the spontaneous petite mutation is very similar to the excision of the lambda prophage from the E. coli chromosome or to the dissociation of a transposon from its host plasmid; in this case, the GC clusters and sequences in the AT spacers play the same role as the insertion sequences delimiting a bacterial transposon, and should also play a role in what can be considered the reverse process, namely the recombination of petite genomes with wild-type genomes or other petite genomes.

The sequence of the repeat unit of a petite mutant

The recent determination of the nucleotide sequence of the repeat unit of the mitochondrial genome of the spontaneous petite mutant $a_{1/1R/Z1}$ ([25]; Fig. 5) and preceding work in Tzagoloff's laboratory [26,27] provided complementary confirmations of several previous results and predictions: that AT spacers are made up of short alternating and non-alternating AT sequences [7,11] and contain repeated sequences and palindromes [12]; that GC-rich clusters are largely contiguous to CCGG sequences and to (CCGG, GGCC) clusters [20]; and that the latter are to some extent endowed with both symmetry and homology [20]. The genome of $a_{1/1R/Z1}$ is of interest in three other respects: (a) it does not contain any gene, and is therefore a clear example of the lethality of petite mutations as far as mitochondria are concerned; (b) it replicates; in fact this is the only function left; since this genome was shown to be made up of a perfect tandem repetition of the basic unit [21], the latter must contain the signal for the initiation of replication; (c) it is excised, in all likelihood, from another petite genome, $a_{1/1R/1}$ (Fig. 4) and not directly from the wild-type genome; the complete sequence of the parental petite genome $a_{1/1R/1}$ should, therefore, provide precise information as to the excision sites involved in the formation of this genome.

Some general issues

From the brief account just presented, two general points are clear. First, the organization of the mitochondrial genome of yeast is typically eukaryotic, in that coding sequences are interspersed with non-coding sequences, the AT spacers and the GC clusters; the existence of split genes in this genome (see [28] for a review) points in the same direction. Second, the non-coding sequences of the mitochondrial genome of yeast are similar to the interspersed repeated sequences and the foldback sequences of the nuclear genome of eukaryotes, in that identical or similar sequences are present in many copies in the genome and that several of these sequences exhibit symmetry.

It is evident that regulatory sequences acting as promotors, operators, sites for the initiation of replication, and sites involved in the processing of transcripts are present in the non-coding sequences of yeast mitochondrial DNA; none of these have so far been identified, although several suggestions concerning the GC clusters have been put forward [20]. Another function of the non-coding sequences is already well documented and has to do with illegitimate, site-specific recombination. The excision of the spontaneous petite genomes just described is an example of these extragenic recombinational events. The same basic mechanism appears, however, to be more general and to account for: (a) the divergence of the mitochondrial genome of wild-type yeast cells; it has been shown [19] that different strains have mitochondrial genomes differing in the length of AT spacers, apparently the result of unequal crossing-overs in the sequences of allelic spacers; (b) similar changes in the mitochondrial genomes of the progeny arising from crosses of different wild-type

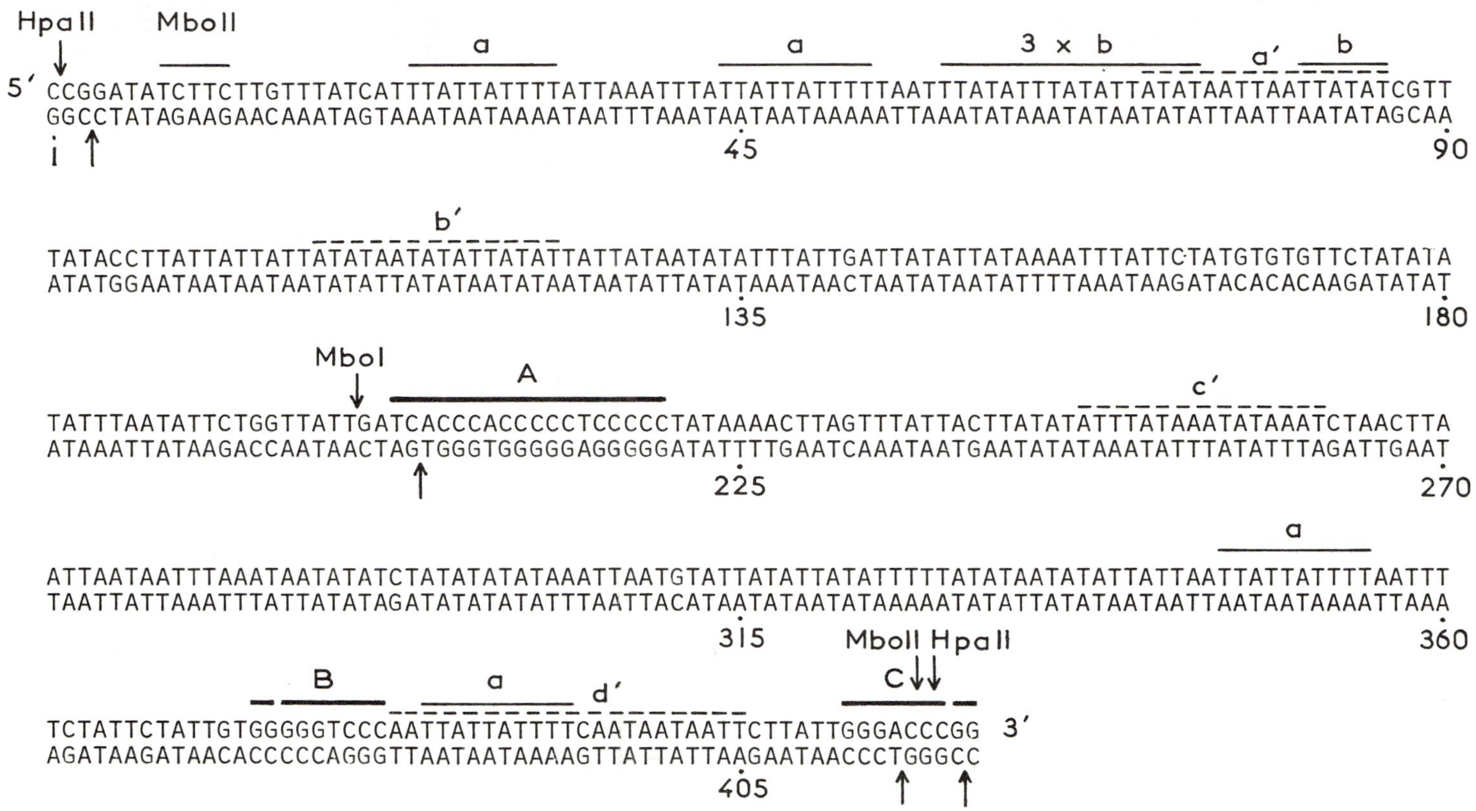

```
Hpall    Mboll
  ↓       ↓        a                      a                    3 x b              a'       b
5' CCGGATATCTTCTTGTTTATCATTATTATTTATTAAATTTATTTTATTTATTTATATATAATTAATTATATCGTT
   GGCCTATAGAAGAACAAATAGTAAATAATTTAAATAATTTAAATAAAATTAATAATAAAATATAATTAATATAGCAA  90
   i  ↑

                b'
     ————————————————————
   TATACCTTATTATTATAATATATTATTATTATAATATATTATTGATTATTATAATAAATTTATTC·TATGTGTTCTATATA
   ATATGGAATAATAATATATATAATAATAATATTATATATATAATAACTAATAATATTATATTTAAATAAGATACACAAGATATAT  180

                     A
            ——————————————————
   TATTTAATATTCTGGTTATTGATCACCCACCCCCTCCCCCTATAAAACTTAGTTTATTACTTATATATTTAAATATAAATCTAACTTA
   ATAAATTTATAAGACCAATAACTAGTGGGTGGGGGAGGGGGATATTTGAATCAAATGAATATATTTATATAAATTTAGATTGAAT  270
   Mbol ↑                            ↑                                c'

                                                             a
                                                       ———————————
   ATTAATAATTTAAATAATATATCTATATAATAAATTAATGTATTATATATATTTTTATATATATTATTATTTTAATTT
   TAATTATTAAATTTATTATATAGATATATTAATTACATAGATATATATAAAAATATATATAATAATAAAATTAAA  360

     B       a       d'             Mboll Hpall
   ———————————————————————————————   C↓    ↓     3'
   TCTATTCTATTGTGGGGTCCCAATTATTATTTCAATAATAATTCTTATTGGGACCCGG
   AGATAAGATAACACCCCCAGGGTTAATAATAAAAGTTATTATTAAGAATAACCCTGGGCC  405
                                                          ↑       ↑
```

Fig. 5. *Nucleotide sequence of the repeat unit of the mitochondrial genome of spontaneous petite mutant* $a_{1/1R/Z1}$ *(see Fig. 4). A, B and C indicate GC-rich clusters; a and b, a repeated decanucleotide and a repeated hexanucleotide, respectively; a', b', c' and d', palindromic sequences in the AT stretches. The restriction sites of Hpa II, Mbo I and Mbo II are indicated by arrows; the recognition site of the latter enzyme is also indicated [25].*

strains [29]; an important point here is that the underlying extragenic unequal crossing-overs seem to be much more frequent than intragenic exchanges.

The organization of the mitochondrial genome of yeast, as it has emerged from our work, is of interest in three additional respects: first, it points to the fact that in eukaryotic genetic systems, where so much of the DNA is non-coding, there is a real need for a molecular approach because approaches based on classical genetics or on the study of gene products suffer from serious intrinsic limitations and are unable to provide an overall picture of the genome; secondly, it disposes of a series of ideas centered about a prokaryotic organization of the mitochondrial genome of yeast; these ideas, which were promoted for many years, can be summarized in the following statements: (1) that the mitochondrial genome of yeast has a unique nucleotide sequence, namely a sequence lacking internal repetition; this view had its origin in a misunderstanding of the renaturation kinetics of mitochondrial DNA; (2) that the mitochondrial genome of yeast has an informational content five times larger than that of animal mitochondrial genomes (which have a unit size of only 10^7); this idea should have been considered with suspicion because there are practically no known gene products encoded in the yeast system and not also in the animal system; (3) that the unit size of the mitochondrial genome decreases in the evolution from unicellular organisms to animals; recent results have shown that some protists have indeed a mitochondrial genome as small as animals; (4) that mitochondria are the remnants of prokaryotic endosymbionts; although this view is almost of a philosophical nature, arguments against it abound (see [30]); finally it suggests that the mitochondrial genome may be a useful, simple model for its big brother in the nucleus.

Conclusions and perspectives

It is a matter of satisfaction to see that the problem of the petite mutation in yeast is now essentially solved at the molecular level and that predictions made ten years ago have proven to be correct. Sequence work currently under way should provide additional details within a short time. It is clear that understanding the molecular mechanism of the petite mutation, has come at least as much from a knowledge of the organization of the mitochondrial genome of wild-type yeast cells as from investigations on petite genomes. The scene is now set to understand the other phenomenon discovered by Ephrussi, suppressivity. The petite mutation and suppressivity seem to be the two faces of the same coin, the former basically consisting of the excision of a petite genome segment from the wild-type genome, the latter having to do with the insertion of a petite genome segment into the wild-type genome. In retrospect, it is evident Ephrussi's work not only opened the field of extrachromosomal genetics, but also provided a fantastic stimulus for the investigations which followed up to this day.

Concerning the future of mitochondrial genetics of yeast, there is ample ground for optimism for three main reasons. The first is that the striking advances of the molecular approach have been accompanied by a rapid progress in the genetic approach following the discovery of mitochondrial recombination of antibiotic resistance markers by Thomas and Wilkie [31] and the discovery of mutants of mitochondrial genes that code for polypeptides by Tzagoloff et al. [32]. The second reason for optimism is that after much tedious but necessary work, detailed physical and genetical maps of the mitochondrial genome have been produced. The third is that our knowledge of the transcription products has progressed very considerably. Under these circumstances, it is not impossible that the mitochondrial genome of yeast will be the first eukaryotic genome to be satisfactorily understood in terms of both structure and function, and this should pave the way to understand the evolution of organelle genomes and the relationships between the latter and the nuclear genomes.

References

1. Ephrussi, B. (1949) in *Unités Biologiques Douées de Continuité Génétique,* Paris, Juin-Juillet, 1948, pp. 165–180, Editions du C.N.R.S., Paris
2. Ephrussi, B. (1952) in *Nucleocytoplasmic Relations in Micro-organisms,* pp. 13–47, Clarendon Press, Oxford
3. Ephrussi, B., Margerie-Hottinguer, H. de and Roman, H. (1955) *Proc. Natl. Acad. Sci. U.S.A.* 41, 1065–1071
4. Bernardi, G., Carnevali, F., Nicolaieff, A., Piperno, G. and Tecce, G. (1968) *J. Mol. Biol.* 37, 493–505
5. Mehrotra, B. D. and Mahler, H. R. (1968) *Arch. Biochem. Biophys.* 128, 685–703
6. Bernardi, G., Faurès, M., Piperno, G. and Slonimski, P. P. (1970) *J. Mol. Biol.* 48, 23–42
7. Bernardi, G. and Timasheff, S. N. (1970) *J. Mol. Biol.* 48, 43–52
8. Campbell, A. M. (1962) *Adv. Genet.* 11, 101–145
9. Bernardi, G., Piperno, G. and Fonty, G. (1972) *J. Mol. Biol.* 65, 173–190
10. Piperno, G., Fonty, G. and Bernardi, G. (1972) *J. Mol. Biol.* 65, 191–205
11. Ehrlich, S. D., Thiéry, J. P. and Bernardi, G. (1972) *J. Mol. Biol.* 65, 207–212
12. Prunell, A. and Bernardi, G. (1974) *J. Mol. Biol.* 86, 825–841
13. Bernardi, G., Prunell, A. and Kopecka, H. (1975) in *Molecular Biology of Nucleocytoplasmic Relationships* (Puiseux-Dao S., ed), pp. 85–90, Elsevier, Amsterdam
14. Locker, J., Rabinowitz, M. and Getz, G. S. (1974) *Proc. Natl. Acad. Sci. U.S.A.* 71, 1366–1370
15. Slonimski, P. P. (1968) in *Biochemical Aspects of the Biogenesis of Mitochondria* (Slater, E. C., Tager, J. M., Papa, S. and Quagliariello, E., eds), pp. 475–476, Adriatica Editrice, Bari
16. Carnevali, F., Morpurgo, G. and Tecce G. (1969) *Science, (N.Y.)* 163, 1331–1333
17. Borst, P. and Kroon, A. M. (1969) *Int. Rev. Cytol.* 26, 107–190
18. Bernardi, G., Prunell, A., Fonty, G., Kopecka, H. and Strauss, F. (1976) in *The Genetic Function of Mitochondrial DNA* (Saccone, C. and Kroon, A. M., eds), pp. 185–198, Elsevier/North-Holland, Amsterdam
19. Prunell, A., Kopecka, H., Strauss, F. and Bernardi, G. (1977a) *J. Mol. Biol.* 110, 17–52
20. Prunell, A. and Bernardi, G. (1977) *J. Mol. Biol.* 110, 53–74
21. Faugeron-Fonty, G., Culard, F., Baldacci, G., Goursot, R., Prunell, A. and Bernardi, G. (1979) *J. Mol. Biol.* (in press)
22. Lewin, A., Morimoto, R., Rabinowitz, M. and Fukuhara, H. (1978) *Mol. Gen. Genet.* 163, 257–275
23. Goldring, E. S., Grossmann, L. I., Krupnick, D., Cryer, D. R. and Marmur, J. (1970) *J. Mol. Biol.* 52, 323–335
24. Nagley, P. and Linnane, A. W. (1970) *Biochem. Biophys. Res. Commun.* 39, 989–996
25. Gaillard, C. and Bernardi, G. (1979) *Mol. Gen. Genet.* (in press)
26. Cosson, J. and Tzagoloff, A. (1979) *J. Biol. Chem.* 254, 42–43
27. Macino, G. and Tzagoloff, A. (1979) *Proc. Natl. Acad. Sci. U.S.A.* 76, 131–135
28. Bernardi, G. (1978) *Nature (London)* 276, 558–559
29. Fonty, G., Goursot, R., Wilkie, D. and Bernardi, G. (1978) *J. Mol. Biol.* 119, 213–235
30. Mahler, H. R. and Raff, R. A. (1975) *Int. Rev. Cytol.* 43, 1–124
31. Thomas, D. Y. and Wilkie, D. (1968) *Biochem. Biophys. Res. Commun.* 30, 368–372
32. Tzagoloff, A., Akai, A., Needleman, R. B. and Zulch, G. (1975) *J. Biol. Chem.* 250, 8236–8242

Giorgio Bernardi is at the Institut de Recherche en Biologie Moléculaire, 2, Place Jussieu – 75005 Paris, France.

The origins of replication of the mitochondrial genome of yeast

Giorgio Bernardi

Recent investigations have provided information on the origin of replication of the mitochondrial genome of yeast and an explanation for the phenomenon of the suppressivity.

Three years ago, I summarized in a *TIBS* review[1] the state of our knowledge on the 'petite colonie' mutation of *Saccharomyces cerevisiae*. As shown by the pioneering work of Ephrussi and his collaborators, this mutation: (a) is characterized by an irreversible loss of respiration and by an extraordinarily high spontaneous mutation rate; and (b) is transmitted to the progeny in a non-Mendelian fashion: crosses of wild-type cells with petite mutants yield either wild-type progeny only, or both wild-type cells and petite mutants in proportions essentially dependent on the particular petite used; in the first case, the petites entering the cross are called neutral, in the second one suppressive. While the molecular basis of the mutation was understood by 1979, its details were not known then, and suppressivity still was the same elusive problem it had been for almost 25 years. The situation is now completely changed and it is possible to give here a full account of the petite mutation.

The excision-amplification process

Previous investigations (see Ref. 1 for a brief review) had established that the first event in the spontaneous cytoplasmic petite mutation is the excision of a segment from one of the 25–50 mitochondrial *genome units* (Fig. 1) of a wild-type cell; excision is then followed by a tandem amplification process in which the excised DNA segment becomes the *repeat unit* of a *defective genome unit* (Fig. 2) which replicates within the parental wild-type cell and segregates into the buds of the latter; further segregation of these defective genome units in the progeny leads to the formation of petite mutants, whose mitochondrial genome is formed by identical units.

The sequence used in the excision process have now been investigated for 17 spontaneous petite genomes[2,3]. In all cases, perfect direct repeats located in the AT spacers or in the GC clusters were found to be used as excision sequences, as predicted a long time ago (see Ref. 1).

These results indicated that sequences of 11–12 base pairs could be used as excision sequences; longer excision sequences were also used and included both canonical and surrogate origins of DNA replication (see below); when shorter sequences were used, they were flanked by regions of patchy homology (Fig. 2).

These results, discussed in more detail elsewhere[2,3], also indicated: (a) that the excision mechanism probably involves unequal, site-specific, crossing-over events within a genome unit and that this process probably is just a special case of the very active recombination processes taking place in the mitochondrial genomes of wild-type yeast cells[4]; (b) that the highest excision rates are associated with sequences capable to form the most stable (longest and/or richest in GC) heteroduplexes; an extreme case involving two *ori* sequences (see below) was described[2]; (c) that the production of petite mutants depends not only upon the excision rate, but also upon the replication rate of the defective genome relative to the parental one, and upon its stability, namely its susceptibility to further excisions, in the cells harboring it[2]; both these parameters favor the production of petite mutants carrying mitochondrial genomes formed by short repeat units containing canonical *ori* sequences (see below). Obviously this trend can be counteracted by genetically selecting petites having a very poor replication efficiency and/or resulting from very rare events; such is the case, for the ethidium-induced petites studied in Tzagoloff's laboratory[5,6]. Here too, direct repeats were used as excision sequences, but they were found to be localized not only in non-coding sequences, but also in the open reading frames of introns and in the exons of mitochondrial genes.

Excision and amplification are general phenomena

It should be stressed that excision and amplification of defective genome units from wild-type mitochondrial genomes are general phenomena not limited to yeast. In the case of obligatory aerobes, however, defective units coexist with wild-type genome units, a certain number of which are required for respiration. The best-known examples[2] are those of senescent cultures of *Podospora anserina*, of 'ragged' mutants of *Aspergillus amstelodami*, of *poky* and *stopper* mutants of *Neurospora crassa*, of male sterile mutants in maize. Furthermore, excision and amplification do not concern only mitochondrial genomes; some bleached mutants of *Euglena gracilis* contain defective chloroplast genomes in which the ribosomal gene region and the origin of replication are preferentially retained; since the chloroplast genome is dispensable in *Euglena*, as is the mitochondrial genome in yeast, these mutants only contain the defective genome[7]. Finally, *Drosophila* and mammalian cells are known to contain extra-

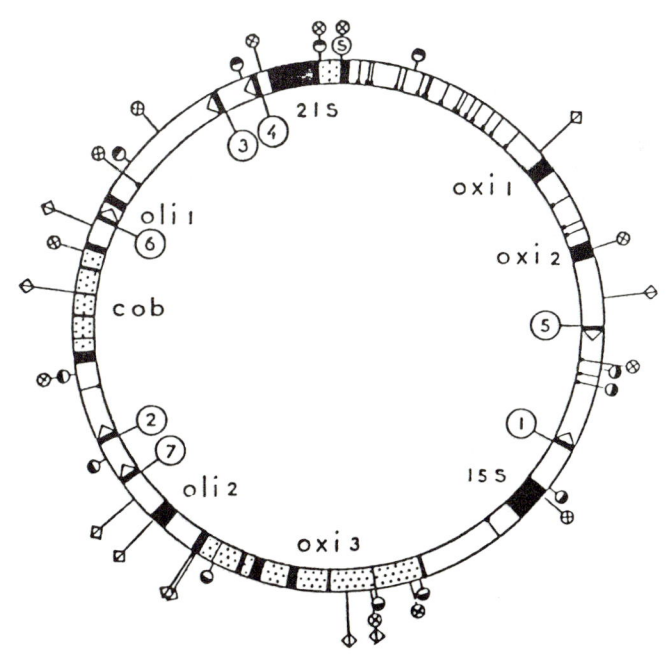

Fig. 1. Physical and genetical map of the mitochondrial genome unit of wild-type yeast (strain A). Some restriction sites are indicated: Hinc II (⊗), Hha I (●), EcoRI (◊) Sal I (Ⓢ). Circled numbers indicate the location of ori sequences 1–7 (arrowheads point in the direction cluster C to cluster A; see Fig. 3). Black and dotted areas correspond to exons and introns of mitochondrial genes, respectively. Thin radial lines indicate tRNA genes. White areas correspond to long AT spacers embedding short GC clusters. (Modified from Ref. 16.)

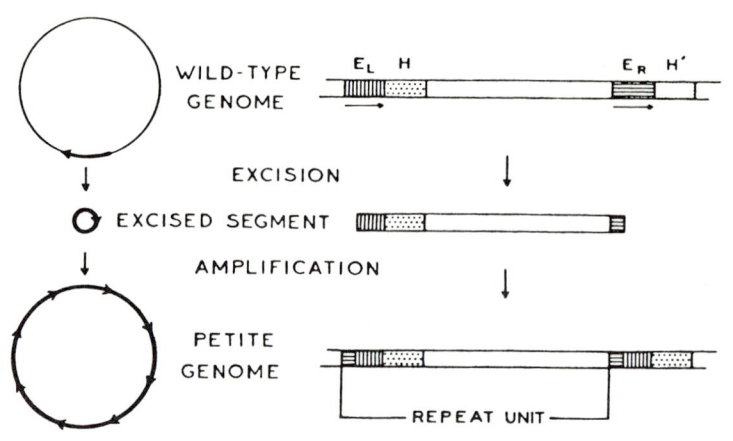

Fig. 2. (A) Scheme depicting the excision–amplification process leading to the formation of the genome of a spontaneous petite mutant. A segment of a unit of a wild-type mitochondrial genome is excised and tandemly amplified into a defective genome unit. This then replicates and segregates into the buds to form the genome of a petite mutant; the petite genome can undergo further excisions leading to the formation of secondary petite genomes. (B) Scheme showing the left and right (E_L, E_R) excision sequences as found on the parental wild-type genome region from which the repeat unit of the petite genome was excised. H, H' indicate sequences flanking the excision sequences and sharing a significant yet imperfect homology (from Ref. 3).

chromosomal circular DNA excised from nuclear DNA[8,9], and double-minute chromosomes are known to be formed in methotrexate-resistant mouse cells[10].

The canonical and the surrogate origins of replication of petite genomes

The mitochondrial genomes of the vast majority of spontaneous petites are exclusively derived from the tandem amplification of a DNA segment excised from any region of the parental wild-type genome[11]. Therefore, either the wild-type genome contains several origins of replication and at least one of them is present on the excised segment[12], or sequences other than the origins of replication of the wild-type genome are used as surrogate origins of replication. In fact, both situations have been found to occur, although with very different frequencies.

Considering that the first explanation was the more likely one, when we first sequenced[13] the repeat units of two petite genomes excised from the same region of the wild-type genome, we looked for an origin of replication in the segment shared by them and found a region characterized by two short GC clusters, A and B, flanking a palindromic AT sequence, p, and a short AT segment, s; and one long GC cluster, C, separated from B by a long AT segment, l (see Fig. 3). The potential secondary structure of the A–B region, the primary structure of cluster C and the general arrangement of the whole region are remarkably similar to those found in other mitochondrial origins of replication (Refs 14 and 15; Fig. 4).

An *ori* sequence like the one just described was found in almost all the mitochondrial genomes of spontaneous petite mutants. Restriction mapping and hybridization of petite genomes on restriction fragments of wild-type genomes[16] provided evidence for the existence of seven such *ori* sequences in the mitochondrial genome of wild-type cells. The primary structure of the *ori* sequences shows that they are extremely homologous, particularly in their GC clusters; some of them, *ori* 4, 6, and 7, contain additional GC clusters, β and γ, identical in sequence and position (Fig. 3). All these *ori* sequences have been

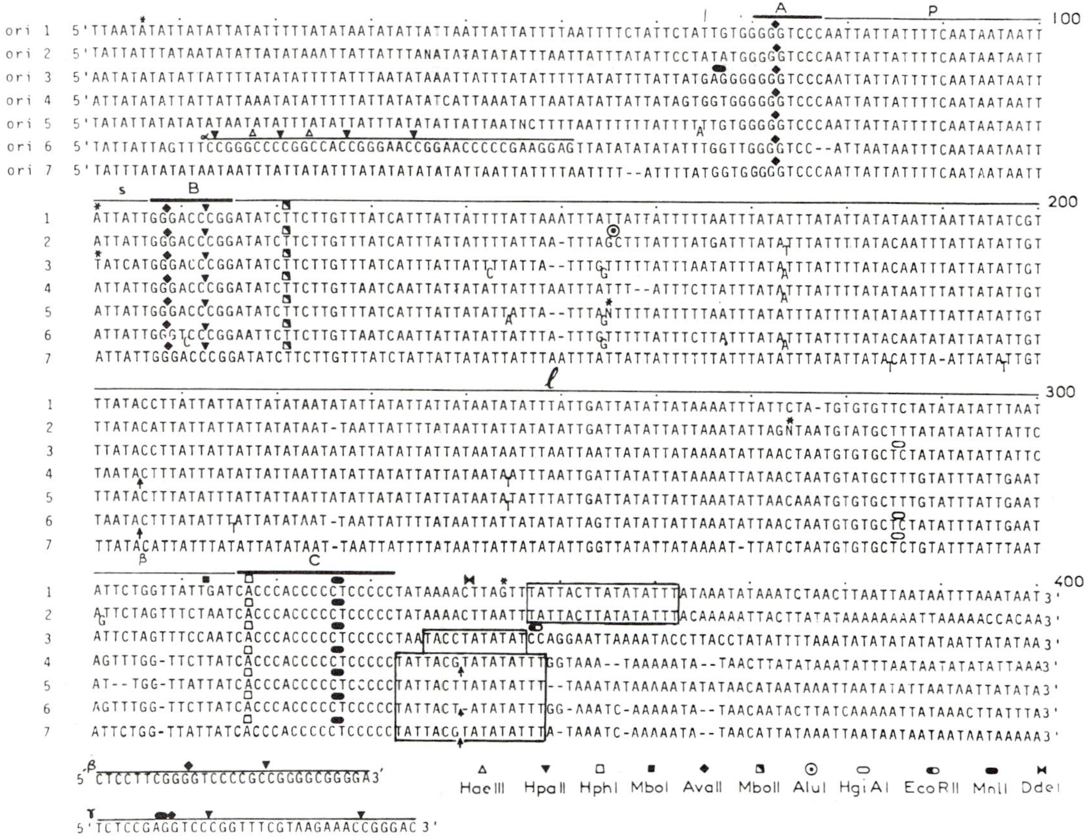

Fig. 3. Primary structure of the ori sequences and their flanking sequences. Thick lines indicate GC clusters A, B, and C, thin lines AT regions p, s and l. The positions and the sequences of extra GC clusters β and γ are given (as well as that of GC cluster α, which is located outside the ori sequence. Restriction sites are indicated by the symbols shown. The sequences homologous to initiation transcription sites at the right of cluster C, are indicated by boxes. (From Refs 16 and 24; G. Baldacci and G. Bernardi submitted, and unpublished results of M. de Zamaroczy.)

precisely localized and oriented on the physical map of the wild-type genome (Fig. 1).

It should be noted: (a) that some *ori* sequences display one orientation on the wild-type genome, and some the opposite; (b) that *ori* 2 and 7 and *ori* 3 and 4 are close to each other and tandemly oriented; (c) that *ori* 4 is absent in a wild-type strain (G. Faugeron-Fonty, personal communication); (d) that *ori* sequences containing the γ cluster have been found only once (*ori* 4) or not at all (*ori* 6, *ori* 7) in extensive screenings of spontaneous petite genomes.

*Ori*⁰ petites, lacking a canonical *ori* sequence, have also been found, although very rarely. An investigation of the mitochondrial genomes of eight such *ori*⁰ petites[17] has revealed that their repeat units contain, instead of canonical *ori* sequences, one or more *ori*ˢ sequences. These 44-nucleotide long surrogate origins of replication are a subset of GC clusters characterized by a potential secondary fold with two sequences ATAG and GGAG inserted in AT spacers; these sequences are followed by two AT base pairs, a GC stem (broken in the middle and in most cases also near the base, by non-paired nucleotides) and a terminal loop (Fig. 5). This structure is reminiscent of that of GC clusters A and B from canonical *ori* sequences (Fig. 4). Like the latter, *ori*ˢ sequences are present in both orientations, are located in intergenic regions and can be used as excision sequences when tandemly oriented. *Ori*ˢ sequences are homologous with many other subsets of GC clusters (one of these subsets, the *ori*ˢ-like sequences, is shown in Fig. 5) some or all of which might perhaps also act as surrogate origins of replication, possibly still less efficiently than *ori*ˢ sequences.

The replication of petite genomes and the phenomenon of suppressivity

A functional evidence that *ori* sequences are indeed involved in the replication of the mitochondrial genome came from crosses of spontaneous petites, characterized in their mitochondrial genome and their suppressivity, with wild-type cells[16,18,19]. Crosses of spontaneous, highly suppressive petites having mitochondrial genomes formed by very short repeat units (400–900 base pairs) with wild-type cells produced diploids which harbored only the unaltered mitochondrial genomes of the parental petite, which was called supersuppressive[18,19]. When petites with different degrees of suppressivity were used in the crosses, the genomes of diploid petite progeny had restriction maps identical to those of the parental haploid petites. A very few exceptions were found and these corresponded to new excision processes affecting one of the parental genomes.

There are two clear correlations between the *ori* sequence of the petite used in the cross and its degree of suppressivity. First, all other things being comparable (namely,

TABLE I. Replication and transcription of petite genomes

A) *Ori*⁺ petites	Suppressivity[a]	Transcription	B) *Ori*⁻ petites	Suppressivity	Transcription
ori 1	>95%	+	*ori* 1 A⁻	80%	+
ori 2	>95%	+	*ori* 1 C⁻	n.d.	−
ori 3	85%	±	*ori* 3 C⁻	< 5%	−
ori 4	b	−	C) *Ori*⁰ petites		
ori 5	90%	+	a-15/4/1/10/3	~ 1%	−
ori 6	b	n.d.	a-3/1/B4	c	−
ori 7	b	n.d.	D) *Ori*ʳ petite	< 5%	n.d.

a Values found for petite genomes having repeat units ~900 (*ori* 1, 2) or 1800 (*ori* 3, 5) base pairs long.
b *Ori* 4 was only found once, *ori* 6 and 7 were never found alone in the extensive screenings of spontaneous petite genomes.
c Diploid.

Fig. 4. (left) Comparison of ori sequences of mitochondrial genomes from yeast[16] and HeLa cells[14]. Homology of potential secondary structure is found for the inverted repeats in cluster A – cluster B region; arrows indicate the base changes found in this region in different petite genomes. Homology of primary structure is found for cluster C. (below) Comparison of the two ori sequences; the arrows indicate the inverted repeats of the A–B region, the broken line corresponding to the looped-out sequence; bp, base pairs. (From Ref. 16.)

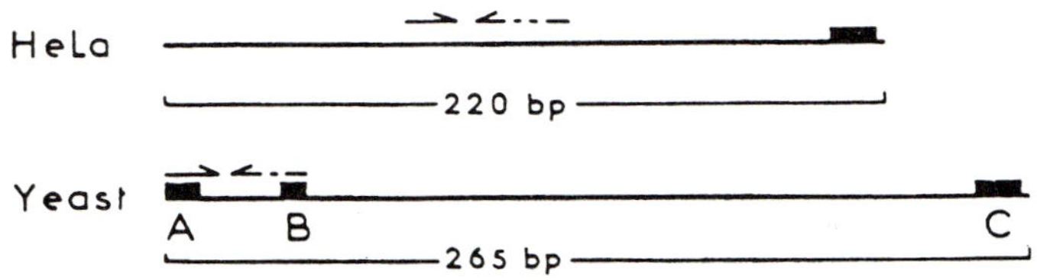

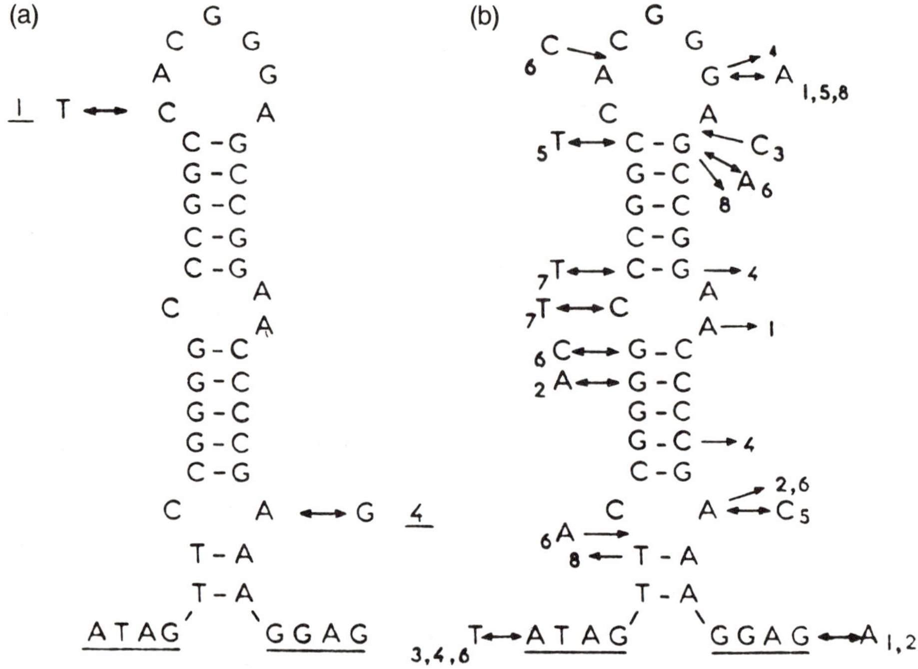

Fig. 5. Potential secondary structure of: (a) the ori^s sequences; and (b) the ori^s-like sequences. All sequences are drawn in the same orientation ATAG → GGAG. Double-headed arrows indicate base exchanges, arrows pointing towards, or away from, the structure indicate insertions and deletions, respectively. Numbers indicate the ori^s, or ori^s-like, sequences presenting these changes. (From Ref. 17.)

the intact state of the *ori* sequence and the total amount of mitochondrial DNA per cell), the lower the overall density of *ori* sequences on the genome units, the lower the supressivity. Second (see Table I), partial or total deletion of the *ori* sequences, or their rearrangement, affects the suppressivity. (a) *Ori*$^-$ petites, in which the *ori* sequence is partially deleted, show a decreased suppressivity relative to *ori*$^+$ petites carrying intact *ori* sequences. The loss of cluster C with its flanking sequence has a much more dramatic effect than the loss of cluster A; (b) *ori*^{r} petites, which show an inverted orientation of two *ori* sequences within the same repeat units (the latter having, in turn, an alternate inverted and tandem orientation) have a very low degree of suppressivity. (c) ori^0 petites, which lack the *ori* sequence altogether but contain *ori*^{s} sequences instead, have low to minimal degrees of suppressivity.

These results provide a molecular basis for a replicative advantage being the explanation for suppressivity. This hypothesis can be traced back to Ephrussi et al.[20], but received a particular attention[21,22] after the work of Mills et al.[23] on the *in vitro* replication of QβRNA. It is quite possible that a replicative advantage also explains amplification of the original monomeric genomes of petites.

The *ori* sequences as transcription initiation sites

Three recent results on the transcription initiation sites in petite genomes are also relevant[24].

(a) Transcription initiation efficiency parallels replication efficiency. Petite genomes containing some canonical *ori* sequences (*ori* 1, 2, 5, and, to a lesser extent, *ori* 3) are transcribed very actively; others, containing *ori* 4, or deleted in their

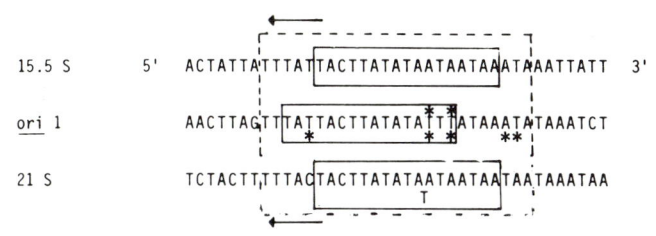

```
15.5 S      5'  ACTATTATTTAT TACTTATATAATAATAAAT AAATTATT   3'

ori 1           AACTTAGTT TATTACTTATATA TT TATAAATA TAAATCT
                         *  *         * *  **

21 S            TCTACTT TTTAC TACTTATATAATAATAA TAATAAATAA
                                  T
```

Fig. 6. Comparison of the transcription initiation sequences of ori 1 and of the 15.5s and 21s rRNA genes[27]. Solid-line boxes indicate the transcription initiation sequences (as read on the coding strand to ensure consistency with Fig. 3); the broken-line box indicates the region of homology among the three sequences; asterisks indicate base differences, the arrows the start of the tRNA transcripts. (From Ref. 24.)

C clusters (but not those deleted in A clusters), or lacking canonical *ori* sequences (*ori*° petites), are not (Table I). Likewise, *ori* sequences containing a γ cluster (*ori* 4, 6 and 7; two of these are in tandem with *ori* 3 and *ori* 2, respectively) are probably not very efficient in DNA replication, as suggested by the fact that they are very rarely or never found in extensive screenings of spontaneous petites and may even be absent (*ori* 4) in some wild-type genomes.

(b) Since transcriptionally active *ori* sequences (see above) are present in both orientations on the wild-type genome, it is very likely that both strands are transcribed, although the non-sense strand appears to be transcribed less accurately or more slowly. Hybridization experiments with separated DNA strands have identified the template strand used in transcription as the strand which contains the oligonucleotide stretch of cluster C.

This conclusion supports previous independent evidence[25-27] and puts the transcription of the mitochondrial genome of yeast in line with that of the animal mitochondrial genome. Similarly, replication might proceed unidirectionally from some *ori* sequences (possibly *ori* 2 and 5 for one strand, and *ori* 1 and 3 for the other). If this is the case, the replication of the mitochondrial genome of wild-type yeast cells would be analogous with the replication of unicircular dimers of the mammalian mitochondrial genome[28].

(c) Transcription initiates next to the oligopyrimidine stretch of cluster C, at a sequence (Fig. 3) very homologous (Fig. 6) to the transcription initiation sequences of rRNA genes[29], and proceeds in from cluster C to cluster A. As already mentioned, the insertion of cluster γ in the middle of this sequence (as in *ori* 4) or the loss of cluster C and of this sequence (as in *ori* C⁻ petite genomes) is accompanied by a loss of transcriptional activity. This suggests that *ori* 2 and *ori* 5 might be among the initiation sites used for transcribing the sense strand, *ori* 1 and *ori* 3 among those used for the transcription of the other strand. A small number of other sequences largely homologous to the transcription initiation sites of tRNA genes have also been found. These might play a role in the multipromotor transcription[30] of the mitochondrial genome of wild-type cells, postulated years ago[12].

In conclusion, the investigations outlined in this review provide answers to the questions raised many years ago by the work of Ephrussi in the petite mutation. In fact, they have done more than this, since they have opened the way to a fine analysis of replication, recombination and expression in the mitochondrial genome of yeast, and have shed some light on the general problem of genome evolution (see Ref. 31 for a recent discussion on the latter point).

References

1 Bernardi, G. (1979) *Trends Biochem. Sci.* 4, 197–201
2 Marotta, R., Colin, Y., Goursot, R. and Bernardi, G. (1982) *EMBO J.* 5, 529–534
3 de Zamaroczy, M., Faugeron-Fonty, G. and Bernardi, G. (1982) *Proc. Natl Acad. Sci. U.S.A.* (submitted)

4 Fonty, G., Goursot, R., Wilkie, D. and Bernardi, G. (1978) *J. Mol. Biol.* 119, 215–235
5 Bonitz, G. S., Coruzzi, G., Thalenfeld, B. E. and Tzagoloff, A. (1980) *J. Biol. Chem.* 255, 11927–11941
6 Thalenfeld, B. E. and Tzagoloff, A. (1980) *J. Biol. Chem.* 255, 6173–6180
7 Heizmann, P., Doly, J., Hussein, Y., Nicolas, P., Nigon, V. and Bernardi, G. (1981) *Biochim. Biophys. Acta* 633, 412–415
8 Flavell, A. J. and Ish-Horowicz, D. (1981) *Nature* 292, 591–595
9 Calabretta, B., Robberson, D. L., Barrera-Saldana, H. A., Lambrou, T. P. and Saunders, F. G. (1982) *Nature* 296, 219–224
10 Kaufman, R. J., Alt, F. W., Kellems, R. F. and Schimke, R. T. (1978) *Science* 202, 1051–1055
11 Faugeron-Fonty, G., Culard, F., Baldacci, G., Goursot, R., Prunell, A. and Bernardi, G. (1979) *J. Mol. Biol.* 134, 493–537
12 Prunell, A. and Bernardi, G. (1977) *J. Mol. Biol.* 110, 53–74
13 Gaillard, C. and Bernardi, G. (1979) *Mol. Gen. Genet.* 174, 335–337
14 Crews, S., Ojala, D., Posakony, J., Nishiguchi, J. and Attardi, G. (1979) *Nature* 277, 192–198
15 Gillum A. M. and Clayton D. A. (1979) *J. Mol. Biol.* 135, 353–368
16 de Zamaroczy, M., Marotta, R., Faugeron-Fonty, G., Goursot, R., Mangin, M., Baldacci, G. and Bernardi, G. (1981) *Nature* 292, 75–78
17 Goursot, R., Mangin, M. and Bernardi, G. (1982) *EMBO J.* 6, 705–711
18 de Zamaroczy, M., Baldacci, G. and Bernardi, G. (1979) *FEBS Letters* 108, 429–432
19 Goursot, R., de Zamaroczy, M., Baldacci, G. and Bernardi, G. (1980) *Current Genet.* 1, 173–176
20 Ephrussi, B., Jakob, H. and Grandchamp, S. (1966) *Genetics* 54, 1–29
21 Rank, G. H. (1970) *Can. J. Genet. Cytol.* 12, 129–136
22 Rank, G. H. (1970) *Can. J. Genet. Cytol.* 12, 340–346
23 Mills, D. R., Peterson, R. L. and Spiegelman, S. (1967) *Proc. Natl Acad. Sci. U.S.A.* 58, 217–224
24 Baldacci, G. and Bernardi, G. *EMBO J.* (in press)
25 Li, M. and Tzagoloff, A. (1979) *Cell* 18, 47–53
26 Coruzzi, G., Bonitz, S. G., Thalenfeld, B. E. and Tzagoloff, A. (1981) *J. Biol. Chem.* 256, 12782–12787
27 Beilharz, M. W., Cobon, G. S. and Nagley, P. (1982) *Nucleic Acid Res.* 10, 1051–1070
28 Clayton, D. A. (1982) *Cell* 28, 693–705
29 Osinga, K. A. and Tabak, H. F. (1982) *Nucleic Acid Res.* 10, 3617–3626
30 Levens, D., Ticho, B., Ackerman, E. and Rabinowitz, M. (1981) *J. Biol. Chem.* 256, 5226–5232
31 Bernardi, G. (1982) in *Mitochondrial Genes* (Slonimski, P. P. *et al.*, eds), pp. 269–278, Cold Spring Harbor, New York

Giorgio Bernardi is at the Laboratoire de Genetique Moleculaire, Institut de Recherche en Biologie Moleculaire, 2, Place Jussieu, 75005 Paris, France.

Organization and expression of the mammalian mitochondrial genome: a lesson in economy-I

Giuseppe Attardi

In this two part review I shall discuss the mitochondrial genetic system of human and of other mammalian cells which exhibits features of remarkable simplicity and economy in its decoding mechanism, in the organization of its genes and in their mode of expression. The gene arrangement in mitochondrial DNA is extremely compact, with the heavy strand sequences which code for the rRNAs, poly(A)-containing RNAs and tRNAs forming a continuum for the quasi-totality of its length. Furthermore, tRNA genes separate from each other nearly all the mitochondrial DNA segments coding for the individual rRNAs and poly(A)-containing RNAs, thus, providing the punctuation in the reading of mitochondrial DNA information.

All eukaryotic cells are endowed with an extrachromosomal genetic system sequestered in the mitochondria, which performs an essential role in the biogenesis of these organelles. The nature of the genetic information contained in mitochondrial DNA (mtDNA) has gradually become apparent in the last fifteen years as a result of investigations carried out in a variety of systems from yeast to man. It is remarkable that, while the essential genetic functions of mtDNA appear to have been preserved in great part in a vast spectrum of organisms at different evolutionary levels in the eukaryotic world, the structure and gene organization of the mitochondrial genome and the mechanisms of its expression have evolved in strikingly diverse ways in different organisms[1-3]. It is the purpose of this review to present an up-to-date account of our knowledge concerning mitochondrial gene organization and expression in mammalian cells. The human mitochondrial genome is used as a model system, and there are references to other systems whenever appropriate.

Role of mtDNA in mitochondriogenesis

Fig. 1 illustrates, in schematic form, the generally accepted role of mtDNA in mitochondriogenesis, based on information derived from several systems[1-5]. mtDNA is represented as a circular molecule attached to the membrane, as it appears to be in HeLa cells[6]. Replication and transcription of mtDNA involve the activity of enzymes which, according to the evidence available for yeast, are apparently coded in nuclear DNA and synthesized in the cytoplasm. The transcription products of mitochondrial DNA include two ribosomal RNA (rRNA) species, which are structural components of mitochondria-specific ribosomes, several mRNAs, which in many organisms, including animal cells, are polyadenylated, and many tRNAs, which in the species analyzed in detail appear to form a complete set for mitochondrial protein synthesis. The two rRNA species combine with ribosomal proteins (which are coded in nuclear DNA and synthesized in the cytoplasm, with one known exception, documented only in yeast and *Neurospora*) to form the large and small ribosomal subunits. These associate with mtDNA coded mRNAs to form polysomes. A considerable amount of evidence indicates that all mRNAs utilized in mitochondria are endogenous. The mtDNA-coded tRNAs, through the activity of specific aminoacyl synthetases, which are also coded in nuclear DNA and cytoplasmically synthesized, are charged by the corresponding amino acids. Mitochondrial polysomes, with the participation of the aminoacyl–tRNA complexes and of nuclear-coded initiation and elongation factors, synthesize a limited set of specific polypeptides. With the exception of the ribosome-associated protein, the known mitochondrial translation products are hydrophobic polypeptides of the inner mitochondrial membrane, which, in association with cytoplasmically synthesized polypeptides, form enzyme complexes of the respiratory chain and the oligomycin-sensitive ATPase. The scheme presented in Fig. 1 illustrates one of the most fascinating aspects of mitochondriogenesis, i.e. the co-operation of nuclear and mitochondrial genomes in the assembly of the protein-synthesizing apparatus and of the enzyme complexes of the inner mitochondrial membrane.

The mammalian mitochondrial genetic code

In the past few years, it has become increasingly clear that the segregation of the mitochondrial genetic system in the special mitochondrial environment has caused it to evolve in such a way that, although not independent of the nuclear genome, is in many respects very distinct from it. The most dramatic evidence of this evolutionary individuality is provided by the recent finding of unusual features in the genetic

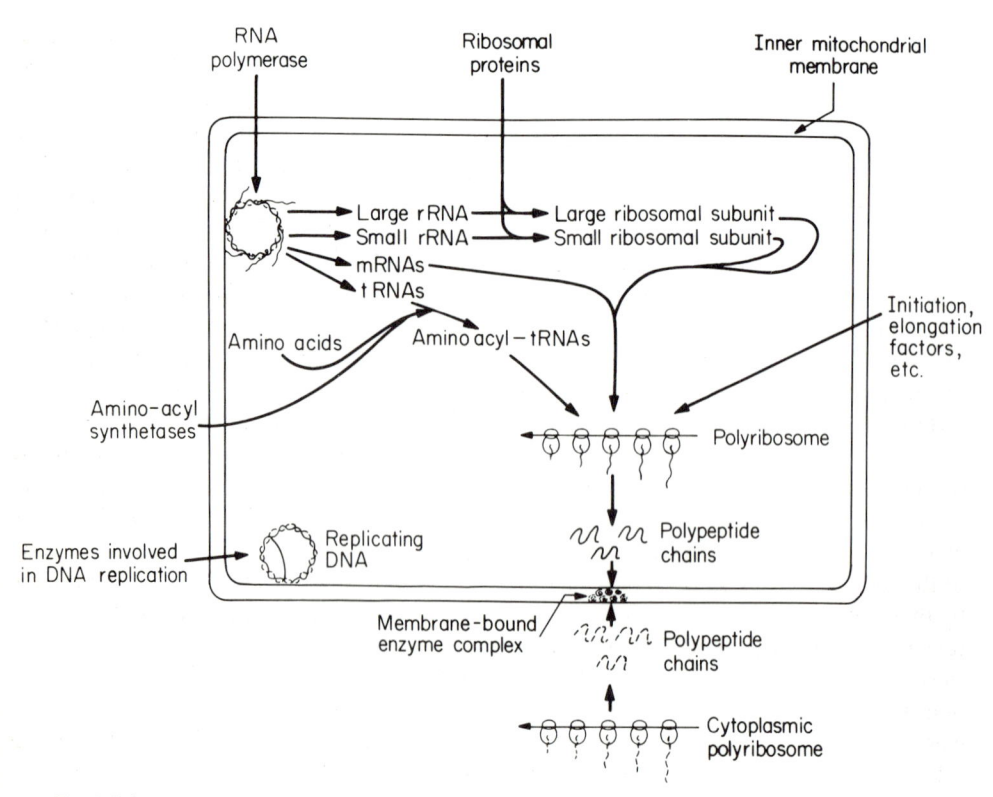

Fig. 1. Schematic representation of the informational role of mitochondrial DNA in mitochondriogenesis.

code and in the codon recognition pattern in mitochondria from mammalian cells, yeast and *Neurospora crassa*. Thus, a comparison of the coding sequence for cytochrome *c* oxidase subunit II (COII) in human mtDNA with the amino acid sequence of the bovine enzyme subunit has unexpectedly revealed that UGA is not used as a stop codon in human mitochondria, but as a tryptophan codon[7] and furthermore, that AUA appears to be read in the human organelles as a methionine codon rather than as isoleucine codon[7]. UGA is also read as a tryptophan codon in yeast[8] and *Neurospora crassa* mitochondria.

As concerns the reading of the genetic code in mitochondria, one intriguing observation, made in several systems, has been the relatively low number of tRNA genes detectable by RNA–DNA hybridization in mtDNA. Thus, 19 tRNA genes were found in human mtDNA[10]; in other systems, the reported number of tRNA genes[22-30] was also lower than the minimum number of 32 tRNAs required to read the genetic code according to the wobble mechanism[11]. Of the three hypotheses put forward to account for this discrepancy, i.e. import of nuclear-DNA-coded tRNAs, restricted codon usage and different pattern of codon recognition by mitochondrial tRNAs, the latter has proved to be the correct one. Very recently, the scanning of the entire human mtDNA sequence has shown the presence of only 22 possible tRNA-coding sequences[12]. A similar set of putative tRNA genes has been found in bovine mtDNA[12]. More significantly, an analysis of the distribution of the tRNAs coded for by these genes among the codon boxes has revealed a striking pattern: namely, that in the eight family boxes with four codons for one amino acid, only one specific tRNA was found, instead of two as in the universal code; in each case, this single tRNA has a U in the first position of the anticodon. It has been speculated that this tRNA is capable of reading all four codons in the family boxes, probably due to the paring of the U with all four bases in the third position of the codons in each family box. A similar conclusion has been reached for yeast[13] and *Neurospora crassa*[9] mitochondria. The decoding mechanism proposed for the family boxes cannot obviously operate in non-family boxes, where two tRNAs are used to read the four codons, because it would lead to misreading. The observation that, in *Neurospora crassa*, mitochondrial tRNA species for the family boxes have an unmodified U in the first position of the anticodon, while tRNAs specific for the two codons ending in purines in the non-family boxes have an unknown modified U in the same position, has led to the proposal that this modification may be the mechanism preventing misreading of the two codons ending in pyrimidines in the non-family boxes[9].

The codon recognition pattern described above could well account for the reduced number of tRNAs utilized for mitochondrial protein synthesis. That a further simplification occurs in the mammalian mitochondrial genetic code is suggested by the fact that no tRNA gene for the arginine codons AG_G^A has been found, and that the codons AGA and AGG do not appear to be present in the significant reading frames of human mtDNA[12]. Thus, it is possible that protein synthesis can proceed in human mitochondria with only 23 tRNA species. (A tRNA gene for the methionine codons AU_G^A, although expected, has not yet been found in human mtDNA.) The proposed codon assignment and codon recognition pattern in mammalian mitochondria are shown in Fig. 2.

The symmetric transcription of HeLa cell mtDNA

One striking feature of the transcription process in HeLa cells is its symmetry[14]. Both strands are completely tran-

UUU ⎫ UUC ⎬ Phe AAG UUA ⎫ UUG ⎬ Leu AAU	UCU ⎫ UCC ⎪ UCA ⎬ Ser AGU UCG ⎭	UAU ⎫ UAC ⎬ Tyr AUG UAA ⎫ UAG ⎬ Ter	UGU ⎫ UGC ⎬ Cys ACG UGA ⎫ UGG ⎬ Trp ACU
CUU ⎫ CUC ⎪ CUA ⎬ Leu GAU CUG ⎭	CCU ⎫ CCC ⎪ CCA ⎬ Pro GGU CCG ⎭	CAU ⎫ CAC ⎬ His GUG CAA ⎫ CAG ⎬ Gln GUU	CGU ⎫ CGC ⎪ CGA ⎬ Arg GCU CGG ⎭
AUU ⎫ AUC ⎬ Ile UAG AUA ⎫ Met UAU ⌐AUG⌐ ⎬ F-Met UAC	ACU ⎫ ACC ⎪ ACA ⎬ Thr UGU ACG ⎭	AAU ⎫ AAC ⎬ Asn UUG AAA ⎫ AAG ⎬ Lys UUU	AGU ⎫ AGC ⎬ Ser UCG AGA ⎫ (Arg) (UCU) AGG ⎬ Ter?
GUU ⎫ GUC ⎪ GUA ⎬ Val CAU GUG ⎭	GCU ⎫ GCC ⎪ GCA ⎬ Ala CGU GCG ⎭	GAU ⎫ GAC ⎬ Asp CUG GAA ⎫ GAG ⎬ Glu CUU	GGU ⎫ GGC ⎪ GGA ⎬ Gly CCU GGG ⎭

Fig. 2. Genetic code and patterns of codon recognition in mammalian mitochondria. In each box, the codons (5'→3') are on the left and the anticodons (3'→5') on the right. The tRNA gene for Met (AUǦ), though expected from the codon usage, has not yet been found; no tRNA gene for Arg (AGǦ), nor any AGǦ codon within the significant reading frames, have been detected.

scribed[15,16]. Pulse-labeling experiments indicate that the light (L)-strand is transcribed at a rate which is two or three times higher than the heavy (H)-strand[17]. However, the L-strand transcripts have a much shorter half-life than that of the H-strand transcripts and do not accumulate in mitochondria to any significant extent[14]. The symmetry of transcription of HeLa cell mtDNA has to be contrasted with the asymmetry of informational content in this DNA. As discussed below, most of the genes identified in mtDNA are localized in the H-strand, while only a few scattered genes are present on the L-strand.

The transcription products of HeLa cell mtDNA

Table I lists the discrete transcripts which have been identified in HeLa cell mitochondria and partially characterized. Quantitatively, the two major classes are represented by the two high molecular weight rRNA components, 16S and 12S RNA, and 4S RNA. Mitochondrial rRNAs of human and animal cells in general, are the smallest known rRNA species, if one excludes the rRNAs of trypanosome mitochondria[18]. Human and hamster mitochondrial rRNAs are methylated[19-21] and in hamster cells, the level of this methylation (only in ribose residues in the large rRNA, in bases in the small rRNA), is considerably lower than that of the cytoplasmic rRNAs[20,21].

The 4S RNA components represent mostly, if not exclusively, mitochondria-specific tRNA species: on the basis of the recently determined mtDNA sequence, the 4S RNA is expected to include 23 species of tRNA specific for all amino acids, with two isoacceptors for serine and two for leucine[12]. Human mitochondrial tRNA sequences, derived from the DNA sequence, reveal only a partial agreement with the pattern, with some striking deviations[7,22]. They are generaly smaller than their cytoplasmic counterparts, varying in size between 59 and 75 nucleotides (Sanger, F. and Barrell, B., personal communication). In human and hamster mitochondrial tRNAs, the level of methylation (all in bases) is about 30% of that of the cytoplasmic tRNAs and, with significant differences in methylation pattern[19,23].

An important heterogeneous class of mtDNA transcripts in HeLa cells is represented by the poly(A) - containing RNAs[24-26]. Mitochondrial poly(A) in HeLa cells has been estimated to consist of stretches corresponding to about 4S$_E$ (~55 nucleotides)[24,25], i.e. considerably shorter than the poly(A) isolated from the mRNA of cytoplasmic polysomes, which is 150–200 residues long. Mitochondrial poly(A) is added post-transcriptionally to the 3'-end of mtDNA-coded RNA species[24]. A large number of poly(A)-containing RNA species have been resolved by electrophoresis through agarose slab gels in the presence of methylmercuric hydroxide as a denaturing agent (Table I)[26]. Among them, components 5, 7, 9, 11–16 and 17, because

TABLE I. Transcripts of HeLa cell mitochondrial DNA

RNA species	Molecular length[a] (number of nucleotides)	Functional assignment	Coding strand
a) Non-poly(A)-containing			
4S RNA	59–75	Includes 22 tRNAs	H; 14 tRNAs L; 8 tRNAs
12S RNA	954	Structural components	H
16S RNA	1559	of ribosomes	H
b) Poly(A)-containing			
1	~10400		L
2	~7070		L
3	~4155		L
4	~2700	Precursor of rRNAs?	H
5	2410		H
6	1938	Precursor of RNA 9?	H
7	1668		H
9	1617	COI mRNA	H
11	1141	Cyt. b mRNA	H
12	1042		H
13	958		H
14	842	ATPase 6 mRNA?	H
15	784	COIII mRNA	H
16	709	COII mRNA	H
17	346		H
18	~215		L

[a] Of non-poly(A) portion, determined from the length of the coding DNA sequences, except for RNAs 2, 3 and 18 (estimated from S1 protection data) and RNAs 1 and 4 (estimated from electrophoretic mobility).

of their relative abundance, their enrichment in partially purified polysomal structures, and their relatively long half-life[27,28], are probably specific mRNAs. Correlation of poly(A)-containing RNA sequences[29], or of reading frames in mtDNA[30], with protein sequence data and with known yeast gene sequences (A. Tzagoloff, personal communication) has recently provided strong support for the mRNA nature of several of these species. Thus, RNAs 9, 16 and 15 appear to be the specific messengers for cytochrome c oxidase subunit I (COI), subunit II (COII) and subunit III (COIII), respectively; RNA 11 is the putative mRNA for cytochrome b and RNA 14 the mRNA for a polypeptide exhibiting partial amino acid sequence homology to the subunit 6 of the yeast oligomycin sensitive ATPase. RNA species 1–4 and 6, because of their presence in only marginal amounts in partially purified polysomal structures and their relatively short half-life[27,28], are presumably either precursors or intermediates in the pathway of maturation of the functional species.

While the presence of poly(A) at the 3'-end of many mitochondrial RNA species represents a definite eukaryotic trait, mitochondrial poly(A)-containing RNAs from HeLa cells lack another eukaryotic attribute, namely, the presence of a 'cap' structure at their 5'ends[31].

Acknowledgements

These investigations were supported by a grant from NIH (GM-11726). I am very grateful to Drs F. Sanger and B. Barrell for communicating to me human mtDNA sequence data prior to publication.

References

1 Borst, P. and Grivell, L. A. (1978) *Cell*, 15, 705–723
2 Cummings, D. J., Borst, P., Dawid, I. B., Weissman, S. M. and Fox, C. F. (eds) (1979) *Extrachromosomal DNA*, ICN-UCLA Symposia on Molecular and Cellular Biology, 15, Academic Press, New York
3 Kroon, A. M. and Saccone, C. (eds) (1980) *The Organization and Expression of the Mitochondrial Genome*, North-Holland, Amsterdam (in press)
4 Borst, P. (1972) *Annu. Rev. Biochem.* 41, 333–376
5 Schatz, G. and Mason, T. L. (1974) *Annu. Rev. Biochem.* 43, 51–87
6 Albring, M., Griffith, J. and Attardi, G. (1977) *Proc. Natl. Acad. Sci. U.S.A.* 74, 1348–1352
7 Barrell, B. G., Bankier, A. T. and Drouin, J. (1979) *Nature (London)* 282, 189–194
8 Macino, G., Coruzzi, G., Nobrega, F. G., Li, M. and Tzagoloff, A. (1979) *Proc. Natl. Acad. Sci. U.S.A.* 76, 3784–3785
9 Heckman, J. E., Sarnoff, J., Alzner-DeWeerd, B., Yin, S. and Rajbhandary, U. L. (1980) *Proc. Natl. Acad. Sci. U.S.A.* 77, 3159–3163
10 Angerer, L., Davidson, N., Murphy, W., Lynch, D. and Attardi, G. (1976) *Cell*, 9, 81–90
11 Crick, F. H. C. (1966) *J. Mol. Biol.* 19, 548–555
12 Barrell, B. G., Anderson, S., Bankier, A. T., deBruijn, M. H. L., Chen, E., Coulson, A. R., Drouin, J., Eperon I. C., Nierlich, D. P., Roe, B. A., Sanger, F., Schreier, P. H., Smith, A. J. H., Staden, R. and Young, I. G. (1980) *Proc. Natl. Acad. Sci. U.S.A.* 77, 3164–3166
13 Bonitz, S. G., Berlani, R., Coruzzi, G., Li, M., Macino, G., Nobrega, F. G., Nobrega, M. P., Thalenfeld, B. E. and Tzagoloff, A. (1980) *Proc. Natl. Acad. Sci. U.S.A.* 77, 3167–3170
14 Aloni, Y. and Attardi, G. (1971) *Proc. Natl. Acad. Sci. U.S.A.* 68, 1757–1761
15 Aloni, T. and Attardi, G. (1971) *J. Mol. Biol.* 55, 251–270
16 Murphy, W. I., Attardi, B., Tu, C. and Attardi, G. (1975) *J. Mol. Biol.* 99, 809–814
17 Cantatore, P. and Attardi, G. (1980) *Nucl. Acids Res.* 8, 2605–2624
18 Borst, P., Hoeijmakers, J. H. J., Frasch, A. C. C., Snijders, A., Janssen, J. W. G. and Fase-Fowler, F. (1980) in Ref. 3 (in press)
19 Attardi, B. and Attardi, G. (1971) *J. Mol. Biol.* 55, 231–249
20 Dubin, D. T. and Taylor, R. H. (1978) *J. Mol. Biol.* 121, 523–540
21 Dubin, D. T., Taylor, R. H. and Davenport, L. W. (1978) *Nucl. Acids Res.* 5, 4385–4397
22 Crews, S. and Attardi, G. (1980) *Cell*, 19, 775–784
23 Davenport, L. W., Taylor, R. H. and Dubin, D. T. (1976) *Biochem. Biophys. Acta*, 447, 285–293
24 Hirsch, M. and Penman, S. (1973) *J. Mol. Biol.* 80, 379–391

25 Ojala, D. and Attardi, G. (1974) *J. Mol. Biol.* 82, 151–176
26 Amalric, F., Merkel, C., Gelfand, R. and Attardi, G. (1978) *J. Mol. Biol.* 118, 1–25
27 Attardi, G., Cantatore, P., Ching, E., Crews, S., Gelfand, R., Merkel, C. and Ojala, D. (1979) in Ref. 2, pp. 443–469
28 Gelfand, R. A. (1980) Ph.D. Thesis, California Institute of Technology, Pasadena, California
29 Attardi, G., Cantatore, P., Ching, E., Crews, S., Gelfand, R., Merkel, C., Montoya, J. and Ojala, D. (1980), in Ref. 3 (in press)
30 Walker, J. E., Anderson, S., Bankier, A. T., deBruijn, M. H. L., Chen, E., Coulson, A. R., Drouin, J., Eperon, I. C., Nierlich, D. P., Roe, B. A., Runswick, M. J., Sanger, F., Schreier, P. H., Smith, A. J. H., Staden, R. and Young, I. G. (1980), in Ref. 3 (in press)
31 Grohmann, K., Amalric, F., Crews, S. and Attardi, G. (1978) *Nucl. Acids Res.* 5, 637–652

Giuseppe Attardi is at the Division of Biology, California Institute of Technology, Pasadena, California 91125, U.S.A.

Organization and expression of the mammalian mitochondrial genome: a lesson in economy-II

Giuseppe Attardi

General organization of HeLa cell mtDNA transcripts

Fig. 1 shows a detailed transcription map of HeLa cell mtDNA, which has been constructed by precisely localizing in the physical map of this DNA by a variety of RNA–DNA hybridization techniques the sequences coding for the rRNA and poly(A)-containing RNA species[1]. In this diagram, the positions and identities of the tRNA genes have been derived from the mtDNA sequence[2], which has been aligned with the physical map. The two rRNA genes, reproducing an arrangement common in the rDNA of bacterial and eukaryotic cells, are close to each other, being separated only by a tRNA gene. In this respect, they differ from the yeast mitochondrial rRNA genes which are far apart from each other. The direction of transcription of the rRNA genes is from the small to the large one, in agreement with the pattern observed in all rDNAs analysed thus far, with the exception of the kinetoplast rDNA[3]. With the exclusion of the D-loop region and of a stretch of about 170 nucleotides between the 5'-end of the H-strand coded tRNA[Phe] and the 5'-end of the L-strand coded RNA 18, the HeLa cell mtDNA sequences are completely saturated by the rRNAs, poly(A)-containing RNAs and tRNAs coded for by the two strands.

No evidence for intervening sequences has been found in the mtDNA segments coding for the rRNAs and poly(A)-containing RNAs; a similar observation has been made in the *Xenopus laevis* mitochondrial genome[4]: this is in contrast with the well-documented occurrence of intervening sequences in the mitochondrial genes of lower eukaryotes[5-7]. With the exception of RNA 4 and RNA 6, which are probably precursors of the two rRNA species and of RNA 9, respectively, there is no apparent overlapping in the H-strand of the sequences coding for the various poly(A)-containing RNA, rRNA and tRNA species. Also the L-strand sequences which specify what are presumably mature RNA products (i.e. RNA 18 and tRNAs) either fall within non-coding regions of the H-strand or correspond to 5'-end or 3'-end proximal segments of H-strand coded poly(A)containing RNAs lying outside the polypeptide coding stretches. The transcription map of human mtDNA is similar to that of *Xenopus laevis* mtDNA[4] and, somewhat less, to that of mouse mtDNA[8].

Mitochondrial rRNA and mRNA coding sequences are immediately contiguous to tRNA genes

The alignment of the positions of the tRNA genes derived from the mtDNA sequence with the transcription map (Fig. 1) has unexpectedly revealed that the sequences coding for the rRNA and poly(A)-containing RNA species on the H-strand are immediately contiguous or very close on one side, and most frequently on both sides, to tRNA coding sequences. These results have been confirmed and refined by recent DNA and RNA sequencing analysis data which have shown that the 5'-end of the coding sequences for 12S rRNA, 16S rRNA and most poly(A)-containing RNAs immediately follows a tRNA gene, as identified from the DNA sequence[1]. Furthermore, preliminary observations on a few poly(A)-containing RNA species indicate that the 3'-terminal nucleotide of these RNAs corresponds to a residue immediately contiguous to the 5'-end of a tRNA gene; likewise the 3'-end of the sequence coding for 12S rRNA (recognized for its homology to the hamster small mitochondrial rRNA sequence[9]) is immediately contiguous to the tRNA^{Val} gene. There are a few exceptions to the rule that each end of a coding sequence for an rRNA or poly(A)containing RNA species is immediately flanked by a tRNA gene. Thus, the RNA 9 coding sequence does not have a tRNA gene on its 5'-side; however, this RNA is probably not a primary transcription product, but is derived from RNA 6 by removal of a 5'-terminal stretch containing sequences complementary to four L-strand coded tRNA species. The coding sequences for RNAs 14 and 15 are apparently not separated by a tRNA gene; the mapping results presented here and recent sequence data indicate that their coding sequences are immediately contiguous on the H-strand, suggesting the possibility of a common processing intermediate giving rise to the two mature species by a precise endonucleolytic cleavage. A similar situation seems to occur for RNAs 5 and 11.

The 'punctuation' model of H-strand gene expression

The mapping data presented above and the sequence data available indicate that the H-strand sequences coding for the rRNA species, the poly(A)containing RNAs and the tRNAs are immediately contiguous to each other, extending continuously from coordinate 2/100 to coordinate 95/100 (relative to the origin taken as 0/100). This arrangement is consistent with a model of H-strand transcription in the form of a single molecule processed by precise endonucleolytic cleavages before and

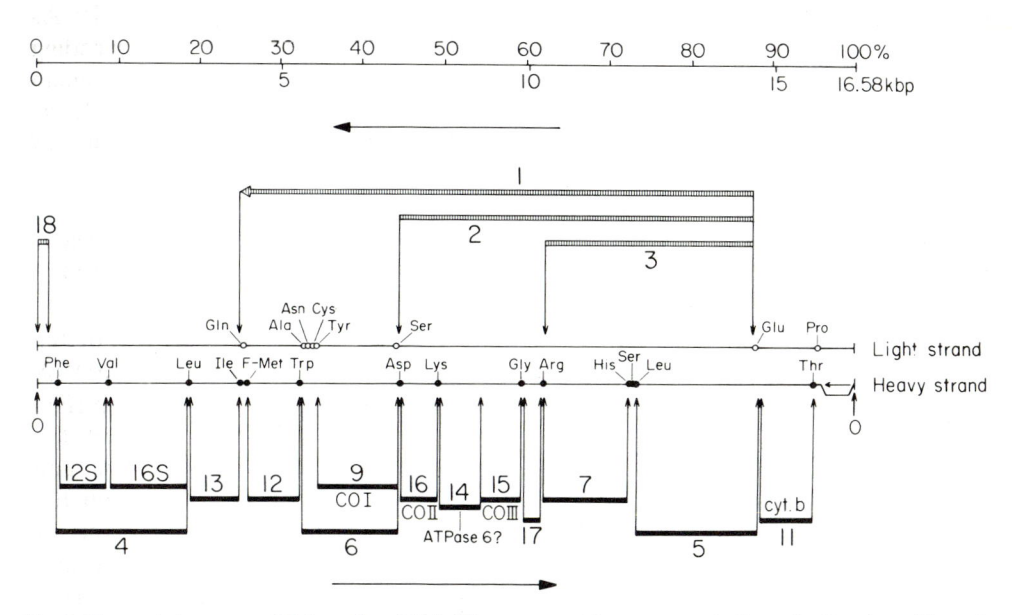

Fig. 1. Transcription map of HeLa cell mtDNA. The upper and lower arrows indicate the direction of L- and H-strand transcription, respectively. See text for details.

after each tRNA sequence to yield the mature products or, in some cases, processing intermediates, like the putative precursors of the rRNA (RNA 4) and of RNA 9 (RNA 6)[1]. A recent mapping study of nascent mitochondrial RNA molecules isolated from transcription complexes of HeLa cell mtDNA[10] has pointed to the existence of an initiation site for H-strand transcription near the origin of replication: such a site may thus represent the promoter of the single large transcripts postulated here.

In the processing of primary transcripts, the tRNA sequences may play an important role as recognition signals, providing the punctuation in the reading of mtDNA information[1,11]. It is conceivable that the processing enzyme(s) recognizes the cloverleaf structure of the tRNA sequence or some portion of it. One interesting observation is that all the RNA species deriving from the initial processing of the primary H-strand transcripts are apparently polyadenylated. This suggests that polyadenylation may be linked in some way to the processing event, independently of the functional role of the products. This may account for the polyadenylation of the putative rRNA precursor (RNA 4) (see also Ref. 12). The observation that many reading frames in human mtDNA apparently lack a stop codon, and that, in these cases, a T or TA follows the last sense codon and immediately precedes a tRNA gene or another reading frame has led to the suggestion that poly(A) addition to T or TA may create the missing stop codon[2].

If a single transcript gives rise to all mitochondrial mature RNA species coded for by the H-strand, premature termination of transcription past the rRNA cistrons and differences in the metabolic stability of the various RNA species may account for the large excess of the rRNA species over mRNAs and for the quantitative differences among the latter[1].

The L-strand transcripts

Although the picture of L-strand transcription is still very preliminary, the available evidence suggests that it may follow the same pattern as the H-strand transcription. In particular, the tRNA sequences may play a role in the processing of L-strand transcription products[1]. The three large L-strand transcripts (RNAs 1, 2 and 3) have an apparently common 5'-end very close to an L-srand tRNA gene (tRNA^{Glu}): this common 5'-end may represent a processing point for primary transcripts having their initiation site near the origin of replication[10]. The 3'-end of RNA 2 maps very close to an L-strand tRNA gene [tRNA^{Ser} (L)], while the 3'-end of RNA 3 appears to correspond to a position in the H-strand very close to the tRNA^{Arg} gene. These transcripts may result from termination of transcription at alternative fixed points or may represent successive steps in a processing scheme. The physiological significance of these large RNAs is not clear. They may be precursors or intermediates in the processing of L-strand coded tRNAs. However, the possibility that these transcripts or their possible derivatives may function as mRNAs, or may serve some other role, possibly related to H-strand gene expression, cannot be excluded.

The L-strand coded RNA 18 has been mapped, on the basis of RNA–DNA hybridization experiments and RNA sequencing analysis, in the region immediately preceding the origin of replication in the direction of L-strand transcription, with its 3'-end corresponding to or very near to this origin[1]. This RNA may be the mRNA of a small polypeptide, or have some other function related to the replication or transcription of mtDNA.

Most human mitochondrial mRNAs start at or very near to the initiator codon

One striking result of the sequencing

UUU } Phe AAG UUC UUA } Leu AAU UUG	UCU UCC } Ser AGU UCA UCG	UAU } Tyr AUG UAC UAA } Ter UAG	UGU } Cys ACG UGC UGA } Trp ACU UGG
CUU CUC } Leu GAU CUA CUG	CCU CCC } Pro GGU CCA CCG	CAU } His GUG CAC CAA } Gln GUU CAG	CGU CGC } Arg GCU CGA CGG
AUU } Ile UAG AUC AUA } Met UAU [AUG] [F-Met UAC]	ACU ACC } Thr UGU ACA ACG	AAU } Asn UUG AAC AAA } Lys UUU AAG	AGU } Ser UCG AGC AGA } (Arg) (UCU) AGG } Ter?
GUU GUC } Val CAU GUA GUG	GCU GCC } Ala CGU GCA GCG	GAU } Asp CUG GAC GAA } Glu CUU GAG	GGU GGC } Gly CCU GGA GGG

Fig. 2. Genetic code and patterns of codon recognition in mammalian mitochondria. In each box, the codons (5'→3') are on the left and the anticodons (3'←5') on the right. The tRNA gene for Met (AU$_G^A$), though expected from the codon usage, has not yet been found; no tRNA gene for Arg (AG$_G^A$), nor any AG$_G^A$ codon within the significant reading frames, have been detected.

analysis of the HeLa cell mitochondrial RNAs has been that almost all putative mRNAs analysed either start directly with an AUG or AUA triplet (a methionine codon in human mitochondria, Fig. 2), or have a few nucleotides (1–8) preceding the AUG or AUA[1]. Sequencing data of bovine and human cytochrome c oxidase subunits and comparison with the yeast cytochrome b and COIII gene sequences have indicated that the AUG of RNA 9 (COI mRNA), of RNA 16 (COII mRNA) and of RNA 11 (cytochrome b mRNA) are initiator codons for the corresponding polypeptides[1,2]. Thus, it seems reasonable to extrapolate from these results and interpret the AUGs of the other mRNAs likewise as initiator codons. There is the possibility that AUA may also function as initiator codon in human mitochondria[4]. A single mRNA (#12), among those analysed, had a long stretch (108 nucleotides) preceding the first AUG.

The observation that most mitochondrial mRNAs either start directly at the initiator codon or have only a few nucleotides (1–8) preceding this codon raises interesting questions concerning the mechanism whereby mitochondrial ribosomes attach to these messengers. Both in eukaryotic and prokaryotic mRNAs, there is a stretch of variable length preceding the initiator codon, and in *E. coli* mRNAs there is evidence that this stretch contains a ribosome-binding site[13,14]. In the case of mitochondria from human cells (and probably from all animal cells), it is reasonable to assume that the special features of their ribosomes make them suitable for binding directly to the initiator codon.

Human mitochondrial genes and gene products

On the basis of a functional analogy with the mitochondrial genetic systems from lower eukaryotes, in particular, yeast, human mtDNA would be expected to code for three subunits of cytochrome *c* oxidase, two to three subunits of oligomycin sensitive ATPase, one subunit of cytochrome bc_1 complex and one protein associated with mitochondrial ribosomes[5-7]. Of these potential genes, as mentioned above, those for the three subunits of cytochrome *c* oxidase, for cytochrome *b* and for a polypeptide presumably homologous to yeast ATPase subunit 6 have already been identified.

In yeast, it is generally believed that the mtDNA gene products listed at the beginning of this section and the corresponding genetic loci, saturate or nearly saturate the yeast mitochondrial genetic map[5-7]. However, in human mitochondria, the situation appears to be more complicated. 13 reading frames longer than 200 nt have been found in the human mtDNA sequence, 12 frames on the H-strand and one on the L-strand[2]; furthermore, 10 H-strand coded mRNAs and at least one potential L-strand coded mRNA have been identified and mapped on human mtDNA. There is an excellent agreement between the map of the H-strand reading frames and that of the putative H-strand coded mRNAs[1,2].

Among the mitochondrial translation products from HeLa cells, defined on the basis of the emetine resistance and chloramphenicol sensitivity of their labeling, 25 discrete components, mol. wt range 3500–51,000 have been identified by bidimensional electrophoresis through an SDS/8 M urea polyacrylamide slab gel and an SDS-polyacrylamide gradient slab gel[1,15]. Unfortunately, there is still very little information on the nature of mitochondrial translation products in human and other animal cells. Among these, only the three subunits of cytochrome *c* oxidase have been identified with any degree of confidence in mitochondria from human cells[16], rat liver cells[17] and *Xenopus* oocytes[18]. It can reasonably be expected that polypeptides corresponding to the other genes already identified in human mtDNA (cytochrome *b*, ATPase 6) will be found fairly soon.

The nature of the remaining polypeptides can only be a matter of conjecture at present. Mitochondrially synthesized ribosomal proteins and other subunits of the ATPase complex already identified in yeast mitochondria are possible candidates. The recent evidence pointing to the synthesis in the yeast organelles of a 'maturase' active in RNA splicing[19] suggests that, in mammalian mitochondria, RNA processing may be under control of the mitochondrial genome, with one or more gene products being involved. The analysis of the nature of the unidentified mitochondrial translation products in mammalian cells and of their relationship represents a formidable challenge for the future, and may reveal unexpected facets of the expression and regulation of the mitochondrial genome in these cells.

mtDNA in other animal cells

A comparison of the bovine mtDNA with the human mtDNA sequence has recently revealed that the detailed gene organization is identical, for the most part, in the genomes of the two species[2]. A similar conclusion is emerging from the available sequence data for mouse[20] and rat mtDNA[21]. Thus, it seems reasonable to extrapolate from these results and predict that mtDNA from all mammals will be found to have essentially the same gene organization and mode of expression, in spite of the fairly rapid sequence divergence which has occurred in the mtDNAs of different species[22]. It also seems likely from the limited available evidence[4,22] that the general features of mitochondrial genome organization have been conserved throughout the animal kingdom, at least among vertebrates.

Evolution to simplicity

We do not know whether the simpler decoding mechanism operating in mitochondria which uses fewer tRNAs than the known prokaryotic and eukaryotic systems, represents the remnant of a primitive mechanism, or a highly evolved form which has emerged in response to particular selective pressures. It is significant, however, that the same codon recognition rules seem to apply to the mitochondrial systems of organisms as far apart in evolution as yeast and man.

In contrast to the apparent uniformity of the mitochondrial decoding mechanism in a large spectrum of eukaryotic cells, the mitochondrial gene organization exhibits an extraordinary diversity in different organisms. At one extreme, there is the loose organization of yeast mtDNA, where the genes are separated by AT-rich stretches and several genes are discontinuous. At the other extreme, there is the extraordinarily compact and lean gene organization of mammalian mtDNA, with its continuous genes mostly butt-jointed to each other and a nearly complete absence of non-coding stretches. The clustered arrangement of the tRNA genes in yeast and *Neurospora crassa* mtDNA likewise contrasts sharply with their scattered distribution in mammalian mtDNA, where they separate the rRNA and structural genes with nearly absolute regularity. This very improbable tRNA arrangement points to a functional role for tRNA sequences related to their position in mtDNA, possibly as recognition signals for RNA processing. It is a plausible idea that the interspersion of tRNA genes with rRNA and structural genes and their immediate juxtaposition have emerged in the course of evolution as a type of organization simplifying the synthesis and processing of mitochondrial RNA and thus the expression of the mitochondrial genes. In the same line of thought, the smaller size of rRNAs and tRNAs, their simpler methylation pattern and the lack of 5' and 3' non-coding stretches in the majority of mRNAs may be viewed as other structural traits marking the evolution to simplicity of mammalian mtDNA.

Acknowledgements

These investigations were supported by a grant from NIH (GM-11726). I am very grateful to Drs F. Sanger and B. Barrell for communicating to me human mtDNA sequence data prior to publication.

References

1 Attardi, G., Cantatore, P., Ching, E., Crews, S., Gelfand, R., Merkel, C., Montoya, J. and Ojala, D. (1980) in Ref. 7 (in press)
2 Walker, J. E., Anderson, S., Bankier, A. T., deBruijn, M. H. L., Chen, E., Coulson, A. R., Drouin, J., Eperon, I. C., Nierlich, D. P., Roe, B. A., Runswick, M. J., Sanger, F., Schreier, P. H., Smith, A. J. H., Staden, R. and Young, I. G. (1980) in Ref. 7 (in press)
3 Borst, P., Hoeijmakers, J. H. J., Frasch, A. C. C., Snijders, A., Janssen, J. W. G. and Fase-Fowler, F. (1980) in Ref. 7 (in press)
4 Rastl, E. and Dawid, I. B. (1979) *Cell*, 18, 501–510
5 Borst, P. and Grivell, L. A. (1978) *Cell*, 15, 705–772
6 Cummings, D. J., Borst, P., Dawid, I. B., Weissman, S. M. and Fox, C. F. (eds) (1979) *Extrachromosomal DNA*, ICN-UCLA Symposia on

Molecular and Cellular Biology, 15, Academic Press, New York
7 Kroon, A. M. and Saccone, C. (eds) (1980) *The Organization and Expression of the Mitochondrial Genome*, North-Holland, Amsterdam (in press)
8 Battey, J. and Clayton, D. A. (1978) *Cell*, 14, 143–156
9 Dubin, D. T. and Baer, R. J. (1980) in Ref. 7 (in press)
10 Cantatore, P. and Attardi, G. (1980) *Nucl. Acids Res.* 8, 2605–2624
11 Attardi, G., Cantatore, P., Ching, E., Crews, S., Gelfand, R., Merkel, C. and Ojala, D. (1979) in Ref. 6, pp. 443–469
12 Cleaves, G. R., Jones, T. and Dubin, D. T. (1976) *Arch. Biochem. Biophys.* 175, 303–311
13 Shine, J. and Dalgarno, L. (1974) *Biochem. J.* 141, 609–615
14 Steitz, J. A. and Jakes, K. (1975) *Proc. Natl. Acad. Sci. U.S.A.* 72, 4734–4738
15 Ching, E. (1979) Ph.D. Thesis, California Institute of Technology, Pasadena, California
16 Hare, J. F., Ching, E. and Attardi, G. (1980) *Biochemistry*, 19, 2023–2030
17 Rascati, R. J. and Parsons, P. (1979) *J. Biol. Chem.* 254, 1594–1599
18 Koch, G. (1976) *J. Biol. Chem.* 251, 6097–6107
19 Jacq, C., Lazowska, J. and Slonimski, P. P. (1980) in Ref. 7 (in press)
20 Van Etten, R. A., Walberg, M. W. and Clayton, D. A. (1980) *Cell* (in press)
21 Saccone, C., Cantatore, P., Pepe, G., Holtrop, M., Gallerani, R., Quagliariello, C., Gadaleta, G. and Kroone, A. M. (1980) in Ref. 7 (in press)
22 Dawid, I. B., Klukas, C. K., Ohi, S., Ramirez, J. L. and Upholt, W. B. (1976) in *The Genetic Function of Mitochondrial DNA* (Saccone, C. and Kroon, A. M., eds), pp. 3–13, North-Holland, Amsterdam

Giuseppe Attardi is at the Division of Biology, California Institute of Technology, Pasadena, California 91125, U.S.A.

Mitochondrial genes and male sterility in plants

Christopher J. Leaver

Cytoplasmically inherited male sterility is of considerable economic importance in the production of hybrid seed for many crop plants. Recent evidence suggests that the genes responsible for this trait are localized on mitochondrial DNA.

It has been recognized since the beginning of this century that the cytoplasm of higher plants is an important source of genetic variation. The basis for this variation was revealed in the early 1960s by the discovery that both mitochondria and chloroplasts contain their own unique DNA. Although considerable progress has been made in our understanding of the structure and role of chloroplast DNA[1], very little is known of the information content and function of mitochondrial DNA in higher plants.

Most of the research on the mitochondrial genetic system has been carried out with the yeast *Saccharomyces cerevisiae*, the mold *Neurospora crassa* and with animal cells. This work has shown that the mitochondrion contains a genetic system capable of replicating, transcribing and translating the genetic information of the mitochondrial DNA (mtDNA)[2,3]. Mitochondrial DNA normally occurs as a covalently closed circular duplex molecule. Its size however varies over a ten-fold range; animal cells contain mtDNA with contour length of 5–6 μm (9–12 × 10^6 mol. wt), while mtDNA of *Saccharomyces* and *Neurospora* has a contour length of about 25 μm (50 × 10^6 mol. wt).

In addition to the mitochondrial RNAs, the mitochondrial genome of fungi and animals codes for 8–10 rather hydrophobic polypeptides; seven of these have been identified as components of three oligomeric enzyme complexes of the inner mitochondrial membrane: three are subunits of cytochrome *c* oxidase, one is apocytochrome *b* (a subunit of the cytochrome bc_1 complex) and two are subunits of the oligomycin-sensitive ATPase complex[4]. One additional mitochondrial translation product (VAR-1) has been identified in yeast and *Neurospora* as a component of the small subunit of mitochondrial ribosomes. The remainder of the mitochondrial proteins are encoded in nuclear DNA, synthesized on cytoplasmic ribosomes and subsequently transferred into the mitochondrial structure. Thus, the respiratory enzyme complexes described above, which are responsible for key steps in the process of oxidative phosphorylation and the generation of ATP, are built up of a combination of mitochondrially and cytoplasmically synthesized polypeptides. The structural and functional integration of each group of subunits into the membrane requires the coordinate synthesis of the other[5].

Mitochondrial DNA of higher plants

The mtDNA of higher plants appears to be an order of magnitude larger than animal mtDNAs and perhaps 2–3 times larger than that of *Saccharomyces*. However there is no general agreement in the literature on the precise size of the plant mitochondrial genome and different values have been obtained depending upon the plant species and method of analysis. Electron microscopic analysis suggests that pea mtDNA consists of a homogeneous population of circular molecules with contour length of 30 μm (70 × 10⁶ mol. wt). More detailed analysis of maize, soybean and sorghum mtDNA revealed the presence of several discrete size classes of circular molecules ranging from less than 5 μm to 30 μm. Digestion of different plant mtDNAs with restriction enzymes yields a surprisingly large number of fragments of differing molar ratios. The pattern obtained is much more complex than predicted from known physical size and suggests that substantial intraspecific as well as interspecific diversity exists among plant species[6]. Furthermore, estimates based on such data suggest molecular weights for mtDNA of between 117 and 200 × 10⁶. The buoyant density of these mtDNAs is in the range 1.705–1.706, indicating a G-C content of c. 47% which together with data on kinetics of mtDNA renaturation, argues against the presence of A-T-rich 'spacer' regions such as those found in yeast mtDNA (G-C content of 18%)[2].

In summary, plant mtDNA appears to be heterogeneous, containing several classes of molecules which may have different sequence arrangements of the same genetic information.

In contrast the circular molecules of chloroplast DNA of a given plant species all have the same contour length (c. 45 μm) which is species specific. Data obtained by a number of techniques suggest that the chloroplast DNA molecules from a single species have the same sequence and that the molecular weights are always in the range 85 to 103 × 10⁶ (ref. 1).

Does plant mtDNA contain additional genes?

Since plant mtDNA is apparently bigger than fungal or animal mtDNA this raises the obvious question, why do plants require so much additional mtDNA to perform the same basic metabolic function of respiration and the generation of energy? Is it possible that plant mtDNA contains additional coding information not found in the mtDNA of other organisms?

Plant mtDNA, like mtDNAs from yeast and animals codes for two specific mitochondrial ribosomal RNAs (rRNA) and about 20 transfer RNAs[7]. Of particular interest has been the identification of a unique mitochondrial 5 S rRNA since such an RNA has not been detected in the mitochondrial ribosomes of fungal and animal cells[8].

In an attempt to determine whether or not plant mitochondria can synthesize (and by extrapolation, contain the genetic information for) additional polypeptides which are not synthesized by mitochondria of other organisms, we have analysed the proteins synthesized *in vitro* by intact mitochondria[9,10]. Mitochondria were isolated from several different plant tissues and incubated in a medium containing [^{35}S]methionine under sterile conditions optimized for the incorporation of amino acids into protein. In Fig. 1 the polypeptides synthesized by mitochondria from dark-grown maize (*Zea mays* L.) shoots are compared with those synthesized by isolated yeast mitochondria. Mitochondria from maize and a range of other higher plants synthesize at least 18 polypeptides compared with the 8–10 synthesized by yeast mitochondria. The translation products of plant mitochondria have molecular weights ranging from 8–54 × 10³ and with only one major exception (a polypeptide of molecular weight 54 × 10³) all are membrane bound. An essentially similar specturm of polypeptides is synthesized by mitochondria *in vivo* when plant cells are pulse labelled with [^{35}S]methionine in the presence of cycloheximide, an inhibitor of cytoplasmic protein synthesis. By immunoprecipitation with monospecific antibodies against purified yeast cytochrome oxidase subunits, we have tentatively identified two of the plant products as subunits I and II of cytochrome oxidase.

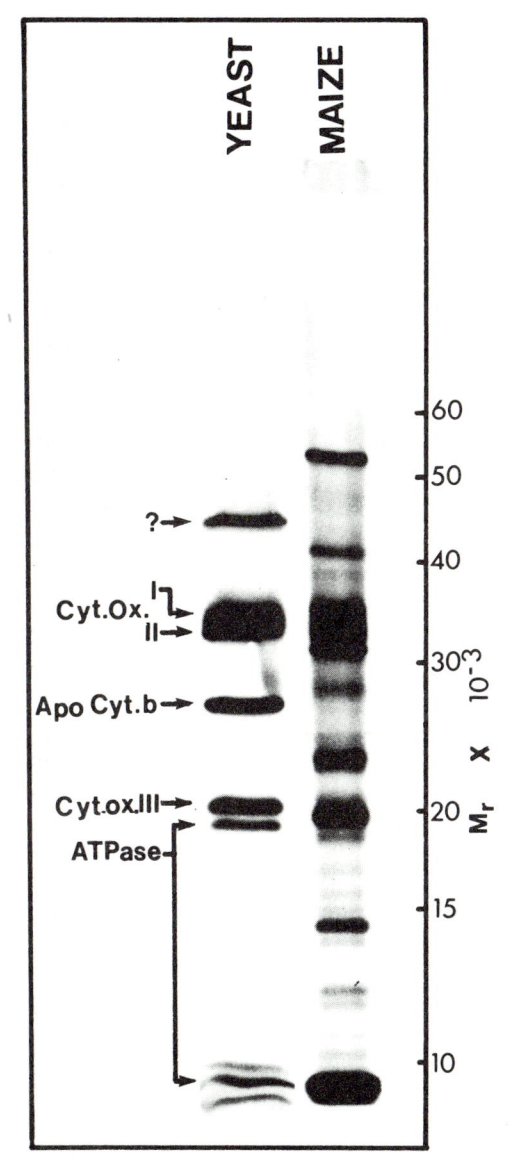

Fig 1. Comparison of polypeptides synthesized by mitochondria from yeast and maize. Mitochondrial translation products were labelled by incubating mitochondria in a medium containing [^{35}S]methionine. The proteins were solubilized in SDS, electrophoresed in 15% (w/v) polyacrylamide slab gels and labelled polypeptides detected by autoradiography.

Even by one-dimensional electrophoresis it is possible to detect a further 10 minor labelled polypeptides besides the 18 major translation products. It is not known whether some or all of these minor polypeptides are the result of aggregation or degradation of other polypeptides or are perhaps precursor proteins.

Nevertheless it seems likely that plant mitochondria do indeed synthesize a larger number of different polypeptides than yeast. If it is assumed plant mtDNA contains one cistron for each of the three rRNAs and 20 tRNAs, and taking into account the estimated number and molecular weight of polypeptides synthesized by plant mitochondria, the total required coding capacity can be estimated to be $c.\ 15 \times 10^6$ daltons; this value is approximately 50% larger than that calculated for yeast mtDNA and still does not account for 80–90% of the potential coding capacity of plant mtDNA[9].

Mitochondrial DNA and cytoplasmic male sterility in higher plants

The impressive progress in mapping genes localized on yeast mtDNA has in large part been made possible by the availability of a wide variety of mutants whose characteristic phenotype results from a mutation in the mtDNA[11]. The unique ability of yeast to live by glycolysis alone, i.e. without functional mitochondria, makes it possible to study mutations which drastically interfere with mitochondrial biogenesis. Genetic analysis of mtDNA of higher plants is much more difficult. Many more of the mutations may be lethal and, unlike many chloroplast mutations, many of the non-lethal mutations in plant mtDNA might not be detected.

However, there is one cytoplasmically inherited trait in higher plants in which the mitochondrion has been strongly implicated. This maternally inherited trait,

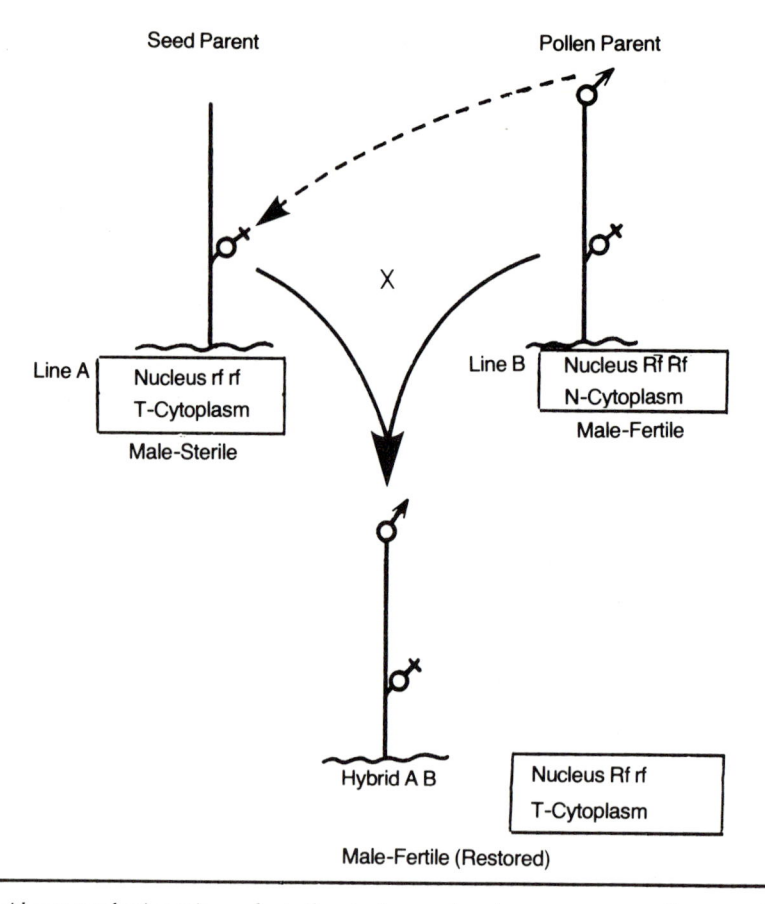

Fig. 2. Hybrid corn production using male-sterile cytoplasm and nuclear restorer genes. N = normal cytoplasm, T = Texas cytoplasm induces pollen sterility, Rf = dominant nuclear gene which restores fertility, rf = recessive nuclear gene with no effect on sterility.

termed cytoplasmic male sterility (CMS), causes pollen to abort in the anthers and thereby prevents self fertilization. Crossing between a CMS plant and a male-fertile partner results in a completely male-sterile progeny, since the cytoplasmic genes responsible for sterility are transferred from the mother to all the offspring via the egg cells; the paternal parent only provides a naked nucleus and does not affect the cytoplasmic constitution[12].

The CMS phenotype has been of major economic importance as the basis for commercial production of hybrid seed varieties, i.e. the crossing of particular inbred lines resulting in F_1 hybrid offspring far superior in productivity, yield and other characteristics to the parents. Today hybrid varieties of many crops are grown commercially (e.g. maize, sorghum, sunflower, sugar-beet) although for a long time hybridization was confined to maize[12]. The production of hybrid varieties in maize was largely facilitated by the occurrence of male and female flowers in separate inflorescences on the same plant. To form hybrid seed, alternating rows of seed lines and pollinator lines were planted. The male flowers (or tassels) of the seed lines were manually removed before they shed their pollen and were then left to be pollinated by the paternal lines. Hand emasculation is however an extremely laborious and time consuming operation and in a crop like sorghum, with numerous, tiny, bisexual flowers grouped in one big inflorescence, almost impossible.

The elimination of this labour intensive and expensive operation of hand emasculation of the seed parent plant by the commercial use of cytoplasmic male sterility began in the production of hybrid maize. This was made possible by the discovery that certain maize lines carry nuclear genes that suppress the male-sterile phenotype and thus restore full pollen fertility. Such genes are called 'restorer genes' or Rf genes[12].

Thus in the production of hybrid maize seed based on cytoplasmic male sterility (see Fig. 2) alternating male-sterile seed and male-fertile pollinator rows are planted next to each other. The female parent (Line A) carries the male-sterile, so called Texas cytoplasm (cms-T) and produces no pollen, therefore eliminating the need for detasseling. The pollen parent (Line B) is homozygous dominant for nuclear restorer alleles (Rf genes). Thus even though the hybrid F_1 generation of seed sold to the farmer carries the male-sterile cytoplasm it will be heterozygous for the Rf genes and hence male-fertile and able to produce corn.

Subsequent genetic analysis has shown that there are three types of male sterility in maize, designated T, C and S, each of which is suppressed (i.e. pollen fertility is restored) by different nuclear Rf genes. The T-source of cytoplasmic male sterility (cms-T) has predominated over other forms, both commercially and in research. It was widely used to produce hybrid corn in America and by 1970 over 80% of the commercially grown hybrid lines carried this cytoplasm. This high degree of genetic uniformity was in retrospect a major factor responsible for the disastrous epidemic of Southern corn leaf blight which swept the United States in 1970, causing losses in maize production of over one thousand million dollars. This epidemic was caused by the fungal pathogen *Helminthosporium maydis* (*Bipolaris maydis*) race T, which preferentially affected plants with T cytoplasm while leaving those with normal (N), 'fertile', cytoplasm virtually untouched. It was soon established that this maternal inheritance of disease susceptibility was in large part due to the production of a host specific toxin (T-toxin) by the fungus.

Several lines of evidence indicate that the genetic determinants that control cytoplasmic male sterility and susceptibility to *H. maydis*, race T-toxin, are carried by the mitochondrion and not the chloroplast. (1) Ultrastructural studies on normal and male-sterile anthers show that mitochondrial degeneration is the first indication of abnormality during pollen development[13]. (2) Restriction endonuclease fragment analysis of chloroplast DNA and mitochondrial DNA has revealed that there are distinct and characteristic differences between the mtDNAs of normal and cms-T plants, although the chloroplast DNAs were indistinguishable[4]. (3) Isolated mitochondria from cms-T lines have been shown to be susceptible to the host-specific toxin of *H. maydis*, race T, whereas N (and C and S) mitochondria are insensitive. The effects of the toxin on cms-T mitochondria include the uncoupling of oxidative phosphorylation and leakage of NAD and other coenzymes from the mitochondria[13]. (4) Mitochondria isolated from normal or cms-T cytoplasm synthesize a virtually identical spectrum of polypeptides. However, one 21,000 M_r polypeptide detected in N mitochondria is missing from cms-T mitochondria and replaced by a polypeptide of 13,000 M_r (see Fig. 3)[9,10]. The synthesis of this variant 13,000 M_r polypeptide is characteristic of cms-T mitochondria in all fifteen nuclear backgrounds so far examined. In addition to being controlled by at least one cytoplasmic gene, the synthesis of the 13,000 M_r polypeptide is controlled by certain nuclear genes. When alleles that restore fertility TRf genes) are introduced into the nuclear background of lines carrying the T cytoplasm there is a marked and specific reduction in the accumulation of the 13,000 M_r polypeptide by isolated mitochondria (see Fig 3), and the polypeptide is no longer detectable in gels stained for protein. However, in none of the restored lines examined was synthesis of the 21,000 M_r polypeptide ever partially recovered[14].

An indication that pollen abortion and susceptibility to the toxin are different expressions of a single genetic defect is given by the finding that mitochondria from TRf lines are significantly less sensitive to H. maydis, race T toxin, than mitochondria from non-restored lines. In addition, selection for resistance to T toxin in tissue cultures derived from cms-T plants has allowed the regeneration of male-fertile, toxin-resistant maize plants[15]. Mitochondria isolated from such spontaneous revertants do not synthesize the 13,000 M_r polypeptide and restriction enzyme analysis of their mtDNA shows that it differs from both the parental cms-T and normal mtDNA.

If it can be demonstrated conclusively, with the help of well-characterized maize genotypes, that suppression of synthesis of the 13,000 M_r polypeptide and reduction in sensitivity to the toxin are under the control of the same nuclear restorer gene(s), this will provide further strong circumstantial evidence that the variant polypeptide is involved in both male sterility and susceptibility to H. maydis, race T toxin.

A model has been proposed by Flavell which suggests that cms-T mitochondria have an altered structure which makes them susceptible to both toxin and to a similar (as yet unidentified) substance produced during the normal course of pollen development[16]. The 13,000 M_r polypeptide synthesized by cms-T mitochondria could be responsible for this altered structure. As a simple working hypothesis, the role of this variant polypeptide in the traits associated with the T cytoplasm could include: (1) providing a binding site for the hypothetical compound and the toxin; (2) altering membrane conformation, so creating an exposing or binding site on another membrane component; or (3) by making the membrane more susceptible to disruption by these substances. There are a number of mechanisms by which nuclear restorer genes could specifically regulate the synthesis of the 13,000 M_r polypeptide. These include direct interaction between the nuclear gene product and the mitochondrial protein synthesizing system, an indirect effect operating through changes in the concentration of an effector molecule, or nuclear control of the processing of mitochondrial mRNA or mitochondrial protein[10].

Other male-sterile cytoplasms in maize

There are two other types of cytoplasmic male-sterility occurring in maize besides the T cytoplasm. These other male-sterile cytoplasms are designated cms-C and cms-S and are distinguished from each other and from T cytoplasm by the different nuclear genes which are required for fertility restoration.

A comparison of the translation products of mitochondria from these cytoplasms show that they too differ from those of N cytoplasm. The cms-C mitochondria synthesize a variant 17,500 M_r polypeptide which replaces one of 15,500 M_r that is synthesized by N mitochondria (and by cms-T and cms-S mitochondria)[9]. Mitochondria from cms-S plants can be distinguished by their ability to synthesize eight additional polypeptides in the molecular weight region 42–85 × 10³. Intriguingly, cms-S mitochondria also possess two mitochondria are encoded by unique DNA species. Alternatively, they may result from a fault in the processing of a primary mRNA transcript or protein precursor.

Prospects

Current research on mitochondrial and chloroplast genome organization and expression has renewed interest in the role of cytoplasmic genes in plant development and crop improvement. Despite the large number of plant breeding programmes exploiting cytoplasmically inherited male-sterility our understanding of the basic biochemical and molecular mechanism(s) underlying CMS is still in its infancy. Further investigation of cytoplasmically inherited male-sterility may lead to an improved understanding of the interaction and coordination of nuclear and mitochondrial genetic systems in organelle biogenesis. It could also lead to the development of chemical gametocides which would cause pollen abortion in normal plants. This would make it possible to produce male-sterile forms of any breeding stock without the need for an expensive and laborious breeding programme to produce male-sterile cytoplasms. Recent work on the two plasmid-like DNA species in mitochondria from the S cytoplasms of maize suggest that they can become integrated into mitochondrial and nuclear genomes. If this is confirmed, the prospect of their use as suitable vectors for transferring desirable genetic information into plant cells is particularly exciting.

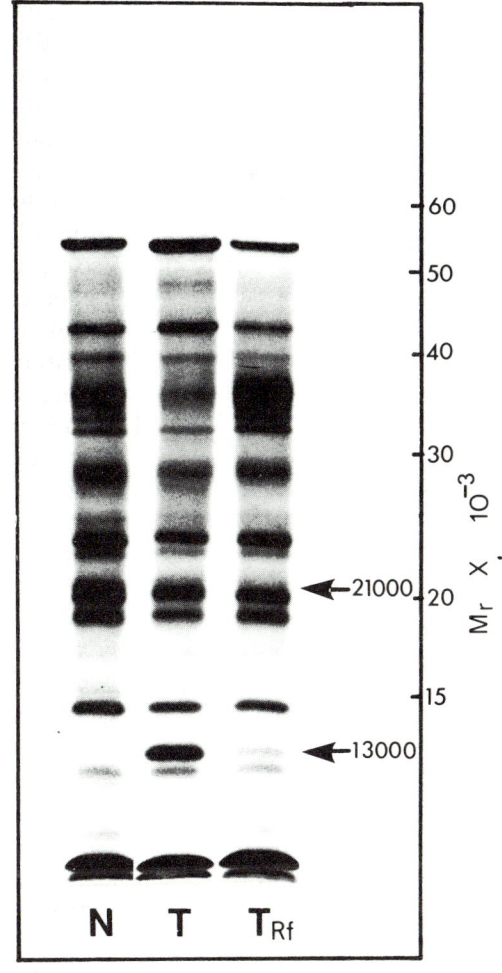

Fig. 3. SDS-polyacrylamide gel electrophoresis of polypeptides synthesized by mitochondria from normal (N), male-sterile (cms-T), and fertility-restored (TRf) lines of maize. The labelled mitochondrial translation products were fractionated and visualized as described in Fig. 1.

linear plasmid-like DNAs with molecular weights of 3.5 and 4.1 × 10^6, and both molecules contain terminal inverted repeats which are 168–196 nucleotides long[6]. Such inverted repeats are frequently involved in insertional events in lower organisms and evidence is mounting that a similar explanation may apply to these molecules. It is tempting to speculate that the additional high molecular weight translation products characteristic of cms-S

References

1 Bogorad, L. (1979) in *Genetic Engineering* Vol. 1 (Setlow, J. K. and Hollaender, S., eds), Plenum, New York
2 Borst, P. and Grivell, L. A. (1978) *Cell* 15, 705–723
3 Borst, P. (1977) *Trends Biochem. Sci.* 2, 31–34
4 Schatz, G. and Mason, T. L. (1974) *Annu. Rev. Biochem.* 43, 51–87
5 Schatz, G. (1979) *FEBS Lett.* 103, 203–211

6 Pring, D. R., Levings, C. S. and Conde, M. F. (1980) in *The Plant Genome* (Davies, D. R. and Hopwood, D. A., eds) pp. 111–120, The John Innes Charity
7 Bonen, L. and Gray, M. W. (1980) *Nucleic Acids Res.* 8, 319–335
8 Leaver, C. J. and Harmey, M. A. (1976) *Biochem. J.* 157, 257–277
9 Leaver, C. J. and Forde, B. G. (1980) in *Genome Organisation and Expression in Plants* (Leaver, C. J., ed.), pp. 407–425, Plenum Press
10 Forde, B. G. and Leaver, C. J. (1980) in *The Plant Genome* (Davies, D. R. and Hopwood, D. A., eds), pp. 131–146, The John Innes Charity
11 Butow, R. A. and Strausbeg, R. L. (1979) *Trends Biochem. Sci.* 110–113
12 Duvick, D. N. (1965) *Adv. Genet.* 13, 1–56
13 Gregory, P. Earle, E. D. and Grace, V. E. (1977) in *Host Plant Resistance to Pests*, A. C. S. Symp. Series No. 62, pp. 90–114, American Chemical Society, Washington D.C.
14 Forde, B. G. and Leaver, C. J. (1980) *Proc. Natl. Acad. Sci. U.S.A.* 77, 418–422
15 Gengenbach, B. G., Green, C. E. and Donovan, C. M. (1977) *Proc. Natl. Acad. Sci. U.S.A.* 75, 3841–3845
16 Flavell, R. (1974) *Plant Sci. Lett.* 3, 259–263

Christopher J. Leaver is at the Department of Botany, University of Edinburgh, Edinburgh EH9 3JH, U.K.

Addendum – *C. J. Leaver*

Since this brief review was written interest in the organization and expression of the plant mitochondrial genome has increased considerably (see Leaver and Gray[17] for review). Recent reports suggest that the plant mitochondrial genome is even larger (mol. wt 190–1600 × 10^6) than previous estimates and that a seven- to eight-fold variation in mitochondrial genome size can exist within a single plant family, the cucurbits[18]. Attempts to obtain a complete restriction map of maize mtDNA are progressing well, and Lonsdale and his colleagues at the Plant Breeding Institute in Cambridge have convincing evidence that the genome comprises at least 650 kb, which can be assigned to two linkage groups (D. Lonsdale personal communication). In mapping the mitochondrial 26S, 18S and 5S rRNA genes[19] these workers made the remarkable observation that a 12 kb DNA sequence present in the maize mitochondrial genome is homologous to part of the inverted repeat of the maize chloroplast genome. In chloroplasts, the sequence contains a 16S rRNA gene, and also the coding sequence for tRNA^{Ile} and tRNA^{Val} (Ref. 20). Preliminary data also suggest the presence of several chloroplast protein-coding gene sequences in the maize mitochondrial genome (Lonsdale, personal communication). The molecular basis of this apparent inter-organellar transport of DNA and its implications in organellar evolutionary studies is obviously of great interest.

The finding that defined yeast mitochondrial gene probes can cross-hybridize under heterologous conditions to homologous sequences in other types of mtDNA, has opened the way for the rapid identification of specific genes on plant mtDNA. The genes for subunits I and II of cytochrome oxidase have been identified and the subunit II gene sequenced. The DNA sequence of the subunit II gene in maize consists of two coding regions separated by a single, centrally located, intervening sequence of approximately 794 bp[21]. The presence of a single intron in the maize gene was unexpected as the corresponding genes in yeast and animals are not interrupted. The sequence data also revealed a further variation on the once 'universal' genetic code. Unlike the fungal and mammalian mitochondrial genomes which use UGA to specify tryptophan instead of stop, maize apparently uses CGG in addition to the usual UGG to specify this amino acid.

The suggestion that cytoplasmic male sterility in maize and sorghum may be due to mutations in the mitochondrial genome, resulting in the synthesis of variant mitochondrial translation products which lead to expression of the CMS phenotype, has been strengthened by several observations[17]. In sorghum four different types of CMS can be distinguished from each other and from the normal fertile counterparts on the basis of specific differences in mitochondrial translation products[22]. In one case, CMS was associated with the synthesis of an altered, higher molecular weight form of subunit I of cytochrome oxidase. In common with mitochondria from the S cytoplasm in maize, the mitochondria from two male-sterile sorghum lines were found to contain equimolar amounts of two plasmid-like, linear DNAs. These DNAs have mol. wts of approx. 3.5 and 4.1×10^6 and possess terminal repeats. Their presence was also correlated with the synthesis of additional characteristic high-molecular-weight mitochondrial translation products[22].

Additional evidence for the origin and involvement of the two linear 'plasmid-like' DNAs (S1 and S2) in the S-type of CMS in maize have been provided by Levings et al.[23]. They have shown that in cms-S plants which have spontaneously reverted to male fertility, due to a genetic change at the cytoplasmic level, the 'free' S1 and S2 DNAs are lost. At the same time the high-molecular-weight mitochondrial polypeptides, which are synthesized by the parental cms-S lines, are not synthesized by the cytoplasmic revertants (C. J. Leaver, unpublished observations). Furthermore Levings et al. have also shown that homology exists between S1 and S2 DNA and sequences in the main band mtDNA from normal fertile plants, parental cms-S plants, and from plants that had cytoplasmically reverted to fertility. However, the organization and arrangement of S1 and S2 related sequences was different in all three genomes. These observations have been interpreted as follows: (1) that S1 and S2 DNAs characteristic of cms-S maize could have arisen as an excision mutation from the mitochondrial genome of normal, fertile lines; (2) that the cytoplasmic reversion to male fertility is due to the reinsertion ('transposition') of S1 and S2 sequences into new locations in the high molecular weight mtDNA with associated loss of the 'free' forms of S1 and S2.

References
17 Leaver, C. J. and Gray, M. W. (1982) *Annu. Rev. Plant Phys.* 33, 373–402
18 Ward, B. L., Anderson, R. S. and Bendich, A. J. (1981) *Cell* 25, 793–803
19 Stern, D. B., Dyer, T. A. and Lonsdale, D. M. (1982) *Nucleic Acids Res.* 11, 3333–3337
20 Stern, D. B. and Lonsdale, D. M. (1982) *Nature* (in press)
21 Fox, T. D. and Leaver, C. J. (1981) *Cell* 26, 315–323
22 Dixon, L. K. and Leaver, C. J. (1982) *Plant Mol. Biol.* 1, 89–102
23 Levings, C. S., Kim, B. D., Pring, D. R., Mans, R. J., Laughnan, J. R. and Gabay-Laughnan, S. J. (1980) *Science* 209, 1021–1023

Genome organization and reorganization

A function for satellite DNA?

Christopher Bostock

Since our present understanding of the function of satellite DNA is inextricably tied to our knowledge of its structure I have summarised some of the salient features of satellite DNA in Fig. 1. Satellite DNAs are isolated as minor components from density gradients. Various reassociation techniques, and more recently restriction endonuclease digestions coupled with gel electrophoresis and DNA sequencing methods, show that a satellite DNA is usually composed of multiple copies of a short nucleotide sequence arranged in tandem arrays in large blocks. Different techniques detect different repeating periodicities in the tandem arrays. For example, the rate of reassociation of mouse satellite DNA suggests that the repeated sequence is about 150 nucleotides long. On the other hand, digestion with the restriction endonuclease EcoRII reveals a main periodicity of 240 nucleotide pairs, whereas direct sequencing methods suggest that the basic sequence is much shorter, perhaps about 18 nucleotides in length [1]. Thus a satellite DNA appears to consist of thousands or millions of copies of very short sequence, which are arranged in tandem, but which have undergone some divergence so that other hierarchies of repeat size can be detected.

Another major property of satellite DNA is its extreme variability. Satellite DNAs have a wide variation in overall base composition ranging from the nearly pure alternating adenine and thymidine satellite of the crab to the guanine and cytosine-rich HS-β satellite of the kangaroo rat. Many species contain several clearly distinguishable satellites; each of these may be related to the other by homology of their basic sequence (e.g. some of the satellites of several *Drosophila* species) or they may be completely unrelated. This wide variation in amount and nucleotide composition and sequence rules out any common coding function. Indeed many satellite sequences would code for very odd proteins, and others contain such a high proportion of translational 'stop' signals that it is unlikely that they could code for any protein. It is unlikely that satellite sequences are ever transcribed, since RNA molecules which are complementary to satellite DNAs have not been found.

Transcriptional inactivity ties in with the third major structural property of satellite DNA, namely its localisation (by *in situ* hybridisation) to constitutive heterochromatin, since this permanently condensed chromatin is also transcriptionally inert. This is not to say that it is without genetic effect; on the contrary, constitutive heterochromatin can have marked effects on the expression of genes in its vicinity (position effects) and on processes related to meiosis. We will return to this later. Constitutive heterochromatin is often located at or around centromeres and secondary constrictions – the sites of nucleolar organisers – and is also found at the end of chromosomes (telomeres) and at intermediate positions (interstitial heterochromatin). The non-centromeric heterochromatin is especially characteristic of plants, but is found in such diverse animals as insects, rodents and whales.

Although there is a strong association between heterochromatin and satellite DNA, we do not know whether small amounts of satellite DNA have gone undetected at other points throughout the length of chromosomes. Conversely, while satellite DNA is clearly a major component of constitutive heterochromatin we do not know if there are other important non-satellite DNA sequences interspersed between the blocks of satellite DNA. It is also clear that not all constitutive heterochromatin is satellite, since the supernumerary chromosomes of some plants and insects do not appear to be enriched in highly repeated satellite-like DNA. Since satellite DNA is almost always associated

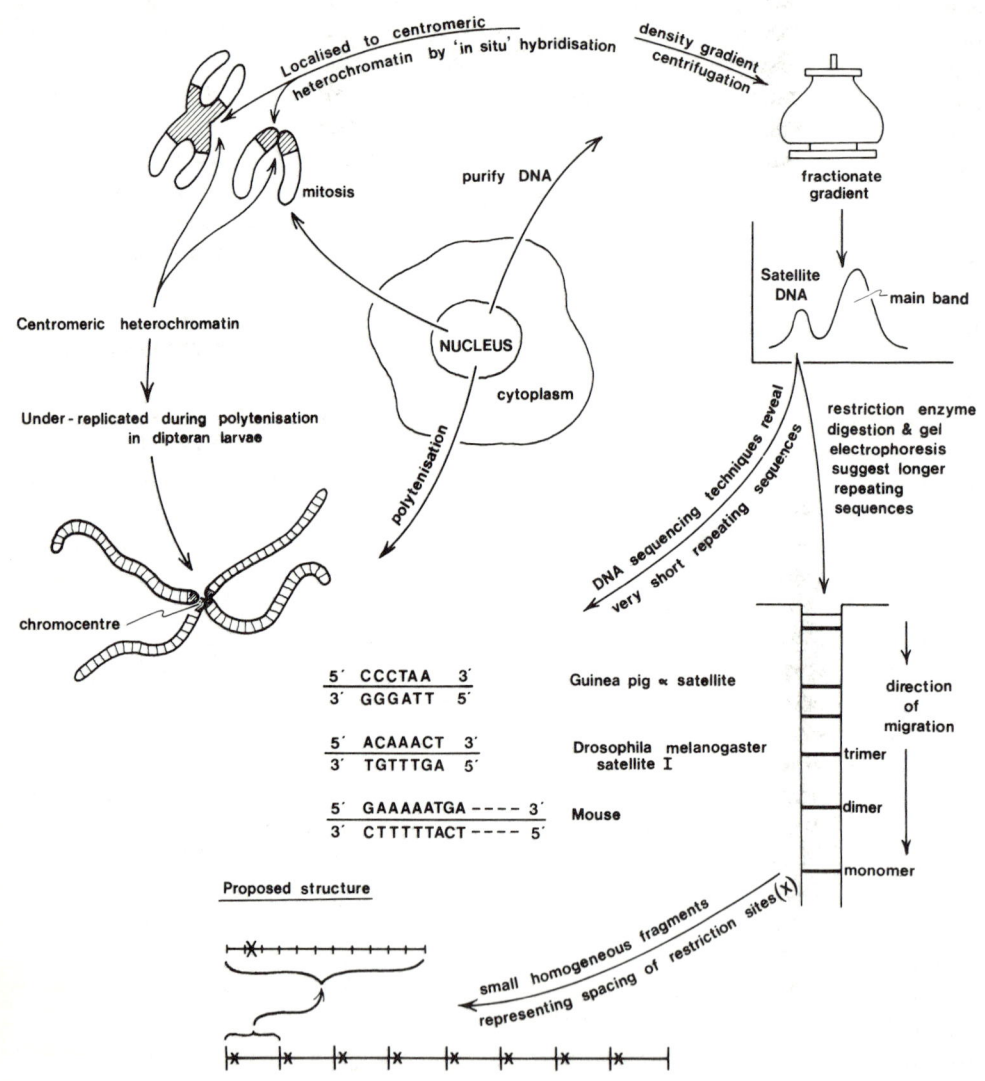

Fig. 1. *A summary of some generalised structural features of satellite DNA. The satellite DNA is located in the nucleus. After purification of nuclear DNA the satellite can be separated from the bulk DNA by centrifugation in a variety of caesium salt density gradients. The purified satellite can be analysed by digestion with restriction endonucleases which produce a series of fragments that can be isolated by gel electrophoresis. The kinetics of digestion and the homogeneity of sizes of fragments show that satellite DNA consists of a series of tandemly repeated sequences. Direct DNA sequencing of satellite DNAs shows that the basic repeating unit is much smaller than that revealed by restriction endonucleases. Purified satellite DNA can be located to the centromeric heterochromatin of mitotic chromosomes by* in situ *hybridisation. This chromatin is under-represented in the large polytene chromosomes of Diptera as well as in some other 'somatic' cell types.*

with a permanently and tightly condensed state of chromatin, there is a good case for saying that the nature of satellite DNA defines a condensed state. Nevertheless, satellite DNA is not a necessary component for permanently condensed chromatin, since facultative heterochromatin does not require the presence of satellite DNA, and not all constitutive heterochromatin is enriched in satellite DNA. However, like satellite DNA, heterochromatin is a highly variable entity, both within and between species. For example, variation in heterochromatin is well described in plants and insects, and in man inherited and stable heteromorphic forms of constitutive heterochromatin have been studied in detail [2].

It is largely this association between heterochromatin and satellite DNA that allows us to speculate about the function of satellite DNA, because we can now call upon cytogenetic and genetic evidence on the role of heterochromatin. Functions that have been proposed for satellite DNA and/or heterochromatin fall into four broad categories: (1) those which have been termed 'housekeeping' functions and which are thought of as being concerned with specifying folding patterns of chromosomes, stabilising the centromere and telomere, and, perhaps, defining individual centromeres; (2) those related to cell metabolism by control of nuclear size and hence determining the rate of cell growth and division; (3) those related to homologous chromosome recognition during pairing; and (4) those related in a rather vague way to evolution and speciation by promoting chromosome rearrangements, or affecting rates and distribution of crossing-over and recombination.

To me, by far the most convincing evidence favours some function for satellite DNA which is related to 'germ line processes'. This view is partly based on a series of observations on diverse organisms that show that satellite DNAs can be eliminated (or drastically reduced) in nuclei with somatic functions, but are retained in full in germ line nuclei. A dramatic example is provided by the hypotrichous ciliates (*Stylonichia* and *Oxytricha*) which form a new vegetative macronucleus (somatic nucleus) from one of the division products of the syncaryon formed during conjugation. Other division products of the syncaryon form the micronuclei (the germ line nucleus). Although the mature polyploid macronucleus contains at least a hundred times the amount of DNA found in a micronucleus, 96% of the DNA sequences of the micronucleus – many of them being satellite-like – are absent from the macronucleus; they are eliminated from the presumptive somatic nucleus during its formation, but are retained throughout vegetative growth in the diploid micronucleus [3]. In the nematode worms, *Ascaris* and *Parascaris*, the well known elimination of heterochromatic chromosomes during early cleavage stages in somatic nuclei is paralleled at the biochemical level by loss of satellite DNA [4]. During polytenization in several tissues of *Drosophila* larvae the satellite DNAs are drastically underreplicated and are barely detectable at the late stages of polyteny, and a similar phenomenon occurs in the polyploid somatic nuclei of adult tissues. In contrast, the diploid nuclei of germ line cells in adult flies retain the full complement of satellite DNAs [5].

What function related to the germ line could it be? Perhaps the particular patterns of arrangement of different blocks of satellite DNA in some way define the precision with which homologous chromosomes recognise each other during pairing. Or maybe the presence of blocks of satellite DNA affects the process of recombination either by exerting an effect on the frequency of crossing-over or on the distribution of chiasmata.

The case for an involvement of satellite DNA in specifying interchromosomal recognition during pairing is best illustrated in *Drosophila melanogaster*. This species has four satellites, clearly distinguishable by their buoyant densities. Each of these has been localised by *in situ* hybridisation within blocks of heterochromatin that surround the centromeres. Three general points emerge. First, all of the satellites are located on most of the chromosomes, but, second, the sizes of the blocks of a particular satellite vary between chromosomes, and third, the juxtaposition of blocks of different satellites varies between chromosomes [6]. Peacock and colleagues argue that such an arrangement could form the basis for recognition between chromosomes during the various types of pairing that they undergo. In larval cells about to undergo polytenization all chromosomes associate at their heterochromatic regions

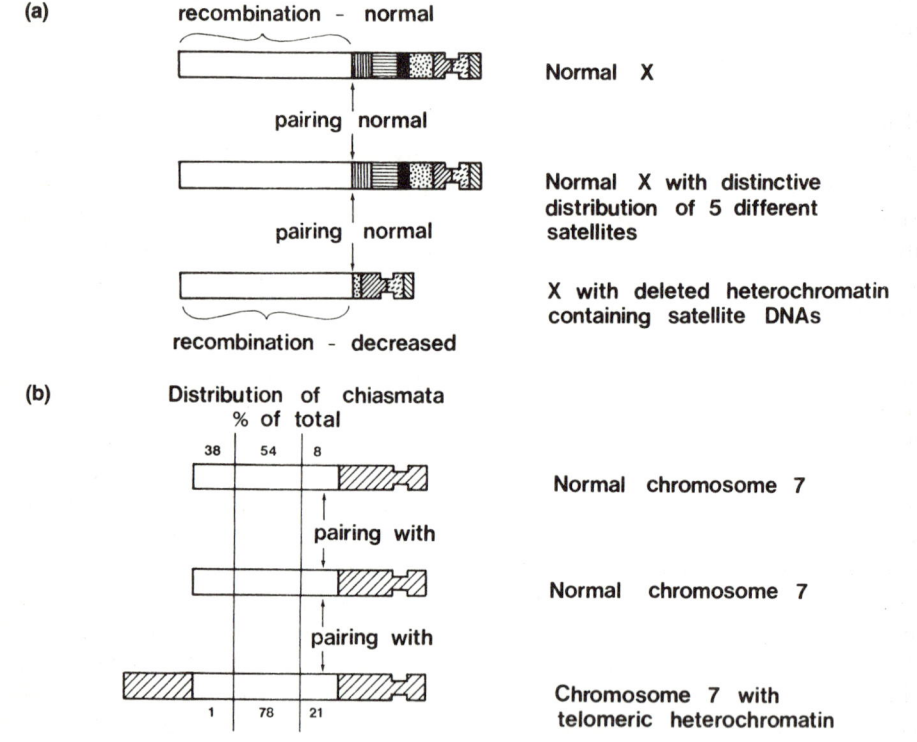

Fig. 2. The involvement of satellite DNA containing heterochromatin in pairing and meiotic recombination. (a) A normal X chromosome of Drosophila melanogaster *has five satellite DNAs distributed in a distinctive pattern within its block of centromeric heterochromatin. X chromosomes have been constructed in which various amounts of this heterochromatin have been deleted. Pairing between normal and deleted X chromosomes appears quite normal, but recombination is decreased in the euchromatic portions when normal and deleted chromosomes are paired together. (b) In* Atractomorpha similis *chromosome 7 is polymorphic for the presence of a telomeric block of heterochromatin. The distribution of chiasmata has been measured when two normal chromosome 7s pair or when a normal chromosome 7 pairs with one containing the telomeric heterochromatin. The presence of the telomeric heterochromatin severely depresses chiasmata formation in the segment of euchromatin most distal to the centromere, even though pairing occurs quite normally. While the evidence presented in (a) and (b) relates to heterochromatins which are known to contain satellite DNAs, it is not possible to distinguish whether it is the satellite DNAs per se which are having the effect, or whether it is simply the condensed state of chromatin, with which satellite DNAs are often associated, which is having an effect on recombination.*
(a) Based on Peacock et al. *[6] and Yamamato and Miklos [7]. (b) Based on Miklos and Nankivell [10].*

to form a chromocentre. Associations such as this, which involve all the chromosomes in the cell, could be mediated via the satellite sequences that are held in common between all chromosomes, even though the satellite sequences are not replicated during subsequent polymerization. Specific pairing between homologues during meiosis on the other hand could be determined by the specific arrangements of different satellites on each of the chromosomes (see Fig. 2).

While such a model is an attractive interpretation of the descriptive structural data, it is nevertheless speculative until tested in some experimental way. Fortunately *Drosophila* is an ideal organism for this purpose since there is a vast library of experimentally manipulated chromosomes at hand. Yamamato and Miklos [7] have studied the behaviour during meiosis of X chromosomes, deficient in up to 80% of their centromeric heterochromatin, and various autosomes with major changes to their heterochromatin. In all cases the pairing and segregation of chromosomes was apparently normal, suggesting that radical alteration to, or deletion of, the patterns of blocks of satellite DNA around the centromere has little immediate effect on specific pairing mechanisms (see Fig. 2a).

It is also difficult to see how the pairing hypothesis can be applied to species which have little or no satellite DNA, or to those that have only one satellite which is present on most of the chromosomes of the complement. The mouse is a well studied example of the latter, although it has to be admitted that even in this case there is still room for chromosomal specificity. While all 38 autosomes of a diploid mouse cell contain satellite DNA, there could be quantitative differences in the amount on each autosome, and the subfractions of minor sequence variants of satellite DNA detected with restriction endonucleases may be located on different autosomes [8].

Differences based on the distribution of restriction endonuclease recognition sites have been demonstrated for human satellite III DNA located on chromosomes 1, 7, 11, 15, 22 and X of man [9]. However, it is difficult to see how these small and subtle differences could determine precise chromosomal recognition required for pairing of homologues, when the vast majority of the sequence present in the satellite DNA is common to several chromosomes. It seems more likely that the differences in satellite structure that are found between chromosomes result from the fact that interchromosomal exchanges in heterochromatin are rare, and thus the satellite sequences are free to diverge independently on each chromosome. This, of course, does not explain how the same original satellite sequence first came to be present on more than one chromosome.

After pairing, homologous chromosomes can undergo genetic exchange by crossing over, which can be visualized at the cytogenetic level by the formation of chiasmata or measured genetically in terms of recombination frequencies between two loci on the same chromosome. Invariably chiasmata do not form in heterochromatic regions, but it is well documented that heterochromatin can have very marked effects upon meiotic recombination in euchromatic segments of chromosomes. Recently Miklos and Nankivell [10] measured both the frequency and the position of chiasmata on chromosome 7 of the grasshopper *Atractomorpha similis* (see Fig. 2b). This chromosome is polymorphic for a telomeric, satellite DNA-containing block of heterochromatin. Chromosomes which lack the telomeric heterochromatin can form chiasmata along the entire distal two-thirds of the chromosome. If a homologue of chromosome 7, which contains the telomeric heterochromatin is present, chiasmata are not formed in the distal one-third of the euchromatin. Thus the pattern of recombination in a chromosome can be altered by the addition of heterochromatin to that chromosome. In *D. melanogaster*, in the example cited earlier, deletion of centromeric heterochromatin (which contains satellite DNA) from the X chromosome, results in a reduction in the frequency of recombination in the euchromatin region immediately proximal to the heterochromatin. Supernumerary chromosome segments, for example B chromosomes and 'knobs' of maize which are heterochromatic, have marked effects on recombination. These can involve both increases or decreases in cross-over frequencies as well as altering the distribution of chiasmata. These effects can operate both within and between chromosomes [11]. In these cases it seems that the heterochromatin is not enriched in satellite or highly repeated DNA, although cytologically it can still be identified as constitutive heterochromatin.

From the point of view of a function for satellite DNA we can see that heterochromatin can have varied but dramatic effects upon recombination in euchromatin, but that these are not always mediated by heterochromatin which contains satellite DNA sequences. It could be that eukaryotic cells have evolved more than one way of creating heterochromatin, the sudden formation of blocks of highly repeated satellite DNAs being just one. In evolutionary terms, an alteration to the rates of recombination in a particular region of a chromosome would have marked effects. Suppression of recombination could protect from change a particularly well-adapted set of genes; alternatively, increasing the rate of recombination might promote variability upon which adaptive evolution is based. In either case the presence of a block of constitutive heterochromatin, which contains satellite DNA, would influence the genetic constitution of the genome which contains it, and would therefore be subject to selection. The selection would not be for any particular satellite sequence, but simply for a DNA structure that is able to maintain the condensed state of heterochromatin. The apparent rapid variability in the amount of satellite DNAs (and heterochromatin) and their chromosomal locations may provide a means of altering the overall genotype of an organism that is faster than, and additional to, that which can be achieved by mutation of structural genes alone. Undoubtedly, many things can affect the behaviour of chromosomes and thus the genotype – mutations to structural genes amongst them – but the possession of an 'auxiliary' system, such as that mentioned above, involving heterochromatin and satellite DNA, may be necessary for eukaryotes (some of which have long generation times) in order that they can adapt fast enough to changing environments.

References

1 Southern, E. M. (1975) *J. Mol. Biol.* 94, 51–69
2 Phillips, R. B. (1977) *Can. J. Genet. Cytol.* 19, 405–413; Hoehn, H., Au, K., Karp, L. E., Martin, G. M. (1977) *Hum. Genet.* 35, 163–168
3 Prescott, D. M., Murti, K. G. and Bostock, C. J. (1973) *Nature (London)* 242, 576 and 597–600
4 Moritz, K. B. and Roth, G. E. (1976) *Nature (London)* 259, 55–57
5 Gall, J. G. and Atherton, D. D. (1974) *J. Mol. Biol.* 85, 633–664
6 Peacock, W. J., Appels, R., Dunsmuir, P., Lohe, A. R. and Gerlach, W. L. (1976) *Int. Cell Biol. Int. Congr. 1st* pp. 494–506
7 Yamamato, M. and Miklos, G. L. G. (1978) *Chromosoma* 66, 71–98
8 Hörz, W. and Zachan, H. G. (1977) *Eur. J. Biochem.* 73, 383–392
9 Beauchamp, R. S., Mitchell, A. R., Buckland, R. A. and Bostock, C. J. (1979) *Chromosoma* 71, 153–166
10 Miklos, G. L. G. and Nankivell, R. N. (1976) *Chromosoma* 56, 143–167
11 Rhoades, M. M. (1978) in *Maize Breeding and Genetics* (Walder, D. B. ed) pp. 641–671, Wiley-Interscience, New York

Christopher Bostock is at the MRC Mammalian Genome Unit, Department of Zoology, University of Edinburgh, Edinburgh EH9 3JT, U.K.

Addendum – C. Bostock

Has anything new been discovered in the last 2 years that alters our view on a possible function for satellite DNA? The short answer to this is probably 'no', but during this interval the notion that satellite DNA might be amongst those DNA sequences with the properties of 'selfishness' could, if correct, make the original question somewhat meaningless [12,13]. As I understand the concept, the essential and only necessary properties of selfish DNA are those that ensure its own replication and avoidance of its deletion from the genome. The need for a DNA sequence to have a selectable effect (i.e. function) on the organism in order for it to be retained is thus eliminated. Once a cellular function is acquired, accidentally or otherwise, the DNA would become subject to maintenance, change or loss through selection on that function and could not be said to be solely 'selfish' anymore. So the question should be broadened to whether satellite DNA has a function at all, and, if so, what it might be.

I should first correct a statement made before, namely that satellite DNA is never transcribed. It has now been clearly shown that both strands of satellite DNA are transcribed, in amphibian oocytes at least [14,15]. It is tempting to suggest that this might be related to a meiotic function, but before jumping to any hasty conclusions it must be realised that this transcription of satellite sequences is probably only a consequence of the failure of normal termination signals within the genes that flank the blocks of satellite. Whether the act of transcribing satellites, or the transcripts themselves, will turn out to have functional significance remains to be seen. There has, however, been another interesting germline related observation. It has been found that satellite DNAs are under-methylated in sperm and eggs, whereas in somatic tissues they are fully methylated [16]. Could this signify a function? Unfortunately we do not even know what the function of methylation itself might be! There are suggestions that the loss of methyl groups might be related to, or the consequence of, gene expression, and it is likely that their presence or absence will affect in some way the structure of the DNA or its ability to interact with proteins. It is known, for instance, that certain proteins bind specifically to satellite DNAs in somatic tissues [17] perhaps determining the condensed state of heterochromatin. Will these same proteins also be found associated with satellite DNA lacking methyl groups in the germ-line?

I want to return to the proposed involvement of satellite DNA in chromosome pairing. A new twist has emerged from Bennett's [18] painstaking analyses of the juxtapositions of mitotic chromosomes of haploid chromosome sets of intergeneric plant hybrids. Bennett showed that interphase chromosomes can be highly ordered in space; the specific order being determined by the arm lengths of chromosomes in a way which maximises the similarity of neighbouring arm lengths. Significantly, adjacent arms from different chromosomes have similar blocks of heterochromatin in equivalent positions. It is unlikely that the satellite DNAs are involved in any sequence-specific recognition for the reasons mentioned in my original article, but their presence and bulk will automatically alter the dimensions of chromosomes and thus influence the position of chromosomes in the spatial order. Are the blocks of satellite DNA in these positions because their chromosome arms are close together and perhaps more readily able to exchange sequences, or are the satellite blocks the instruments of change causing perturbations to the spatial order? If Bennett's model is correct, the translocation or *de novo* formation of a block of satellite DNA would result in a major reshuffling of chromosomes, bringing together in space new combinations of genes on different chromosomes. What effect, if any, this would have on the cell remains to be determined.

In conclusion, it seems that we are no nearer to deciding between the various propositions for the function of satellite DNA. It is likely that this impasse will remain until more attempts are made to experimentally manipulate blocks of satellite DNA within the chromosomal environment, and to examine the consequences of these against our existing knowledge and hypotheses about satellite DNA.

References

12 Orgel, L. E. and Crick, F. H. C. (1980) *Nature (London)* 284, 604–607

13 Doolittle, W. F. and Sapienza, C. (1980) *Nature (London)* 284, 601–603

14 Diaz, M. O., Barsacchi-Pilone, G., Mahon, K. A. and Gall, J. G. (1981) *Cell* 24, 649–659

15 Varley, J. M., Macgregor, H. C., Nardi, I., Andrews, C. and Erba, H. C. (1980) *Chromosoma* 80, 289–307

16 Storm, K. S. and Taylor, J. H. (1981) *Nucl. Acids. Res.* 9, 4536–4546

17 Hsieh, T. and Brutlag, D. L. (1979) *Proc. Natl Acad. Sci. U.S.A.* 76, 726–730

18 Bennett, M. D. (1982) in: *Gene Evolution* (Dover, G. A. and Flavell, R. B., eds), pp. 239–261, Academic Press, London

The human *Alu* family of repeated DNA sequences

Elisabetta Ullu

The structural characterization of the Alu *family of human interspersed repeated sequences, and the study of its relationship with cellular RNAs may give some indications of the possible function of eukaryotic repeated sequences.*

Nearly 15 years ago Britten and Kohne[1] discovered that eukaryotic DNAs contain a high proportion of sequences which are present in up to a million copies in the haploid genome; these are called repeated sequences. In lower eukaryotes, such as yeast, repeated sequences are usually few while in higher eukaryotes they may constitute up to 80% of the entire genome. Some of these sequences are organized in clusters of tandemly repeated units, others are interspersed throughout the genome. Satellite DNA and some repeated genes represent examples of the clustered arrangement. Interspersed sequences will be the subject of this article.

Interspersed repeated sequences vary in length from 150–300 bp to a few kilobases and can be subdivided into different families. The members of a family cross-hybridize and most have similar but not identical nuleotide sequences. In higher eukaryotes several families are usually present within a genome and their distribution seems to be random. Repeated DNA elements can be intermingled with unique sequences or with members of other repeated families.

The ubiquity and abundance of repetitive sequences in eukaryotic genomes have stimulated many discussions about the common functions they might fulfil. Since repeated sequences are found down or upstream from genes coding for proteins it has been suggested that they might play a regulatory role in the process of gene expression. In 1969, Davidson and Britten[2] proposed that the transcripts of these elements could regulate the transcription of functionally related genes. On the other hand, Doolittle and Sapienza[3] and Orgel and Crick[4] have suggested that the major function of repetitive sequences is 'survival within genomes' such that 'no phenotypic or evolutionary function need be assigned to them'. At present, neither of the two hypotheses has been proven. A detailed structural analysis of families of repeated sequences and study of their transcription products might help to clarify their possible function, if any, in the cell. For this reason, several families of repeated sequences have recently been cloned and analysed.

The best characterized to date is the *Alu* family, first isolated from the human genome. I will summarize the results of these studies and try to outline the features of the *Alu* family which might indicate its biological role.

Identification of the *Alu* family

Human DNA contains short interspersed sequences, approximately 300 nucleotides long, which represent at least 5% of the entire genome and can be present in either orientation[5]. About 60% of these sequences have a common cleavage site for the restriction enzyme *Alu* I, and, on this basis, they have been regarded as a family, the *Alu* family[5]. Subsequent cloning and sequencing of this DNA has confirmed that the *Alu* family consists of closely related sequences

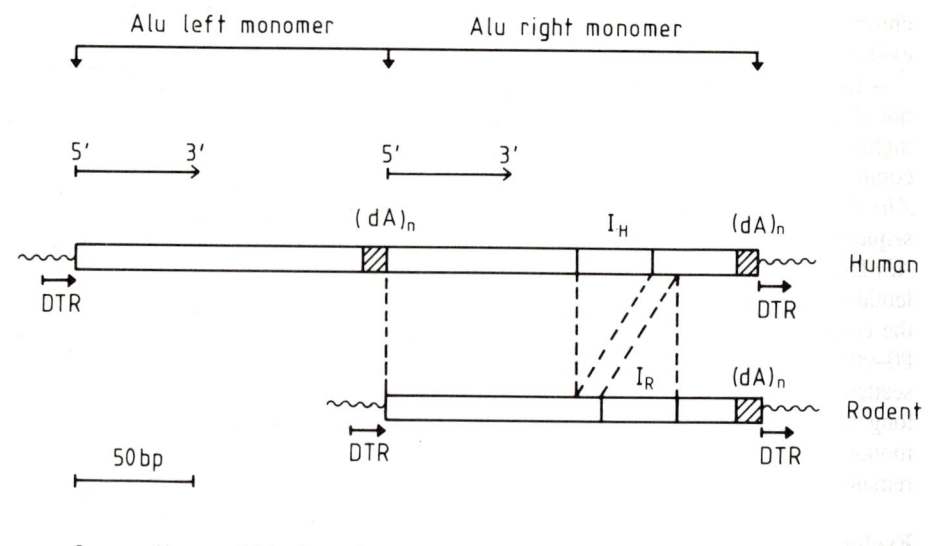

Fig. 1. Structure of human (12) and rodent (17) Alu DNAs.

and is the most abundant repeated DNA component within the human genome[6]. More than 300,000 Alu sequences, 300 bp long, are dispersed in a genome of 2.7×10^9 bases. Assuming that they are randomly distributed, Alu sequences should be found every few thousand nucleotides. The analysis of cloned genomic fragments has confirmed this expectation. Members of the Alu family have been mapped in the neighbourhood of all the structural genes isolated so far[7-9]. In particlar, two Alu sequences have been mapped in the beta-like globin gene cluster within regions which are believed to be involved in the regulation of expression of the fetal globin in adults[10]. Alu sequences have also been found in two different intervening sequences (introns) of a human onc gene[11], suggesting that they can be part of transcriptional units.

Structure of Alu sequences

Several Alu sequences have been characterized by nucleotide sequence analysis. A 300 bp long Alu sequence has the following structure[12]: two similar sequences, approximately 130 bp long, are arranged as a head to tail dimer (Fig. 1) and represent the left and right monomer of the sequence. The right half contains a 31 bp insert which is usually, but not always, absent in the left half[9]. A stretch of 10–20 Å residues is located at the 3' end of each monomer. In addition, in their genomic context, Alu sequences can be flanked by short direct repeats, 5–20 bp long[7-9].

Jelinek et al.[12] have found homologies between a 14 bp sequence near the origin of replication of BK, SV40 and polyoma viruses and Alu DNA and have suggested that Alu sequences might therefore function as origins of replication in mammalian chromosomes. However, there is no direct evidence for this.

Although members of the *Alu* family are not identical, their sequence appears to be highly conserved. Deininger et al.[13] have compared the nucleotide sequence of ten *Alu* clones and have derived a consensus sequence for the left and right monomer of *Alu* DNA. The homology between the nucleotide sequence of individual clones and the consensus sequence is on the average 80–90%. Although the base changes are scattered throughout the sequence, a 40 nt long region, close to the 5' end of each monomer, is more conserved than the remainder[14].

Evolution of the *Alu* family

Repeated sequences very close in nucleotide sequence to human *Alu* DNA have been found in the genome of mouse (B1 family,[15]), monkey[16] and Chinese hamster (CHO-*Alu*-equivalent families, types 1 and 2[17]). The dimeric structure, typical of human *Alu* DNA, is conserved in the monkey, while in mouse and Chinese hamster most of the *Alu*-equivalent sequences seem to be monomers. A comparison of the structure of human and rodent *Alu* DNAs is shown in Fig. 1. The best homology between human and rodent *Alu* DNAs is found with the human *Alu* right monomer[17]. Rodent *Alu* DNA contains a 32 bp insert which is different in nucleotide sequence from the 31 bp insert present in human *Alu* right monomers and is located nine nucleotides downstream from the insertion point of the human 31 bp insert. Excluding the insert regions, the homology between human and rodent *Alu* sequences is approximately 80%. Possibly the dimeric *Alu* sequence of primates originated by tandem duplication of an ancestral *Alu* monomer which is still the prevalent structure in the genome of rodents.

Recently I have found *Alu*-like sequences in the genome of much more distantly related organisms such as slime molds, echinoderms, amphibians and birds (Ullu, E., Esposito, V. and Melli, M. L., in preparation). Such sequences crosshybridize to human *Alu* sequences under stringent conditions which require at least 80% total homology in blocks of more than 20 bp. The extreme evolutionary conservation of the *Alu* family could be due to selective pressures related to the function of these sequences. Alternatively, *Alu* DNA might not have any 'phenotypic' function[3,4] and its evolutionary conservation may have resulted from a selective advantage of the sequence as such.

Relationship between *Alu* DNA and cellular RNAs

At least some *Alu* sequences are transcribed. Repetitive sequences are well represented in heterogeneous nuclear RNA (HnRNA) isolated from human cells and the majority of them belong to the *Alu* family[13]. *Alu* sequences in HnRNA are present in either orientation, as they are arranged in the genome, and can be found both in poly (A)$^+$ and poly(A)$^-$ HnRNA[14] and in high as well as low molecular weight HnRNA molecules (Fig. 2A). In addition, two low molecular weight RNAs, human 7S[18] and rodent 4.5S[17] RNAs, have sequence homologies with *Alu* DNA. 7S RNA is a cytoplasmic RNA species, approximately 300 bp long (Fig. 2B), which was first found, along with other small cellular RNAs in the virions of mouse and avian oncornaviruses, but is coded by the cell genome. The structure of human 7S RNA is composite (Fig. 3): a non-*Alu* sequence of 140 bp is inserted within an *Alu* right monomer[19]. The 5' and 3' regions of this RNA are more than 80% homologous to the consensus sequence of the human *Alu* right monomer. The 100 nucleotide long 4.5S RNA, found in rodents but not in humans, is hydrogen bonded to nuclear and cytoplasmic poly(A)$^+$ RNA and has approximately 70% total homology with the rodent *Alu* consensus sequence[17] (Fig. 3). In particular, a 50 nucleotide long internal region of this RNA is 85% homologous to *Alu* DNA. Since both these RNAs are extensively homologous to the *Alu* sequ-

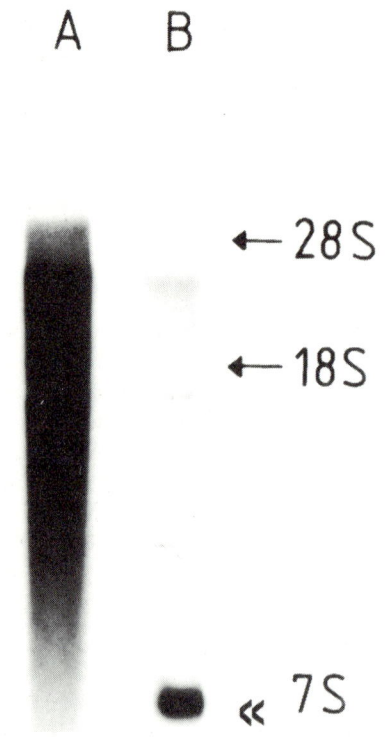

Fig. 2. *Hybridization of human* Alu *DNA to HnRNA (A) and to total cytoplasmic RNA (B) of HeLa cells. 20 μg were electrophoresed on a 1.2% agarose gel containing 2.2 M formaldehyde, transferred to a nitrocellulose filter and hybridized to cloned* Alu *DNA(19) labelled with* ^{32}P *by nick translation. Arrows indicate the positions of 28S and 18S ribosomal RNAs as molecular weight markers. The cytoplasmic RNA species, distinct from 7S RNA, which hybridize to* Alu *DNA probably represent HnRNA species contaminating the cytoplasmic RNA fraction.*

ence, their genes might be regarded as subsets of the *Alu* family.

Calabretta et al.[20] have found that mRNA isolated from human polysomes, contains sequences of 200–300 bp which can form stable R-loops with cloned *Alu* DNA. The region of complementarity can be located internally or at one end of the RNA molecule. They suggest that the transcription of *Alu* sequences might be tissue specific since the RNA isolated from human placenta does not hybridize to *Alu* probes.

In summary, *Alu* sequences are found in different types of RNA each of which have distinct functions. The *Alu* transcripts in HnRNA and mRNA might only reflect the widespread distribution of this DNA in the genome. Their presence in intervening sequences and in mRNA shows that *Alu* DNA can be transcribed as part of larger primary transcripts. And, if its presence does not imply a function, at least it means that the *Alu* sequence is sufficiently neutral not to interfere with transcription, RNA processing and mRNA translation.

Alu sequences also code for the homogeneous small 7S and 4.5S RNA species. Although the function of these RNAs is unknown, their abundance, their specific cellular localization and, at least in the case of the 7S RNA, the conservation in birds and mammals[21], strongly suggest a structural role. It has been proposed that the 7S RNA may participate in the transport of mRNA from the nucleus to the cytoplasm[22], while the 4.5S RNA could be involved in the processing of HnRNA[17]. In both cases *Alu* sequences would be responsible for specific functions.

In vitro transcription of *Alu* sequences

Some, but not all, cloned *Alu* sequences can be efficiently transcribed *in vitro* by RNA polymerase III[14] the enzyme which transcribes 7S RNA *in vivo*[22] and the sense strand of *in vitro* Alu transcripts and of 7S RNA is the same[19].

Transcription of *Alu* DNA *in vitro* starts almost precisely at the 5′ end of the left hand *Alu* monomer, but termination depends on the presence of a T-rich region downstream from the *Alu* dimer[14]. Therefore the length of the transcripts obtained varies from clone to clone.

Does the fact that some cloned *Alu* sequences can be efficiently transcribed *in vivo* by RNA polymerase III mean that these sequences are also transcribed *in vivo* by this enzyme? Recently, Haynes and Jelinek[23] have identified *Alu*-like low molecular weight RNA species, distinct from 7S and 4.5S RNAs, in the nuclei of Chinese hamster cells which might represent *in vivo* *Alu* transcripts and might be homologous to those obtained *in vivo*. Fingerprint analysis of these RNAs suggest that they are heterogeneous in sequence and, therefore, specified by similar but not identical *Alu* sequences.

How did the *Alu* family originate?

Nothing is known about the origin of repeated sequences and, particularly, about the molecular mechanisms which have led to their abundance and interspersion.

Prokaryotes and eukaryotes contain transposable elements several kilobases long which can move from one place to another in the genome. In general, these elements are flanked by short 5–10 bp direct repeats which are produced by duplication of the DNA at the site of insertion. The presence of short direct repeats flanking some members of the *Alu* family has led to the suggestion that *Alu* sequences might be mobile[14,24]. If so, this would explain their abundance and interspersion and would make them the shortest transposable elements known. Van Arsdell et al.[24] and Jagadeeswaran et al.[14] have speculated about a mechanism for the transposition of *Alu* sequences which involves the reverse transcription of an RNA polymerase III dependent RNA transcript of the *Alu* sequence and the insertion of the complementary DNA molecule in the genome.

Alternatively, it is also possible that members of the *Alu* family are, or were, part of larger transposable elements; they might be equivalent to the long terminal repeats, approximately 300 bp long, which flank eukaryotic transposable elements[9]. Sometimes the excision of a transposable element is incomplete and a long terminal repeat may be left behind. This has been described for the Tyl transposon of yeast[9]. Such a process, repeated a sufficient number of times would produce a family of repeated sequences.

In conclusion, many structural and transcriptional features of the *Alu* family of repeated sequences have been elucidated and several suggestions about its function

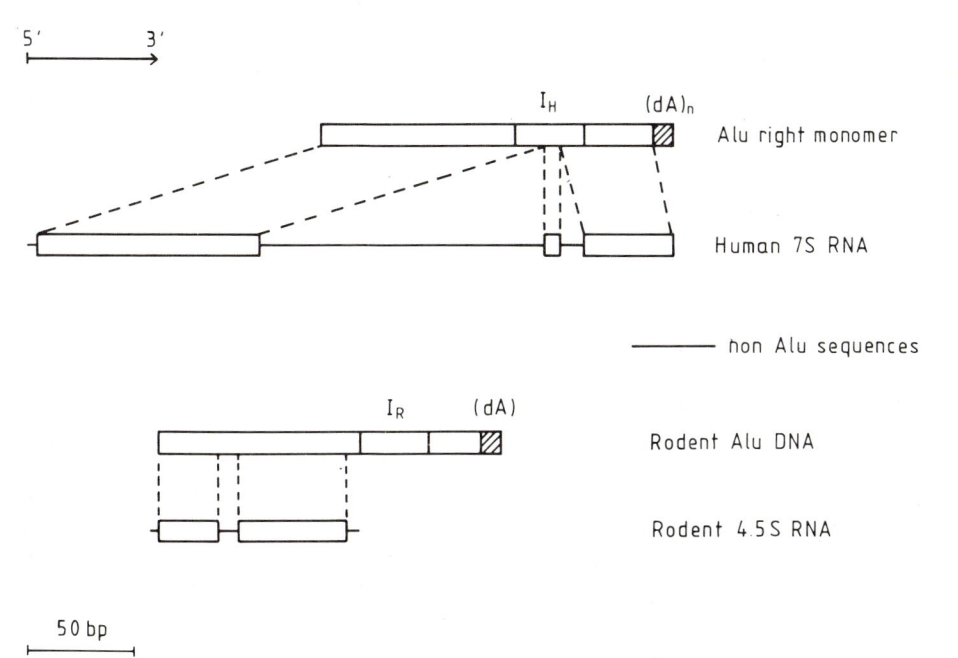

Fig. 3. Alu sequences in human 7S (19) and rodent 4.5S (17) RNAs.

have been put forward. However, the biological role of the *Alu* family is still unknown. If *Alu* sequences are mobile or degenerate descendants[3] of larger transposable elements, their location in the genome might be accidental and not related to their function. As I mentioned above, *Alu* sequences are found adjacent to structural genes, within intervening sequences, as part of messenger RNA and, in addition, represent a major portion of the primary structure of 7S and 4.5S RNAs. The location of *Alu* DNA at different positions in respect to coding regions and its presence in functionally distinct RNA species might suggest that subsets of the *Alu* family fulfil different functions in the cell. However, it is also possible that the majority of *Alu* sequences have no function and might be regarded as parasitic DNA. According to Doolittle and Sapienza[3] and to Orgel and Crick[4], such DNA would not contribute to the phenotype of the organism and would be only 'a slight burden to the cell that contains it'. The concept of parasitic DNA does not exclude the possibility that some members of the *Alu* family have acquired a specific function during evolution. The 7S and 4.5S RNAs, which are homologous to *Alu* DNA and probably have a function, might represent examples of such an evolutionary process.

Acknowledgements

I am grateful to M. L. Melli and to V. Pirrotta for helpful discussions and critical reading of the manuscript. The author holds an EMBL postdoctoral fellowship and is on leave of absence from the Istituto di Biologia Generale of the University of Rome.

Note added in proof

The nucleotide sequence of rat 7S RNA is almost identical to that of human 7S RNA (Li, W., Reddy, R., Henning, D., Epstein, P. and Busch, H., in press).

References

1 Britten, R. J. and Kohne, D. (1968) *Science* 161, 529
2 Davidson, E. H. and Britten, R. J. (1969) *Science* 204, 1052
3 Doolittle, W. F. and Sapienza, C. (1980) *Nature (London)* 284, 601
4 Orgel, L. E. and Crick, F. H. C. (1980) *Nature (London)* 284, 604
5 Houck, M. C., Rinehart, F. P. and Schmid, C. W. (1979) *J. Mol. Biol.* 132, 289
6 Rubin, C. M., Houck, M. C., Deininger, P. L., Friedmann, T. and Schmid, C. W. (1980) *Nature (London)* 284, 372
7 Bell, G., Pictet, R. and Rutter, W. J. (1980) *Nucl. Acids Res.* 8, 4091
8 Baralle, F. E., Shoulders, C. C., Goodbourn, S., Jeffreys, A. and Proudfoot, N. J. (1980) *Nucl. Acids Res.* 8, 4393
9 Duncan, C. H., Jagadeeswaran, P., Wang, R. R. C. and Weissman, S. (1981) *Gene* 13, 185
10 Fritsch, E. F., Lawn, R. M. and Maniatis, T. (1980) *Cell* 19, 959
11 Dalla Favera, R. D., Gelmann, E. P., Gallo, R. C. and Wong-Staal, F. (1981) *Nature (London)* 292, 31
12 Jelinek, W. R., Toomey, T. P., Leinwand, L., Duncan, C. H., Biro, P. A., Choudary, P. V., Weissman, S. M., Rubin, C. M., Houck, C. M., Deininger, P. L. and Schmid, C. (1980) *Proc. Natl Acad. Sci. U.S.A.* 77, 1398
13 Deininger, P. L., Jolly, D. J., Rubin, C. M., Friedmann, T. and Schmid, C. W. (1981) *J. Mol. Biol.* 151, 17
14 Jagadeeswaran, P., Forget, B. G. and Weissman, S. M. (1981) *Cell* 25, 205
15 Krayev, A. S., Kramerov, D. A., Skryabin, K. J., Ryskov, A. P., Bayev, A. A. and Georgiev, G. P. (1980) *Nucl. Acids Res.* 8, 1201
16 Grimaldi, G., Queen, C. and Singer, F. M. *Nucl. Acids Res.* 9, 5553
17 Haynes, S. R., Toomey, T. P., Leinwand, L. and Jelinek, W. R. (1981) *Mol. Cell. Biol.* 1, 573
18 Weiner, A. M. (1980) *Cell* 22, 209
19 Ullu, E., Murphy, S. and Melli, M. L. (submitted for publication)
20 Calabretta, B., Robberson, D. L., Maizel, A. L. and Saunders, G. F. (1981) *Proc. Natl Acad. Sci. U.S.A.* 78, 6003
21 Erikson, E., Erikson, R. L., Henry, B. and Pace, N. R. (1973) *Virology* 53, 40
22 Zieve, G. (1981) *Cell* 25, 296
23 Haynes, S. R. and Jelinek, W. R. (1981) *Proc. Natl Acad. Sci. U.S.A.* 78, 6130
24 Van Arsdell, S. W., Denison, R. A., Bernstein, L. B., Weiner, A. M., Manser, T. and Gesteland, R. F. (1981) *Cell* 26, 11

Elisabetta Ullu is at the European Molecular Biology Laboratory, Postfach 102209, D-6900 Heidelberg, F.R.G.

Structure and regulation of a collagen gene

Benoit de Crombrugghe and Ira Pastan

The alpha 2(I) collagen gene is 38 kilobases in length and contains at least 52 exons. Many of the exons are 54 base pairs in length suggesting the gene evolved from a small primordial gene.

The recent advances in recombinant DNA techniques have opened the way for the study of molecular regulation of specific animal genes. Until recently, such studies were only feasible with genes from bacteria or bacteriophages. We have chosen to study the structure and regulation of the collagen genes for several reasons. They are a family of developmentally regulated genes coding for a class of interesting proteins which constitute the major protein component of the extracellular matrix in animal tissues. Their principal function is probably to provide a structural scaffold around which cells can attach. In higher vertebrates and in man there are at least five genetically distinct collagen types with as many as nine different polypeptide chains[1]. Because the levels of these different collagen types vary greatly from tissue to tissue, their synthesis is probably controlled by differentiation programs that are tissue specific and developmentally regulated. The differences in primary structure between the collagen chains probably reflect tissue-related functional differences.

In cultured chick embryo fibroblasts (CEF) the synthesis of type I collagen is severely inhibited when these cells are transformed by Rous sarcoma virus[2]. This inhibition is selective because the synthesis of most other cellular proteins is unchanged. Type I collagen, which consists of two alpha 1 chains and one alpha 2 chain, is the principal collagen made in these cells. Although the effect of p60^{src} on collagen synthesis is almost certainly indirect, there is good evidence that it is mediated by a transcriptional control mechanism[3-7]. Because the product of the transforming gene of RSV, p60^{src}, perturbs a normal regulatory signal, RSV transformation provides a tool with which to examine the developmental regulation of the genes for type I collagen. A similar inhibition of type I collagen synthesis is found in rodent fibroblasts transformed by other retroviruses[8].

To study the control of collagen gene expression in appropriately reconstituted *in vivo* and *in vitro* systems, one of the collagen genes expressed in chick embryo fibroblasts had to be isolated; we chose the gene coding for the alpha 2 chain of type I collagen.

Structure of the collagen proteins

The polypeptide chains of type I collagen as well as of most other collagen types are synthesized as precursor molecules. They contain, beginning at the NH$_2$ terminal end: a signal peptide; a N-terminal propeptide; the major helical part of the molecule; and a C-terminal propeptide. As with other secreted proteins, the signal peptide is cleaved off in the rough endoplasmic reticulum, around the time when the chains associate. After hydroxylation of many lysine and proline residues and the addition of carbohydrate sidechains, the molecule is secreted. Following the removal of the N and C propeptides by specific enzymes, the collagen molecules assemble in highly ordered microfibrils and fibrils.

The central helical part of most collagen species has unique structural features.

These include: (1) the presence of glycine residues every third amino acid; (2) an abundance of proline and lysine residues, many of which are hydroxylated; (3) a characteristic configuration of the molecule composed of three subunits which interact with each other to form a triple helical structure; (4) the presence of many inter- and intra-molecular crosslinks which result in a higher order structural organization. Except for the glycine residues, it is difficult to find extensive sequence homologies between different portions of the same chain or between different collagen chains.

Structure of the alpha 2(I) collagen gene

H. Ohkubo and G. Vogeli[9,10], isolated the gene for the alpha 2 chain of type I collagen from a library of chick genomic DNA fragments (generously provided by J. Dodgson, D. Engel and R. Axel). They used as a probe a cloned cDNA which had previously been constructed in our laboratory. The gene is 38,000 base pairs (bp) in length and is more complex than any gene so far isolated. Its coding information is subdivided into at least 52 exons. The introns which interrupt these exons vary greatly in size and range from about 80 to more than 2,000 bp (Fig. 1).

The existence of more than 50 introns in the alpha 2(I) collagen gene implies that the conversion of the primary transcript of the gene to mature translatable collagen mRNA includes more than 50 precise splicing events. By virtue of its unusual complexity, the collagen gene, more than any other gene, illustrates some of the essential properties of the splicing reactions[11]. It is clear that the cellular splicing machinery needs to

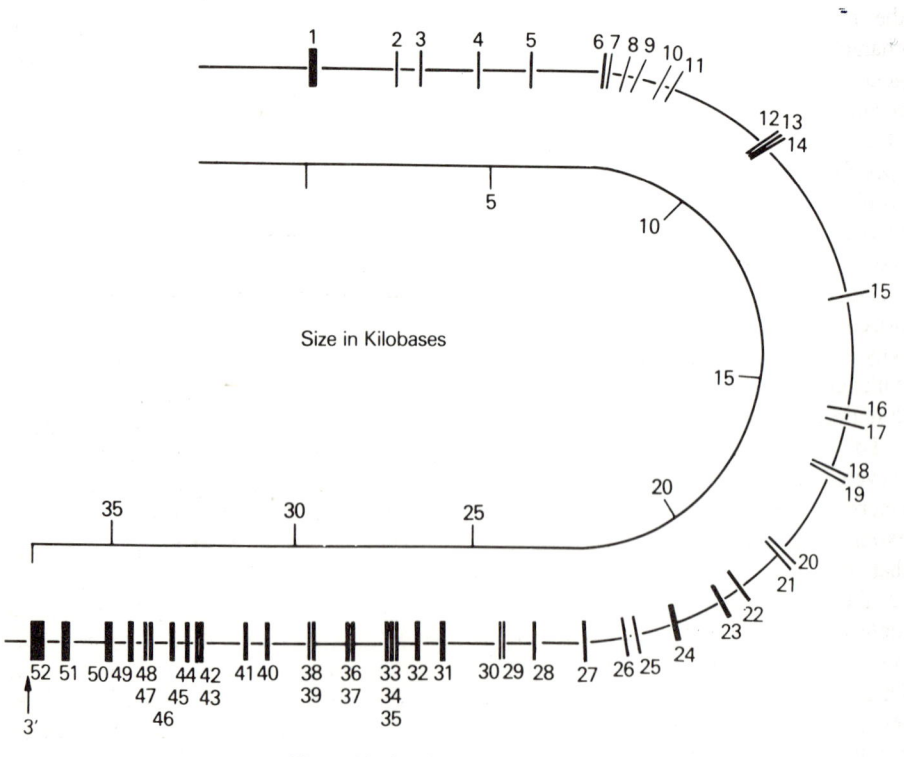

Chick α-2 (type I) Collagen Gene

Fig. 1. Chick α-2 (type I) collagen gene

ASSEMBLY OF THE ANCESTRAL COLLAGEN GENE

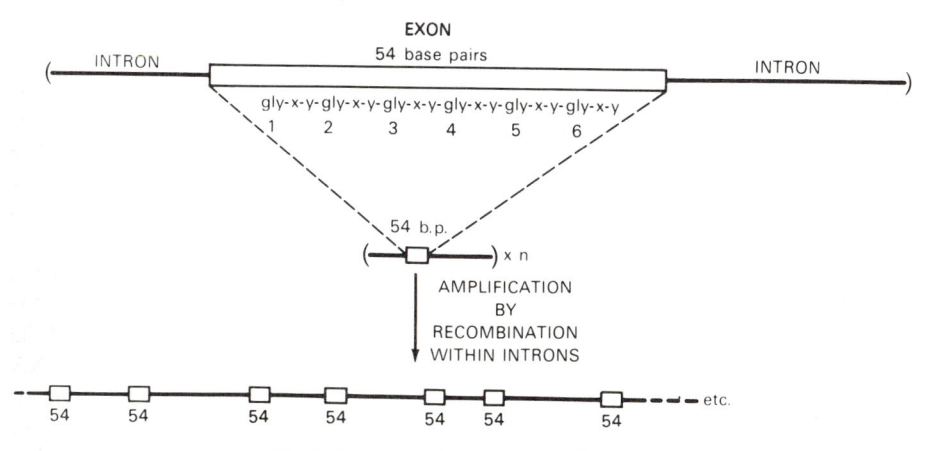

Fig. 2. Assembly of the ancestral collagen gene

be endowed not only with a high degree of precision, so that the nucleotides at the ends of exons are spliced together, but also with an error-free mechanism that will enable it to splice the 52 exons in unique order.

The most attractive hypothesis to explain the multi-exon organization which is a characteristic feature of many eucaryotic genes, postulates that these genes arose by recombination between functional DNA blocks[12]. Each exon would code for a specific protein domain. Recombination would occur within introns and so probably increase the rate of recombination between these functional DNA blocks. During the isolation of the collagen gene, we were surprised to find so many exons, and wondered why a molecule as uniform and regular as collagen would be subdivided into so many domains.

To investigate this question, E. Avvedimento, Y. Yamada and M. Mudryj determined the size of a small number of exons in three different parts of the gene that correspond to three different segments of the helical portion of the collagen molecule[13]. Later some additional exons were sequenced in collaboration with L. Dickson, B. Olsen and P. Fietzek[4] of Rutgers University Medical School. The results show a remarkable conservation of the size of the exons. Nine out of twelve exons which were examined had an identical length of 54 bp, two exons contained 108 bp (2 x 54) and one had 99 bp. H. Boedtker and her colleagues at Harvard University[15] have also sequenced exons in the same gene and found, in addition to a number of 54 bp and 108 bp exons, a few other exons which had a different length but all were multiples of nine. It is important to note that, although most exons have the same length (54 bp), the sequences within these exons vary except for the glycine codons (GGX) which appear every third triplet. The 108 bp exons probably arose by fusion of two adjacent 54 bp exons and the precise deletion of an intervening intron. The number of bp in those exons which are different from 54 or 108 bp diverge by nine or a multiple of nine. To preserve the continuity of the triple helix only deletions of nine bp or multiples of nine can be tolerated. Indeed, the basic gly-x-y repeat of the collagen molecule is encoded by nine bp.

The finding that most exons of the gene for the alpha 2 chain of type I collagen have an identical length of 54 bp has important implications for the evolutionary assembly of this gene. It suggests that the ancestral gene for collagen was assembled by multiple duplications of a single genetic unit containing an exon of 54 bp (Fig. 2). Later the exon sequences evolved by successive point mutations but also, in some cases, by deletions or additions of nine bp or multiples of nine bp. The 54 bp exon structure is probably common to all collagen genes found in nature. Y. Yamada and M. Mudryj have recently sequenced two exons in another chick collagen gene; one exon contained 54 bp, the other 108 bp.

Although the exon sequences vary, they present enough similarities to suggest that they are derived from each other. We speculate that the primordial exon for the collagens may have encoded a glycine-proline-proline tripepetide repeated six times. Such a glycine-proline peptide of 18 amino acids probably had the minimal length needed to form a stable triple helical structure.

The C$_1$Q component of the complement system contains a triple helical collagen tail[16]. This finding raises the possibility that the original function of the primordial collagen exon peptide may have been different from its current role in collagen. A short triple helical segment could have helped stabilize the tertiary structure of a three-subunit protein in which each subunit contains a single collagen extension. This same primordial exon may later have been utilized to construct the precursor of the collagen gene family.

Regulation of the alpha 2(I) collagen gene and structure of its promoter

There are a number of findings that support the notion that the effect of p60^{src} on collagen synthesis, although indirect, is mediated by a transcriptional control mechanism. First, our co-workers S. Adams, M. Sobel and B. Howard[3-5] showed that the decrease in type I collagen synthesis in RSV transformed chick embryo fibroblasts was due to a reduction of more than tenfold in the levels of fully processed, cytoplasmic mRNA. Later, Avvedimento[6] demonstrated a severe reduction in the levels of nuclear intron-specific RNA for alpha 2(I) collagen. Recently, P. Bornstein and colleagues[7] have shown that the rate of synthesis of type I collagen RNA is six- to eight-fold lower with nuclei of CEF transformed by RSV than with nuclei from normal CEF. Finally, in cells infected with a mutant of RSV in which the *src* gene codes for a thermosensitive p60^{src}, the levels of type I collagen and of type I col-

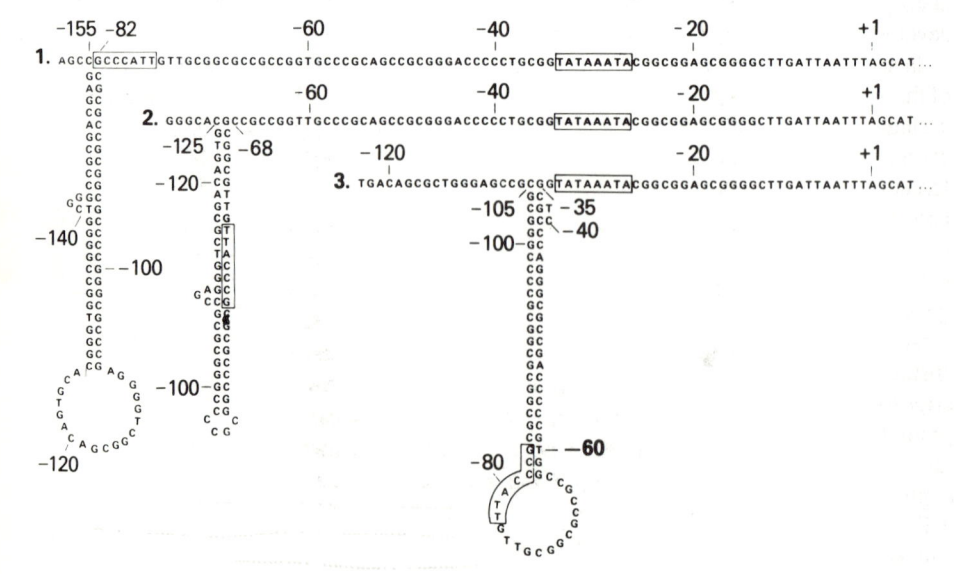

Fig. 3. Dyads of symmetry in the α2 collagen promoter

lagen RNA are low at the permissive temperature and elevated at the non-permissive temperature[5]. If the temperature is shifted from a permissive to a non-permissive temperature, a typical induction curve for collagen and collagen RNA is obtained[5]. All these experiments strongly suggest that the control of collagen synthesis occurs at the level of transcription.

Because studies on bacteria and bacteriophages show most transcriptional control factors operate at a promoter region located at the 5' end of the gene, we have characterized the structure of the 5' region (promoter) of the alpha 2(I) collagen gene[17]. The promoter region of this gene contains two short characteristic sequences which are found at approximately the same distance from the start site of transcription in a number of eucaryotic genes. One is called the Goldberg-Hogness sequence, the other the 'CAT' box sequence. In addition the promoter region of the alpha 2(I) collagen gene contains another unique feature. There are three large inverted repeats preceding the start site of transcription which overlap each other and which could therefore form mutually exclusive stem and loop structures (Fig. 3). We suspect that these structures could be binding sites for regulatory proteins and thus play a role in the developmental regulation of this gene.

In summary, our studies on the structure of the alpha 2(I) collagen gene indicate how a single small genetic unit containing an exon of 54 bp, which may have had a different function in a more primitive cell, was used to construct the ancestor for a family of genes, encoding a set of abundant extracellular proteins. The structure of this gene also illustrates that an error-free succession of splicing reactions is essential to ensure the conversion of the primary transcript to translatable messenger RNA. The promoter of this gene has a unique structure. Examination of the control of the expression of the promoter for the alpha 2(I) collagen gene in appropriately reconstituted *in vivo* and *in vitro* systems should help us understand how the gene is developmentally regulated and how oncogenic growth factors alter its expression.

References

1 Bornstein, P., and Sage, H. (1980) *Ann. Biochem.* 49, 957
2 Levinson, W., Bhatnagar, R. S. and Liu, T.-Z. (1975) *J. Natl Cancer Inst.* 55, 807
3 Adams, S. L., Sobel, M. E., Howard, B. H., Olden, K., Yamada, K. M., de Crombrugghe, B. and Pastan, I. (1977) *Proc. Natl Acad. Sci. U.S.A.* 74, 3399
4 Howard, B. H., Adams, S. L. Sobel, M. E., Pastan, I. and de Crombrugghe, B. (1978) *J. Biol. Chem.* 253, 5869
5 Sobel, M., Yamamoto, T., de Crombrugghe, B. and Pastan, I. (1981) *Biochemistry* 20, 2678
6 Avvedimento, E., Yamada, Y., Lovelace, E., Vogeli, G., de Crombrugghe, B. and Pastan, I. (1981) *Nucleic Acids Res.* 9, 1123
7 Sandmeyer, S., Gallis, B. and Bornstein, P. (1981) *J. Biol. Chem.* 256, 5022
8 Hata, R. and Peterkofsky, B. (1977) *Proc. Natl Acad. Sci. U.S.A.* 74, 2933
9 Ohkubo, H., Vogeli, G., Mudryj, M., Avvedimento, V. E., Sullivan, M., Pastan, I. and de Crombrugghe, B. (1980) *Proc. Natl Acad. Sci. U.S.A.* 77, 7059
10 Vogeli, G., Avvedimento, E. V., Sullivan, M., Maizel Jr., J. V., Lozano, G., Adams, S. L., Pastan, I. and de Crombrugghe, B. (1980) *Nucleic Acids Res.* 8, 1823
11 Lewin, B. (1980) *Cell* 22, 324–326
12 Gilbert, W. (1979) *Nature (London)* 271, 501
13 Yamada, Y., Avvedimento, V. E., Mudryj, M., Ohkubo, H., Vogeli, G., Irani, M., Pastan, I. and de Crombrugghe, B. (1980) *Cell* 27, 887
14 Dickson, L. A., Ninomiya, Y., Bernard, M. P., Pesciotta, D. M., Parsons, J., Green, G., Eikenberry, E. F., de Crombrugghe, B., Vogeli, G., Pastan, I., Fietzek, P. P. and Olsen, B. R. (1980) *J. Biol. Chem.* (in press)
15 Wozney, J., Hanahan, D., Morimoto, R., Boedtker, H. and Doty, P. (1981) *Proc. Natl Acad. Sci. U.S.A.* 78, 712
16 Porter, R. R. and Reid, K. B. M. (1978) *Nature (London)* 275, 699
17 Vogeli, G., Ohkubo, H., Sobel, M., Yamada, Y., Pastan, I. and de Crombrugghe, B. (1981) *Proc. Natl Acad. Sci. U.S.A.* 78, 5334

Benoit de Crombrugghe and Ira Pastan are at the Laboratory of Molecular Biology, Division of Cancer Biology and Diagnosis, National Cancer Institute, National Institutes of Health, Bethesda MD 20205, U.S.A.

Corticotropin-β-lipotropin precursor – a multi-hormone precursor – and its gene

Shosaku Numa and Shigetada Nakanishi

The characteristic repetitive structure of the common precursor of the pituitary hormones corticotropin and β-lipotropin has been elucidated by determining the nucleotide sequence of its cDNA. The gene encoding this multi-hormone precursor provides an attractive system for studying the structure, evolution and regulation of hormonally controlled eukaryotic genes.

Endocrine peptides are initially synthesized as larger precursors, which are then converted to their respective peptides by post-translational cleavage. The common precursor of the pituitary hormones corticotropin (ACTH) and β-lipotropin (β-LPH) provides a typical example of a precursor protein containing multiple, biologically-active peptides. This review deals with the elucidation of the whole primary structure of this precursor by way of the nucleotide sequence of its cDNA and with the structural organization and regulation of its gene. The possible biological significance of the multi-hormone precursor is discussed.

Cell-free translation and purification of the mRNA coding for the ACTH–β-LPH precursor

The initial precursor of ACTH was demonstrated by translating bovine pituitary poly (A)-containing RNA in a cell-free protein-synthesizing system derived from wheat germ or rabbit reticulocytes[1]. It exhibited a molecular weight about seven times larger than that of ACTH itself and was identified by specific immunoprecipitation with antibody to ACTH, followed by electrophoretic analysis of the dissociated immunoprecipitate. Further evidence for the identity of the precursor protein was provided by demonstrating a common peptide in the chymotryptic digests of the translation product and authentic ACTH. Subsequent studies revealed that the same translation product was immunoprecipitated with antibody to β-endorphin, indicating the presence of a common precursor of ACTH and β-LPH[2]. A similar ACTH–β-LPH precursor was found in cultured mouse pituitary tumor cells[3,4]. Thus, this precursor protein contains a number of peptides with different biological activities, i.e. ACTH and its component peptides, α-melanotropin (α-MSH) and corticotropin-like intermediate lobe peptide (CLIP), as well as β-LPH and its component peptides, γ-lipotropin (γ-LPH), β-melanotropin (β-MSH) and endorphins.

The fact that ACTH and β-LPH together account for only about half of the molecular weight of the precursor protein raised the question: what peptides and what biological functions are in its remaining portion? A promising approach to the answer seemed to be to deduce the whole amino acid sequence of the ACTH–β-LPH precursor by determining the nucleotide sequence of cloned DNA complementary to its mRNA. This mRNA is a major mRNA species in the intermediate lobe of the pituitary – its translation product amounting to almost one-third of the products encoded by total translatable mRNA[5]. This unexpected abundance aided purification[6] and suggests that the intermediate lobe may perform a highly specialized function in producing a large amount of ACTH–β–LPH precursor. RNA from membrane-bound polysomes of bovine pituitary neurointermediate lobes was subjected to chromatography on oligo (dT)-cellulose and on poly (U)-Sepharose and then fractionated twice by sucrose density gradient centrifugation. The final mRNA preparation was homogeneous as shown by analysis of its translation product, electrophoresis on polyacrylamide gel in the presence of formamide and analysis of the kinetics of hybridization with its cDNA[6].

Nucleotide sequence of cDNA for the ACTH-β-LPH precursor and assignment of its amino acid sequence

Bacterial plasmids were constructed using purified bovine mRNA coding for the ACTH-β-LPH precursor[7]. Double-stranded cDNA species were synthesized and cloned in *Escherichia coli* χ 1776 by inserting them into the *Pst*I endonuclease cleavage site of the plasmid pBR322 with the use of poly(dG)·poly(dC) homopolymeric extensions.

The nucleotide sequence of a cloned cDNA insert was determined[8] by the method of Maxam and Gilbert. The sequence of the remaining 15 nucleotide residues at the 5' extremity of the mRNA, not included in the cDNA clone used, was determined by sequencing the reverse transcript formed by elongation of a DNA primer hybridized to a segment of the 5'-untranslated region of the mRNA, as well as by RNA sequencing[9]. The mRNA coding for the bovine ACTH-β-LPH precursor is 1,098 nucleotides long, including the 5'-untranslated region with 128 nucleotides, the protein-encoding region with 795 nucleotides and the 3'-untranslated region with 175 nucleotides (excluding the poly(A) tail with an average length of 68 nucleotides[6]). It is suggested that a 'cap' structure is present at the 5' terminus of the mRNA[9].

The complete amino acid sequence of the ACTH-β-LPH precursor was deduced from the nucleotide sequence of its cDNA[8]. The structure of the precursor protein thus elucidated is schematically represented in Fig. 1. The amino acid residues 1–39 and 42–134 correspond precisely to the sequences determined for bovine ACTH and β-LPH, respectively, except that two additional amino acid residues, Ala–Glu, are present between positions 35 and 36 of the generally accepted structure of bovine β-LPH determined by protein sequencing. The nucleotide residues encoding the glutamine at the carboxyl terminus of β-LPH are followed by the termination codon UGA, indicating that the carboxyl terminus of β-LPH corresponds to that of the precursor protein. The amino terminus of β-LPH is connected with the carboxyl terminus of ACTH by the paired basic amino acid residues, Lys–Arg. The amino terminus of ACTH is likewise joined with the adjacent peptide by a Lys–Arg sequence.

In assigning the amino acid sequence of the cryptic portion of the ACTH-β-LPH precursor, the reading frame of amino acid codons was determined by the frame corresponding to the amino acid sequences of ACTH and β-LPH. The location of the translational initiation site at the methionine residue at position −131 was verified by determining the partial amino acid sequence of the amino-terminal region of the ACTH-β-LPH precursor (prepro-hormone) synthesized in a cell-free translation system[10]. This study, as well as amino acid sequence analysis of the amino-

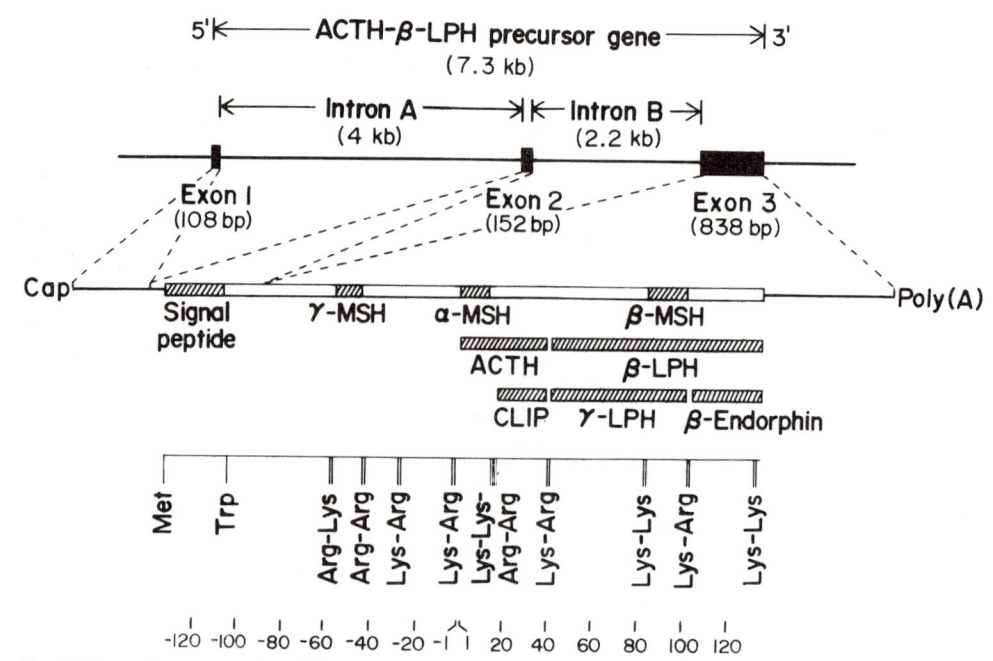

Fig. 1. Schematic representation of the structure of the bovine ACTH–β-LPH precursor and its gene. The structural organization of the gene is shown above; the lengths of the gene, exons (indicated by closed boxes) and introns are given in kilobase pairs (kb) or base pairs (bp). The topology of the mRNA is displayed below the gene, the thick portion representing the protein-encoding sequence. The shaded bars on and under the mRNA stand for the regions coding for the component peptides. The initiator methionine residue, the amino-terminal tryptophan residue of the prohormone, the paired basic amino acid residues and the amino acid numbers are shown underneath; amino acid residues are numbered from the amino terminus to the carboxyl terminus, beginning with the first residue of ACTH, and the residues on the amino-terminal side of ACTH are indicated by negative numbers.

terminal region of the precursor protein (prohormone) or its fragment isolated from intact cells (reviewed in Refs 11 and 12) confirmed the protein structure predicted from the nucleotide sequence of the cDNA. The latter studies also located the amino terminus of the prohormone at the tryptophan residue at position −105, indicating that the first 26 amino acid residues starting with the initiator methionine, which include a large number of hydrophobic amino acids (16 non-polar residues including 7 leucine residues), constitute a signal peptide characteristic of secretory proteins.

The cryptic portion of the ACTH–β-LPH precursor contains an amino acid sequence (residues −55 to −44) strikingly similar to the sequences of α-MSH and β-MSH. This newly found peptide fragment, named γ-melanotropin (γ-MSH), shares tyrosine and methionine residues with the known MSHs, and has at equivalent positions the characteristic tetrapeptide sequence His–Phe–Arg–Trp, which is required for melanotropic activity. γ-MSH is also flanked on both sides by paired basic amino acid residues. There is evidence for the presence of the γ-MSH sequence in bovine and human pituitaries as well as in ectopic ACTH-producing human tumors[13] (see also Ref. 12 and refs therein).

Another peptide fragment that exhibits some structural similarity to the MSHs is that comprising residues −111 to −105, most of which are contained in the signal peptide. It is also noteworthy that the peptide fragment composed of residues −104 to −73 has some homology with the hormone calcitonin.

On the basis of these findings, the ACTH–β-LPH precursor molecule is thought to be composed of three repetitive units, each containing an MSH sequence, plus a signal peptide (Fig. 1). The repetitive units and their peptide components are each bounded by paired basic amino acid residues. This structure of the precursor protein implies that each of its units is destined to be separated from its neighbour by proteolytic cleavage and that further processing occurs within the units to yield smaller peptides. Furthermore, the presence of an unusually large number of repeated nucleotide sequences within the ACTH–β-LPH precursor mRNA supports the view that the structural gene for the precursor protein has evolved by a series of genetic duplications[8].

Also noteworthy is the fact that the 5'-untranslated region of the bovine ACTH–β-LPH precursor mRNA contains three segments of potential secondary structure which partially overlap, so that it can exist in a number of alternative base-paired configurations[9]. However, its interaction with the 3'-terminal segment of 18S rRNA at the site of maximal complementarity would fix the mRNA configuration in such a way as to bring the possible site of ribosome binding near to the initiation codon. It seems attractive to speculate that alterations in the RNA configuration are involved in the regulation of translational initiation or possibly in that of transcription termination. An analogous shift between alternative RNA configurations is associated with the regulation of transcription termination at the attenuators of various operons of enteric bacteria (reviewed in Ref. 14).

Structural organization of the ACTH–β-LPH precursor gene

In view of the recent knowledge about the organization of eukaryotic genes, it would be particularly interesting to investigate the relationship between the repetitive structure of the ACTH–β-LPH precursor containing different functional components and the structural organization of its gene. Using the lambda cloning system, the entire bovine ACTH–β-LPH precursor gene was isolated as a set of overlapping genomic DNA fragments extending over a length of approximately 17 kilobase pairs[15,16]. Restriction mapping of these fragments and nucleotide sequence analysis of the whole mRNA-encoding segments and their surrounding regions established the structural organization of the gene (schematically represented in Fig. 1). The bovine ACTH–β-LPH precursor gene is approximately 7.3 kilobase pairs long and codes for the mature mRNA of 1,098 nucleotides. The mRNA-encoding sequence is divided by two large introns into three exons which are arranged linearly. Exon 1 encodes all but the distal 20 nucleotides of the 5'-untranslated sequence and is separated from exon 2 by intron A of approximately four kilobase pairs. The distal portion of the 5'-untranslated region and the amino-terminal 44 amino acids including the signal peptide sequence are encoded by exon 2, which is separated from exon 3 by intron B of approximately 2.2 kilobase pairs. Exon 3 encodes most of the protein structure, including the three repeated MSH sequences, other biologically active component peptides and whole 3'-untranslated sequence. Thus, there is no apparent correspondence between the repetitive structure of the precursor protein and the structural organization of its gene.

The isolation of human and rat genomic DNA segments containing an uninterrupted sequence equivalent to exon 3 of the bovine ACTH–β-LPH precursor gene has been reported[17,18]. The entire human ACTH–β-LPH precursor gene has been isolated recently, and its structural organization is essentially the same as that of the bovine gene (unpublished results). The human gene is approximately 7.6 kilobase pairs long, and its introns A and B, approximately 3.6 and 2.9 kilobase pairs long, respectively, are inserted at exactly the same positions as the bovine counterparts. Nucleotide sequence analysis of the whole exons has revealed the complete mRNA and amino acid sequences of the human ACTH-β-LPH precursor. Comparison of the bovine and human gene sequences shows that three regions are highly conserved, i.e. the region extending from the signal peptide to γ-MSH, the ACTH region and the β-MSH/β-endorphin region. This suggests that the peptide(s) in the amino-terminal region of the prohormone, including γ-MSH, may be of physiological importance, as is the case for ACTH and β-endorphin.

According to the hypothesis of exon shuffling advocated by Gilbert and Tonegawa (reviewed in Ref. 19), an intron is derived from the flanking regions of a translocated primordial gene in the process of producing a new gene structure. Assuming that the repetitive structure of the ACTH-β-LPH precursor gene has evolved by intron-mediated joining of primordial DNA segments, the absence of introns between the three MSH-encoding regions may have resulted from the loss of introns that have once existed. Such a loss of an intron has been proposed for the region encoding the C-peptide of the rat preproinsulin I gene[20]. Alternatively, the repetitive structure of the present-day ACTH-β-LPH precursor gene may have evolved in such a way that a primordial DNA segment was duplicated at a position adjacent to itself and that the translational reading frame of the duplicated segments was recovered by addition or deletion of nucleotides. The structural organization of immunoglobulin genes is considered to reflect their evolution by intron-mediated assembly of different structural and functional domains[21]. On the other hand, the partly homologous repetitive units of the ACTH-β-LPH precursor are destined to be separated from one another to exhibit biological functions, and it seems possible that the gene for such a hormone precursor has evolved by duplication of an ancestral gene at adjacent positions. To test the different possibilities of evolution one would need to compare the ACTH-β-LPH precursor genes of different species.

Blot hybridization analysis of bovine pituitary nuclear RNA indicates that the entire ACTH-β-LPH precursor gene is transcribed into a primary hnRNA product, which is then spliced to form the mature mRNA[16]. The message strands at both ends of the two introns are complementary to the 5'-terminal sequence of U1 small nuclear RNA[16]. It is proposed that the base-pairing interaction due to complementarity between U1 small nuclear RNA and the consensus sequences at RNA splice sites is involved in the processing of hnRNA (reviewed in Ref. 22).

Using blot hybridization analysis and DNA sequencing, the interspersed repetitive sequences, located in the 5'-flanking region and within the introns of the ACTH-β-LPH precursor gene have been identified (Ref. 16 and unpublished results). The repetitive sequences in the human ACTH-β-LPH precursor gene region are the *Alu* family sequences, and the bovine repetitive sequences are partly homologous to the human counterparts. It is suggested that the interspersed repetitive DNA sequences characteristic of eukaryotic genomes are involved in the replication or transcription of genes or in the processing of their transcripts[23].

Regulation of expression of the ACTH-β-LPH precursor gene

Studies with the rat and a mouse pituitary tumor cell line have shown that the cellular contents of the mRNA and the hnRNA encoding the ACTH–β-LPH precursor are depressed by glucocorticoids (Refs 24, 25 and unpublished results). The effects of various steroids on the amount of mRNA correlate well with the binding specificity of the glucocorticoid receptor. This suggests that the expression of the ACTH–β-LPH precursor gene, which represents a unique system in view of its negative control by glucocorticoids, is regulated via the glucocorticoid receptor.

Comparison of the 200-base-pair sequence in the 5'-flanking region of the bovine ACTH–β-LPH precursor gene with those of the mammalian α-globin and β-globin genes reveals sequence similarities[16]. This is of interest in view of the fact that globin synthesis in Friend leukemia cells induced by dimethyl sulfoxide is depressed by glucocorticoids[26], although it is not possible yet to determine whether the homologous sequences noted have occurred by chance or represent common functional sites shared by these evolutionarily distant genes. In contrast, no impressive sequence similarities are observed between the 5'-flanking region of the ACTH–β-LPH precursor gene and those of the chicken ovalbumin and conalbumin genes[16], which, unlike the ACTH–β-LPH precursor gene, are positively regulated by steroid hormones.

Biological significance of the common precursor of multiple hormones

The fact that the different biologically active component peptides of the ACTH–β-LPH precursor are encoded by a single gene suggests that their functions may be related to or co-ordinated with one another. It is intriguing to speculate that these peptides perform co-ordinate functions in the defense mechanism of the living organism. ACTH increases the production of glucocorticoids, which elevate blood glucose, and β-LPH stimulates lipolysis in adipose tissue, both mobilizing stored energy. CLIP has been reported to enhance insulin secretion for efficient utilization of blood glucose, and β-endorphin exhibits analgesic activity. Melanocyte stimulation by MSHs could be a rudimentary mechanism for self-protection. Furthermore, MSH-related peptides are implicated in brain functions such as the enhancement of the arousal state and increase in attention and motivation (reviewed in Ref. 27). The newly found γ_3-MSH (amino acid residues -55 to -29) has been shown to potentiate the steroidogenic action of ACTH synergistically[28] and to exhibit striking hypothermic and behavioral effects including transient behavioral hyperactivity[29]. In view of the fact that the component peptides of the ACTH–β-LPH precursor are present in various regions of the central nervous system, it seems attractive to postulate that they may act co-ordinately as modulators of neural functions.

The amino-terminal fragment of the ACTH–β-LPH precursor represents the first example of the discovery of new endocrine peptides using the nucleotide sequence of the DNA encoding them. It seems probable that analogous precursors containing multiple peptides with co-ordinate functions occur rather commonly. This approach should be effective in unravelling the structure of other possible multi-hormone precursors.

References

1 Nakanishi, S., Taii, S., Hirata, Y., Matsukura, S., Imura, H. and Numa, S. (1976) *Proc. Natl. Acad. Sci. U.S.A.* 73, 4319–4323
2 Nakanishi, S., Inoue, A., Taii, S. and Numa, S. (1977) *FEBS Lett.* 84, 105–109
3 Mains, R. E., Eipper, B. A. and Ling, N. (1977) *Proc. Natl. Acad. Sci. U.S.A.* 74, 3014–3018
4 Roberts, J. L. and Herbert, E. (1977) *Proc. Natl. Acad. Sci. U.S.A.* 74, 5300–5304
5 Taii, S., Nakanishi, S. and Numa, S. (1979) *Eur. J. Biochem.* 93, 205–212
6 Kita, T., Inoue, A., Nakanishi, S. and Numa, S. (1979) *Eur. J. Biochem.* 93, 213–220

7 Nakanishi, S., Inoue, A., Kita, T., Numa, S., Chang, A. C. Y., Cohen, S. N., Nunberg, J. and Schimke, R. T. (1978) *Proc. Natl. Acad. Sci. U.S.A.* 75, 6021–6025
8 Nakanishi, S., Inoue, A., Kita, T., Nakamura, M., Chang, A. C. Y., Cohen, S. N. and Numa, S. (1979) *Nature (London)* 278, 423–427
9 Inoue, A., Nakamura, M., Nakanishi, S., Hidaka, S., Miura, K. and Numa, S. (1981) *Eur. J. Biochem.* 113, 531–539
10 Nakamura, M., Inoue, A., Nakanishi, S. and Numa, S. (1979) *FEBS Lett.* 105, 357–359
11 Herbert, E., Roberts, J., Phillips, M., Allen, R., Hinman, M., Budarf, M., Policastro, P. and Rosa, P. (1980) in *Frontiers in Neuroendocrinology* (Ganong, W. F. and Martini, L., eds). Vol. 6, pp. 67–101, Raven Press, New York
12 Imura, H. (1980) *Adv. Cancer Res.* 33, 39–75
13 Tanaka, I., Nakai, Y., Jingami, H., Fukata, J., Nakao, K., Oki, S., Nakanishi, S., Numa, S. and Imura, H. (1980) *Biochem. Biophys. Res. Commun.* 94, 211–217
14 Yanofsky, C. (1981) *Nature (London)* 289, 751–758
15 Nakanishi, S., Teranishi, Y., Noda, M., Notake. M., Watanabe, Y., Kakidani, H., Jingami, H. and Numa, S. (1980) *Nature (London)* 287, 752–755
16 Nakanishi, S., Teranishi, Y., Watanabe, Y., Notake, M., Noda, M., Kakidani, H., Jingami, H. and Numa S. (1981) *Eur. J. Biochem.* 115, 429–438
17 Chang, A. C. Y., Cochet, M. and Cohen, S. N. (1980) *Proc. Natl. Acad. Sci. U.S.A.* 77, 4890–4894
18 Drouin, J. and Goodman, H. M. (1980) *Nature (London)* 288, 610–613
19 Crick, F. (1979) *Science* 204, 264–271
20 Lomedico, P., Rosenthal, N., Efstratiadis, A., Gilbert, W., Kolodner, R. and Tizard, R. (1979) *Cell* 18, 545–558
21 Sakano, H., Rogers, J. H., Hüppi, K., Brack, C., Traunecker, A., Maki, R., Wall, R. and Tonegawa, S. (1979) *Nature (London)* 277, 627–633
22 Lewin, B. (1980) *Cell* 22, 324–326
23 Jelinek, W. R., Toomey, T. P., Leinwand, L., Duncan, C. H., Biro, P. A., Choudary, P. V., Weissman, S. M., Rubin, C. M., Houck, C. M., Deininger, P. L. and Schmid, C. W. (1980) *Proc. Natl. Acad. Sci. U.S.A.* 77, 1398–1402
24 Nakanishi, S., Kita, T., Taii, S., Imura, H. and Numa, S. (1977) *Proc. Natl. Acad. Sci. U.S.A.* 74, 3283–3286
25 Nakamura, M., Nakanishi, S., Sueoka, S., Imura, H. and Numa, S. (1978) *Eur. J. Biochem.* 86, 61–66
26 Lo, S.-C., Aft, R., Ross, J. and Mueller, G. C. (1978) *Cell* 15, 447–453
27 Reith, M. E. A., Schotman, P., Gispen, W. H. and de Wied, D. (1977) *Trends Biochem. Sci.* 2, 56–58
28 Pedersen, R. C., Brownie, A. C. and Ling, N. (1980) *Science* 208, 1044–1046
29 Henriksen, S. J., Benabid, A. L., Madamba, S., Bloom, F. E. and Ling N. (1980) *Soc. Neurosci. Abstr.* 6, 681

S. Numa and S. Nakanishi are at the Department of Medical Chemistry, Kyoto University Faculty of Medicine, Yoshida, Sakyo-ku, Kyoto 606, Japan

Gene amplification during development
by R. Chisholm

Cellular differentiation often requires the precisely timed appearance of new gene products. The production of globin in reticulocytes, or of ovalbumin in chick oviduct occurs by the synthesis of stable messages which accumulate over relatively long periods. Other gene products, such as *Drosophila* chorion proteins are also needed in significant amounts, but must accumulate during a relatively short period of synthesis. Recent reports from Allan Spradling, Anthony Mahowald and their collaborators have described an unusual mechanism used during *Drosophila* oogenesis to produce the required amount of egg shell proteins in the relatively short time during which they are synthesized. They have shown that the copy number of specific chorion genes is dramatically increased by specific amplification of the chromosome segments containing them.

The *Drosophila* chorion consists of several protein-rich layers and is synthesized by ovarian follicle cells over a period of about 5 h. To study the expression of these chorion protein genes, Spradling and Mahowald produced cDNA clones complementary to egg chamber RNA[1]. One of these clones was shown to encode the predominant 36,000 dalton egg shell protein, s36. When this clone was used to probe restriction digests of egg chamber DNA, sequences complementary to the cloned s36 gene were found to be more abundant than sequences encoding non-chorion genes. However, when this cloned chorion sequence was used to probe DNA prepared from non-ovarian tissue, the chorion sequence was not over-represented. Interestingly, a mutation resulting in female sterility (ocelliless) known to map near the s36 gene and reduce s36 synthesis, caused a reduction in the over-representation of the s36 coding sequences to 25% of that seen in wild-type flies. Spradling and Mahowald concluded that the s36 coding sequences were being specifically amplified during *Drosophila* oogenesis. They also showed that a genetically unlinked chorion gene, encoding an abundant 18,000 dalton protein (s18) was also specifically over represented in ovarian DNA, and that it was not amplified in ocelliless mutants. Thus, the specific amplification of chorion protein genes occurs at more than one chromosomal location.

Amplification of particular DNA sequences has been observed before, and has been best characterized for ribosomal RNA genes during *Xenopus* oogenesis[2]. Here, the ribosomal RNA genes are amplified as extrachromosomal elements, probably by a process of rolling circle replication. Another example is the amplification of the dihydrofolate reductase gene in methotrexate-resistant mammalian cells. This gene is specifically amplified, apparently as tandem duplications of a large chromosomal region, to produce as many as 200 tandem copies[3]. Such amplification has also been observed in human tumors that do not respond to methotrexate.

Spradling and Mahowald[4,5] have now described experiments which suggest that chorion genes are amplified by a third mechanism: bi-directional replication, initiated at a discrete chromosomal location, followed by random termination to produce a structure which has been likened to an 'onion skin' (see diagram,

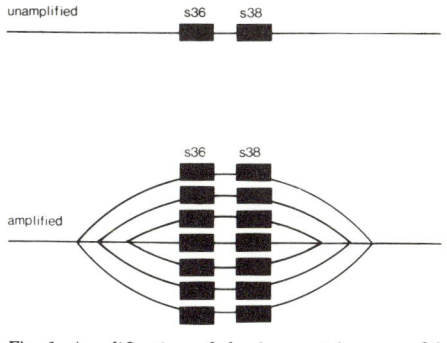

Fig. 1. Amplification of chorion protein genes s36 and s38.

Ref. 6). Using the s36 cDNA clone as a probe, 100 kb region of the *Drosophila* X-chromosome, centered on the s36 gene has been isolated as overlapping lambda clones. When these cloned restriction fragments are used to determine the relative abundance of different segments of the amplified region, the sequences containing the s36 gene were the most over represented, and the relative over representation of the adjoining sequences decreased the further they were from the s36 gene. In all, a region of about 90 kb was found to be amplified. This pattern, along with the absence of new restriction fragments in the amplified DNA, suggests amplification results from DNA replication originating near the s36 gene and terminating at random locations over about 45 kb in each direction.

The ocelliless mutation is a small chromosomal inversion with one of its breakpoints lying near the s36 gene. Careful restriction analysis of this region in the ocelliless fly suggested that the inversion had occurred such that the site of initiation of DNA replication responsible for the amplification had been moved to near the other breakpoint. Using cloned probes for the non-chorion sequences now adjacent to the end of the inversion, the relative abundance of these non-chorion sequences in ovarian DNA was determined. Indeed, these sequences were now amplified, and the normally amplified chorion sequences at the other end of the inversion were not over represented. Thus, it appears that chorion gene amplification occurs by initiation of DNA replication at a specific chromosomal site adjacent to the chorion genes. This also suggests that clustering of the chorion protein genes observed at the s36 protein gene location has biological significance. Activation of a single origin of replication can result in the amplification of several genes whose products must all be synthesized during a relatively short period of time.

Recent experiments suggest that the amplification of genes as a mechanism for ensuring that adequate concentrations of a gene product can be synthesized quickly may be a more widespread phenomenon. In experiments to be published in *PNAS*, Ronald Schwartz and his collaborators report the amplification of actin gene sequences during chick myogenesis. Only time will tell how many more examples of gene amplification during development will be found.

References

1 Spradling, A. C. and Mahowald, A. P. (1980) *Procl Natl Acad. Sci. U.S.A.* 77, 1096–1100
2 Rochaix, J. D., Bird, A. and Bakken, A. (1974) *J. Mol. Biol.* 87, 473–487
3 Nunberg, et al. (1978) *Proc. Natl Acad. Sci. U.S.A.* 75, 5553–5556
4 Spradling A. C. (1981) *Cell* 27, 193–202
5 Spradling A. C. and Mahowald, A. P. (1981) *Cell* 27, 203–210
6 Botchan, M., Topp, W. and Sambrook, J. (1978) *Cold Spring Harbor Symp. Quant. Biol.* 43, 709–720

R. CHISHOLM

Department of Biology, Massachusetts Institute of Technology, Cambridge, MA 02139, U.S.A.

The rearrangements of immunoglobulin genes

Nicholas Gough

Antibodies are encoded by a complex gene system that apparently undergoes unprecedented rearrangements during lymphocyte development. The immunoglobulin gene system has evolved novel and flexible strategies for dealing with large amounts of genetic information.

Antibody (immunoglobulin) molecules are composed of two identical *light* polypeptide chains (mol. wt ~23,000) and two identical *heavy* chains (mol. wt ~53,000–70,000) joined by disulphide bonds (see Fig. 1 and Ref. 1 for review). There are five major classes of immunoglobulin molecule (IgG, IgM, IgA, IgD and IgE) which mediate different biological effector functions (see Table 1 for examples). Counterparts of each of these immunoglobulin classes are found in the human, the rabbit and the mouse. Some of these classes can be further divided into subclasses, the nature and number of which varies between different species. In the mouse there are four IgG subclasses, IgG1, IgG2a, IgG2b and IgG3. Each class or subclass is distinguished by the nature of its component heavy chain, which, as indicated in Table I, is designated by the corresponding Greek letter (γ1, γ2a, γ2b, γ3, μ, α, δ, ϵ). Light chains, which are of two types, kappa (κ) and lambda (λ), occur in all immunoglobulin classes. Whether an antibody molecule with a κ light chain has different functions to one with an λ chain is not known.

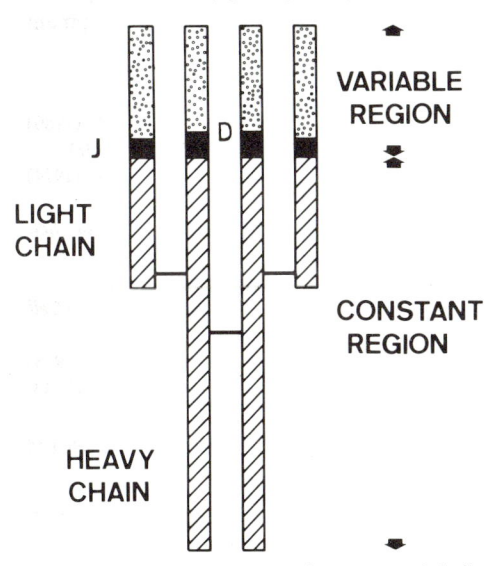

Each heavy and each light chain (Fig. 1) has an N-terminal region of ~110 amino acids, the *variable* (V) region, that exhibits extensive sequence diversity, and a C-terminal portion, the *constant* (C) region that exhibits negligible diversity[1]. The antigen-binding specificity is determined by the particular amino acid sequence within the variable regions of the heavy and light chains.

The division into variable and constant regions reflects the functionally bipartite nature of immunoglobulins – the diverse V regions are able to react with an enormous array of different foreign molecules, whereas C regions mediate the same range of effector functions. This fundamental duality is reflected in the organization of immunoglobulin genes.

The arrangement of immunoglobulin genes in germline DNA

Immunoglobulin genes are arranged in three multigene families (see Fig. 2) one

Fig. 1. Schematic representation of an immunoglobulin molecule.

TABLE I
Immunoglobulin classes

Class	Heavy chain	Light chain	Examples of functions
IgG	γ	κ or λ	Predominant serum antibody. Activates complement.
IgM	μ	κ or λ	Cell-surface receptor. Serum antibody early in immune response. Activates complement.
IgA	α	κ or λ	Predominant antibody in saliva, colostrum and intestinal fluids.
IgD	δ	κ or λ	Cell-surface receptor on immature B lymphocytes.
IgE	ε	κ or λ	Releases histamine from mast cells. Anti-parasite immune response.

for each of the κ and λ light chains and one for all of the heavy-chain classes[2]. These families are on three separate chromosomes. As originally proposed by Dreyer and Bennet[3], V and C regions are encoded by gene segments that are distant from each other in germline DNA. Each different constant region (the two types of light chains and eight classes of heavy chains) is encoded by a unique gene. The order and spacing of heavy-chain constant region (C_H) genes within the heavy-chain locus[4,5,6] is shown in Fig. 2. Whereas the separation between most C_H genes is large, the C_μ and C_δ genes are quite close to each other, possibly to allow co-transcription of this pair[5] (see below).

The genome contains multiple V_κ[7,8] and V_H[8,9] genes, which can be divided into families of closely related sequences. There are probably between 100–300 V_κ[8] genes and possibly a similar number of V_H genes. Closely related V_H genes, from at least two different families, are clustered in tandem arrays with the same spacing (14–16 kilobases) between genes (Fig. 2)[8,9]. V_κ genes are somewhat more complicated; the distance between related V_κ genes varies from 14 to 34 kilobases (J. Gorski, R. Cattaneo, C. Gertsch and B. Mach, in preparation). In contrast there is probably only one, or at most a few, V_λ genes in the mouse[10]. The reason for this paucity is uncertain, but presumably explains the low abundance of λ chains in mouse serum.

Somatic rearrangements of immunoglobulin genes

During the differentiation of an antibody-producing lympocyte (B lymphocyte) two functionally and mechanistically distinct rearrangements of immunoglobulin genes occur. In the first (*V gene translocation*) a V gene is selected and associated with its C gene, generating a functional immunoglobulin gene and determining the antigen specificity of that lymphocyte and its progeny. This rearrangement occurs independently of contact with antigen. In a particular cell only one of the light-chain genes (κ or λ) is expressed along with a heavy-chain gene. The choice between κ and λ gene expression will be discussed in a future *TIBS* article. In the second type of rearrangement (*heavy-chain class switching*) the expressed V_H gene is reassociated with a different C_H gene. Thus, the class (and hence biological activity) of the antibody produced is changed, but the antigen specificity determined by the first rearrangement is retained. This rearrangement, which presumably occurs after antigenic stimulation, allows flexibility in the use of an antigen binding site.

Translocation of light chain V genes

Light-chain V genes are actually encoded by two distinct gene segments. Most of a V region is encoded by the V gene proper but the last 13 amino acids are

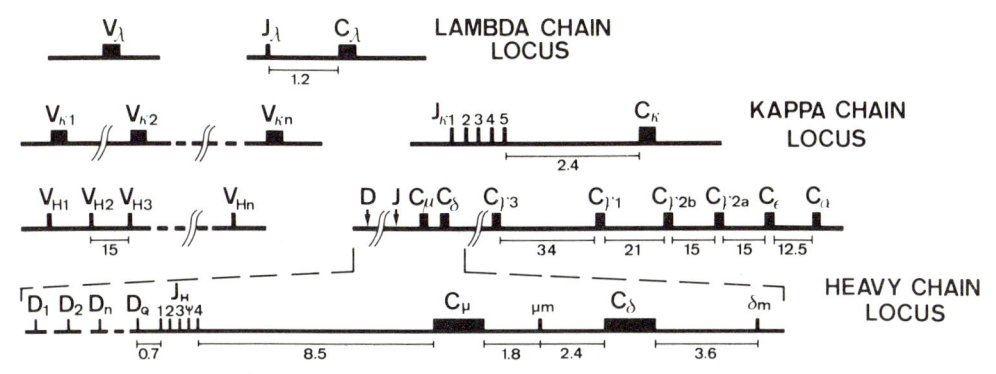

Fig. 2. The arrangement of immunoglobulin genes in mouse germline DNA. The kappa and lambda gene families are drawn to the same scale, whereas the heavy chain family is on a twenty-fold larger scale. The region of the heavy chain locus shown enlarged, has been expanded ten-fold. The distances given are in kilobases. JHψ is a pseudo-J segment[19]. DQ is the DQ52 gene segment identified recently[20].

encoded by a *joining* (J) gene segment[11-13] (Fig. 1). These two genetic elements must be joined prior to expression. In the germline (see Fig. 2), the J segments are actually closer to their corresponding C gene than V. There is a single J_λ segment, 1.2 kilobases 5' to the C_λ gene[10] and five J_κ segments located 2.4 kilobases 5' to the C_κ gene[12,13].

A V gene joins a J segment during lymphocyte development[10,14] and the germline DNA between V and J is deleted. Thus, in the example shown in Fig. 3, the second V_κ gene is joined to the fourth J_κ segment and $V\kappa 3$–$V\kappa n$ and $J\kappa 1$–$J\kappa 3$ are deleted from the genome. The intervening sequence between VJ and C, which is retained during V/J joining, is transcribed and removed from the pre-mRNA by excision and splicing[15,16]. The J gene segments and their flanking DNA sequences are thus involved in two key processes essential for the expression of an immunoglobulin gene: V/J joining and VJ–C RNA splicing.

A complete V$_H$ gene is generated by fusion of three gene segments

In contrast to the bipartite structure of a light-chain V gene, a complete heavy-chain V gene is formed by joining three DNA segments – V$_H$, D$_H$ and J$_H$ (Refs 17, 18). There are four J$_H$ genes located 8.4 kilobases 5' to the Cμ gene (Fig. 2) which encode the final 15 or 17 amino acids of the classical V$_H$ region[17-19]. The D$_H$ gene segments encode a short portion of the heavy-chain polypeptide, of variable length and sequence, that immediately precedes the J$_H$ segment (Fig. 1). Recently several D$_H$ gene segments have been identified by molecular cloning and, in some cases, nucleotide sequence analysis[20,21]. One D$_H$ segment is located only 0.7 kilobases 5' to the J$_H$ locus (D$_Q$ in Fig. 2), but the location of the other segments has not been established. Based on the known D$_H$ amino acid sequences[19] and the identified D$_H$ gene segments[20,21], it is estimated that there are at least 20 D$_H$ gene segments, many of which are closely related. Thus, the formation of a complete V$_H$ gene involves two joining events, V to D and D

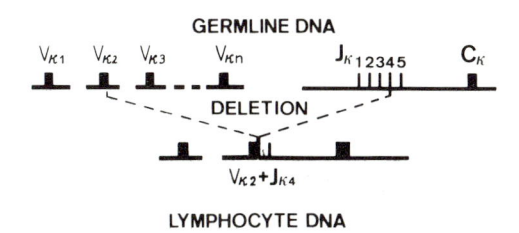

Fig. 3. Translocation of a kappa variable region gene during lymphocyte development.

to J; both occur by deletion of intervening DNA[20].

Two short blocks of nucleotides (7 and 10 bases long) that occur immediately 5' to all J segments appear in the inverse orientation 3' to V gene segments[11-13,18,19] and also, flank D$_H$ segments[20,21]. These highly conserved sequences are attractive candidates for mediating the V(/D)/J recombination event, either by way of DNA:DNA base pairing or more probably as recognition sites for recombination enzymes.

In addition to playing a key role in the activation of an immunoglobulin gene, V(/D)/J recombination plays a crucial role in expanding the repertoire of antigen specificities inherited in the germline. (See page 145.)

Heavy chain class switching

The B-lymphocyte lineage moves through an orderly series of early developmental stages[22]. Immature B-lymphocytes, which synthesize IgM class antibody as a cell-surface antigen receptor, differentiate to cells bearing both IgM and IgD. On a given cell these two classes share a common V region. Subsequently, after antigenic stimulation, these cells differentiate to become plasma cells which either actively secrete IgM or change the type of immunoglobulin synthesized to one of the other classes (IgG, IgA, IgE). Throughout these switches the antigen specificity is retained. How is one V$_H$ gene expressed simultaneously with two C$_H$ genes?, or sequentially with different C$_H$ genes? How is the switch from membrane-bound to secreted IgM achieved?

Deletions of genes accompanies C$_H$ class switching

Since J$_H$ segments occur at only one locus, 5' to the C$_\mu$ gene (Fig. 2), it seems obligatory that C$_\mu$ expression accompany the initial V gene translocation event and thus IgM is the first class synthesized. To

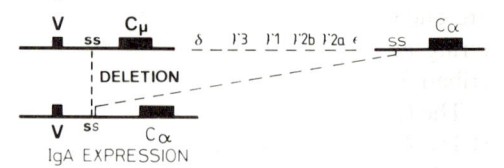

Fig. 4. Switch in the class of heavy chain expressed, from μ to α.

switch expression from IgM to, for example, IgA (Fig. 4) a large segment of DNA spanning most of the C$_H$ locus is deleted[4,23], relocating the V$_H$ gene next to the C$_\alpha$ gene. The DNA sequences involved in this relocation or switching are distinct from those involved in V(/D)/J recombination. Those regions 5' to each C$_H$ gene in which *switch recombination* occurs span several kilobases of DNA and are composed of tandemly-repeated homologous sequences[24-26] which contain multiple *switch-sites* (SS in Fig. 4). The multiple switch-sites ahead of the different C$_\gamma$ genes bear reasonable homology[25,26]. However, the C$_\mu$ and C$_\gamma$ switch sites neither bear homology with each other nor with the different C$_\gamma$ sites[24] and thus it is unlikely that switch recombination, at least involving the C$_\mu$ and C$_\alpha$ genes, occurs by homologous recombination. In an IgM secreting plasmacytoma, a fully differentiated lymphocyte that has not switched expression to another immunoglobulin class, a large segment of DNA 5' to the C$_\mu$ gene that encompasses the C$_\mu$ switch-sites has been deleted[27]. Deletion of switch-sites may therefore represent a mechanism for 'freezing' a lymphocyte clone at the stage of IgM expression[27].

mRNAs for secreted and membrane μ chains are produced from a single Cμ gene by alternative RNA processing

The heavy chains of membrane-bound IgM are a little larger and more hydrophobic than those of secreted IgM. These differences are confined to the

C-terminus. These two forms of the μ chain are encoded by two distinct mRNAs, differing at their 3' termini, that are transcribed from the same C_μ gene[16,28].

The C_μ gene (Fig. 5) has four large exons (CH1–CH4) that encode the four domains of the polypeptide[29]. Contiguous with CH4 is a coding segment (μ sec) that specifies the hydrophilic μ-secreted C-terminus and 3'-noncoding region. At 1.8 kilobases downstream are two further exons (red in Fig. 5) specifying the μ-membrane C-terminus and 3'-noncoding region[28]. The amino acid sequence encoded for the μ-membrane C-terminus has hydrophobic characteristics consistent with its being a transmembrane polypeptide.

It has been proposed[28] that if the synthesis of a precursor RNA terminates immediately 3' to the μ-sec exon (first example in Fig. 5) then this exon is retained in the mature mRNA and a secreted μ chain is encoded. If the transcript proceeds beyond this putative stop signal and the μ-mem exons are included (second example), RNA splicing will remove the μ-sec exon and fuse the μ-mem exons to CH4. Thus, the relative ratio of secreted to membrane-bound μ mRNA synthesized could be controlled by the efficiency of termination of transcription adjacent to the μ-sec exon.

Dual expression of IgM and IgD probably also occurs by alternative RNA processing

DNA deletion is unlikely to account for the dual expression of IgM and IgD, since certain lymphoma cell-lines continue to produce both classes after long periods in culture, eliminating the possibility of a long-lived μ mRNA persisting after the switch to IgD expression. Indeed, in two such IgM + IgD expressing lymphomas recently examined[30], the $C\mu$ gene has not been deleted as it is in cell-lines expressing other C_H genes[4].

An attractive model is analogous to that explaining the balance between μ-secreted and μ-membrane production. If transcription terminates immediately 3' to the μ-mem exons RNA processing will generate a mature μ mRNA. If transcription proceeds beyond this point and includes the C_δ gene then RNA splicing may preferentially join the V and C_δ regions, generating a mature δ mRNA. Thus, as a lymphocyte differentiates to become an IgM-secreting plasma cell, transcription is terminated prior to the μ-mem exons (above) and therefore the C_δ gene is switched off.

Acknowledgements

I am grateful to Dr Bernard Mach for providing data prior to publication and Dr Susumu Tonegawa for manuscripts. The author holds an Australian N.H. and

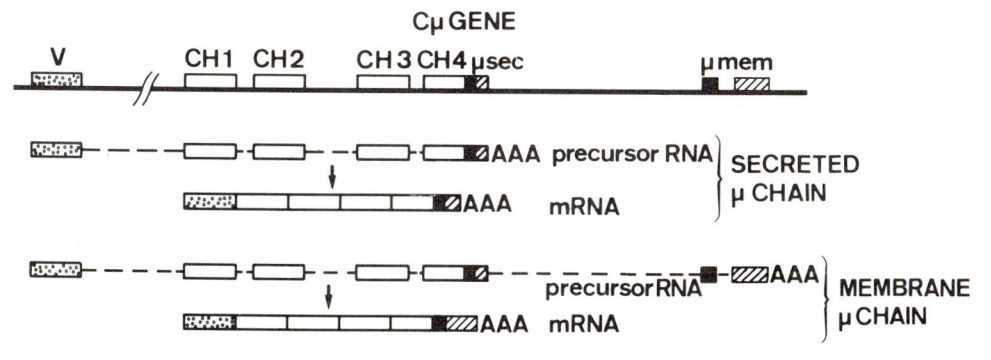

Fig. 5. Model for the formation of two different μ mRNAs, one encoding a secreted μ chain and the other a membrane-bound μ chain, from a single μ gene.

M.R.C. 'C. J. Martin' post-doctoral fellowship.

References

1. Gally, J. A. (1973) in *The Antigens* (Sela, M., ed.), vol. 1, pp. 162–298, Academic Press, New York
2. Gally, J. A. and Edelman, G. M. (1972) *Ann. Rev. Genet.* 6, 1–46
3. Dreyer, W. J. and Bennet, J. L. (1965) *Proc. Natl. Acad. Sci. U.S.A.* 54, 864–869
4. Cory, S., Jackson, J. and Adams, J. M. (1980) *Nature (London)*, 285, 450–456
5. Lui, C. P., Tucker, P. W., Mushinski, J. F. and Blattner, F. R. (1980) *Science*, 209, 1348–1353
6. Shimizu, A., Takahashi, N., Yamawaki-Kataoka, Y., Nishida, Y., Kataoka, T. and Honjo, T. (1981) *Nature (London)*, 289, 149–153
7. Seidman, J. G., Leder, A., Edgell, M. H., Polsky, F., Tilghman, S. M., Tiemeier, D. C. and Leder, P. (1978) *Proc. Natl. Acad. Sci. U.S.A.* 75, 3881–3885
8. Cory, S., Kemp, D. J., Bernard, O., Gough, N., Gerondakis, S., Tyler, B., Webb, E. and Adams, J. M. (1981) *ICN-UCLA Symposium – Immunoglobulins and Idiotypes* (in press)
9. Kemp, D. J., Cory, S. and Adams, J. M. (1979) *Proc. Natl. Acad. Sci. U.S.A.* 76, 4627–4631
10. Brack, C., Hirama, M., Lenhard-Schuller, R. and Tonegawa, S. (1978) *Cell*, 15, 1–14
11. Bernard, O., Hozumi, N. and Tonegawa, S. (1978) *Cell*, 15, 1133–1144
12. Sakano, H., Huppi, K., Heinrich, G. and Tonegawa, S. (1979) *Nature (London)*, 280, 288–294
13. Max, E. E., Seidman, J. G. and Leder, P. (1979) *Proc. Natl. Acad. Sci. U.S.A.* 76, 3450–3454
14. Seidman, J. G. and Leder, P. (1978) *Nature (London)*, 276, 790–795
15. Schibler, U., Marcu, K. B. and Perry, R. P. (1978) *Cell*, 15, 1495–1509
16. Wall, R. (1980) *Trends Biochem. Sci.* 5, 325–327
17. Early, P. W., Huang, H., Davis, M., Calame, K. and Hood, L. (1980) *Cell*, 19, 981–992
18. Sakano, H., Maki, R., Kurosawa, Y., Roeder, W. and Tonegawa, S. (1980) *Nature (London)*, 286, 676–683
19. Gough, N. M. and Bernard, O. (1981) *Proc. Natl Acad. Sci. U.S.A.* 78, 509–513
20. Sakano, H., Kurosawa, Y., Weigert, M. and Tonegawa, S. (1981) *Nature (London)*, 290, 562–565
21. Kurosawa, Y., von Boehmer, H., Haas, W., Sakano, H., Trauneker, A. and Tonegawa, S. (1981) *Nature (London)*, 290, 565–570
22. Strober, S. (1977) *Prog. Immunol.* 3, 183–191
23. Davis, M. M., Calame, K., Early, P. W., Livant, D. L., Joho, R., Weissman, I. L. and Hood, L. (1980) *Nature (London)*, 283, 733–739
24. Davis, M. M., Kim, S. K. and Hood, L. E. (1980) *Science*, 209, 1360–1365
25. Tyler, B. M. and Adams, J. M. (1980) *Nucleic Acids Res.* 8, 5579–5598
26. Kataoka, T., Miyata, T. and Honjo, T. (1981) *Cell*, 23, 357–368
27. Cory, S., Adams, J. M. and Kemp, D. J. (1980) *Proc. Natl Acad. Sci. U.S.A.* 77, 4943–4947
28. Early, P., Rogers, J., Davis, M., Calame, K., Bond, M., Wall, R. and Hood, L. (1980) *Cell*, 20, 313–319
29. Gough, N. M., Kemp, D. J., Tyler, B. M., Adams, J. M. and Cory, S. (1980) *Proc. Natl Acad. Sci. U.S.A.* 77, 554–558
30. Moore, K. W., Rogers, J., Hunkapiller, T., Early, P., Nottenburg, C., Weissman, I., Bazin, H., Wall, R. and Hood, L. E. (1981) *Proc. Natl Acad. Sci. U.S.A.* 78, 1800–1804

Nicholas Gough is at the European Molecular Biology Laboratory, Postfach 102209, 6900 Heidelberg, F.R.G.

Gene rearrangement can extinguish as well as activate and diversify immunoglobulin genes

Nicholas Gough

The longstanding hypothesis that immunoglobulins are encoded by two distinct gene segments that become fused in the genome of an antibody-producing cell has proven to be essentially correct. This translocation event is not only a mechanism for gene activation but is also a means by which immunoglobulin genes can be switched off and, in addition, it plays a major role in expanding the repertoire of inherited antibody gene sequences.

Antibodies, or immunoglobulins, are multi-chain proteins composed of two light and two heavy chains. Since only one light-chain gene and one heavy-chain gene is active in any particular antibody-producing cell, in an individual antibody molecule the two light chains are identical, as are the two heavy chains. The C-terminal portions of the different immunoglobulin chains, the constant (C) regions, are responsible for mediating the biological effector functions of antibody molecules, whilst the N-terminal region of about 110 amino acids, the variable (V) region, which exhibits extensive amino acid sequence diversity, is responsible for binding antigens (see Ref. 1 for review).

Amino acid sequence variation does not occur randomly throughout the V region, but is concentrated into three regions of particularly high variability[2]. These so-called *hypervariable regions* of both the heavy and light chains, which are in close spatial proximity, form the antigen-binding site of the antibody molecule[3]. Thus, the antigen-binding specificity of an antibody is determined by the particular amino acid sequences of the hypervariable regions of its constituent heavy and light chains. Two types of alterations in hypervariable regions markedly influence the antigen-binding site. Amino acid substitutions that alter the charge or size of a particular residue can greatly change the binding affinity or specificity. Secondly, insertions and deletions within the hypervariable regions probably play a major role in the diversification of the antigen-binding site, determining whether the hypervariable regions form a deep cavity, suitable for binding a small antigen, or a shallow groove appropriate for a linear antigen (see Ref. 3 for review).

The arrangement of genes for immunoglobulins and the rearrangements that they undergo during lymphocyte development were reviewed in a recent *TIBS* article[4]. Briefly, immunoglobulin genes are arranged in three genetically unlinked families, one each for the two different light chains (κ and λ) and one for all of the different heavy chains. Within each family V and C regions are encoded by different gene segments that are distant from each other. A functional immunoglobulin gene is formed during lymphocyte development by a recombination event (*V gene translocation*) that associates a particular V gene with its C gene. This review looks at the role V gene translocation plays in generating antibody diversity and in switching immunoglobulin genes off.

V gene translocation as a mechanism for generating antibody diversity

V genes themselves are actually composed of distinct gene segments. Most of a variable region is encoded by the V gene proper and the last 13–17 amino acids are encoded by a *joining* (J) gene segment[5–9]. For heavy-chain V regions, a segment that immediately precedes the J region in the heavy-chain polypeptide is encoded by a third gene segment, the *diversity* (D) segment[7,8,10]. The D$_H$ region and a portion of the J$_H$ region constitutes one of the three hypervariable regions (antigen-binding sites) of the heavy-chain. V gene translocation joins a V gene with a J (and D) segment located close to the corresponding C gene thus forming a complete, active immunoglobulin gene.

The mouse germline contains multiple V genes, which represent the contribution to antibody diversity generated during evolution. There are probably between 100 and 300 V$_\kappa$ genes[11] and 30–80 V$_H$ genes[11,12], although this estimate for the number of V$_H$ genes is possibly low. Further antibody diversity is generated by the recombination events (V gene translocation) that occur during ontogeny. There are four active J$_\kappa$ segments clustered close to the C$_\kappa$ gene which encode the final 13 amino acids of the kappa V region[6] (Fig. 1). The first amino acid residue encoded by the different J$_\kappa$ segments is different in each case and thus joining of any one V$_\kappa$ gene segment to each of the four J$_\kappa$ segments will generate four different V genes. Since the first J$_\kappa$ residue is part of a hypervariable region, and hence potentially part of the antigen-binding site, each of these four V genes will possibly have different antigen-binding properties. Thus, by combination, the 100–300 germline V$_\kappa$ genes and four J$_\kappa$ segments can be expanded to some 400–1200 different V genes.

In contrast to the J$_\kappa$ segments, which are all of the same length, Fig. 1 shows that the four J$_H$ segments[7,8,9] differ in length as well as sequence. Whereas for the J$_\kappa$ segment only the first residue is part of a hypervariable region, the first four residues of J$_{H2}$ and J$_{H3}$ and first six of J$_{H1}$ and J$_{H4}$ are within the hypervariable region. Within this portion of the J$_H$ sequences there are no invariant residues. Because each J$_H$ gene contributes a markedly different sequence to the hypervariable region, each presumably has a unique effect upon the nature and configuration of the antigen-binding site. Thus, combinations of J$_H$ genes with different V$_H$ genes will contribute signifi-

J$_\lambda$			Trp	Val	Phe	Gly	Gly	Gly	Thr	Lys	Leu	Thr	Val	Leu	Gly
J$_{\kappa 1}$			Trp	Thr	Phe	Gly	Gly	Gly	Thr	Lys	Leu	Glu	Ile	Lys	Arg
J$_{\kappa 2}$			Tyr	Thr	Phe	Gly	Gly	Gly	Thr	Lys	Leu	Glu	Ile	Lys	Arg
J$_{\kappa 4}$			Phe	Thr	Phe	Gly	Ser	Gly	Thr	Lys	Leu	Glu	Ile	Lys	Arg
J$_{\kappa 5}$			Leu	Thr	Phe	Gly	Ala	Gly	Thr	Lys	Leu	Glu	Leu	Lys	Arg
J$_{H1}$	Tyr	Trp	Tyr	Phe	Asp	Val	Trp	Gly	Ala	Gly	Thr	Thr	Val	Thr	Val	Ser	Ser
J$_{H2}$		Tyr	Phe	Asp	Tyr	Trp	Gly	Gln	Gly	Thr	Thr	Leu	Thr	Val	Ser	Ser
J$_{H3}$		Trp	Phe	Ala	Tyr	Trp	Gly	Gln	Gly	Thr	Leu	Val	Thr	Val	Ser	Ala
J$_{H4}$	Tyr	Tyr	Ala	Met	Asn	Tyr	Trp	Gly	Gln	Gly	Thr	Ser	Val	Thr	Val	Ser	Ser
			Hypervariable Region											

Fig. 1. Amino acid sequences encoded by the different mouse J region segments. Variant residues and residues within hypervariable regions are highlighted.

TABLE I.
A selection of D_H region amino acid sequences

Protein	D_H Region
MOPC173	Pro
J558	Arg
Hdex6	Ser-His
Hdex7	Ala-Asp
Hdex10	Val-Asn
HPC76	Gly-Val-Pro
MOPC315	Asn-Asp-His-Leu
HOPC1	Gly-Glu-Pro-Pro-Tyr
S107	Tyr-Tyr-Gly-Ser-Ser
X44	His-Tyr-Tyr-Gly-Tyr-Asn

cantly to the final repertoire of antigen-binding sites.

An even more important contribution to diversification of the heavy-chain antigen-binding site is made by the third genetic element of the V_H gene – the D_H segment[7,8,10]. The residues encoded by the D_H segments constitute the remainder of the third hypervariable region of the heavy chain and, as indicated in Table I, D_H region amino acids are particularly diverse in both sequence and length. Since any V_H gene is free to join to any D_H segment and in turn to any J_H segment, the 30–100 V_H genes, at least 30 D_H and four J_H segments can be expanded to 3600–12,000 distinct V genes.

Sequence variation resulting from such combinations can be augmented further by flexibility in the exact position at which a J segment is joined to a V or D segment. Thus, for J_κ segments, additional sequence variation is introduced at the first J codon (that is, just within a hypervariable region) by recombination within that codon[6,13]. Fig. 2 shows the possible variant junction sequences generated by the fusion of one particular germline V_κ gene with $J_{\kappa 1}$. In example 1 (the 'normal' junction) the 95th codon of the V gene is joined directly to the first J codon, generating the junction amino acid sequence Pro-Trp-Thr. In the subsequent examples, the first 1, 2 or 3 bases of the '95 + 1th' codon of the germline V gene are combined with the last 2, 1 or 0 bases of the first J codon, generating variant junction sequences Pro-Arg-Thr and Pro-Pro-Thr. All three of these amino

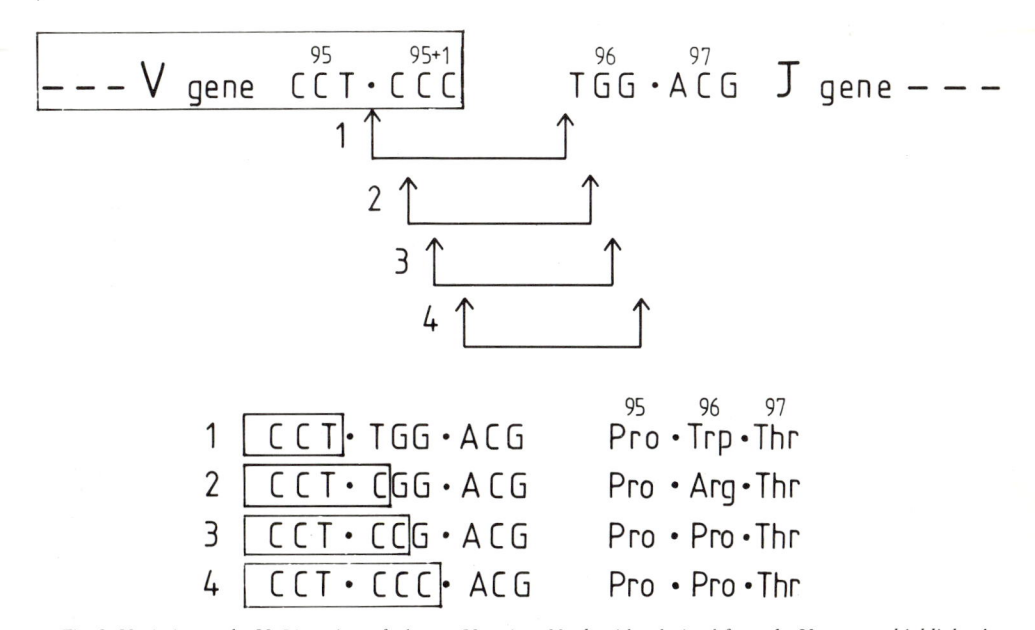

Fig. 2. Variation at the V–J junction of a kappa V region. Nucleotides derived from the V gene are highlighted.

acid sequences have been observed in polypeptides encoded by this pair of V and J genes[6]. Such junctional variation appears to play a significant role in diversifying V_κ sequences, since approximately 25% of mouse kappa chains have variant V/J junctions[13].

For heavy-chain J segments junctional variation is even more marked, involving, in addition to the formation of such hybrid codons, the deletion of one or two J_H codons[7,14]. Such codon deletion and variation is very common in V_H sequences: of 27 known V_H amino acid sequences, only nine include the entire J_H sequence, 12 have deleted all or part of the first J codon and six lack the first two codons[9]. Thus, random selection of D_H and J_H segments (which differ in both sequence and length) and variation in their joining allows an extremely large number of different hypervariable regions (antigen-binding sites) to be created during V_H translocation.

Mutations occurring in V genes during the somatic development of antibody-producing cells must also contribute significantly to the final repertoire of antibody diversity, since it is clear that in certain V gene 'families' there are insufficient germ-line V genes to encode all of the polypeptide sequences known to belong to that family[5,11].

V gene translocation and probabilistic models of allelic exclusion

The expression of immunoglobulin genes, unlike that of other autosomal genes, is confined to one allele[15]. Thus, in a particular antibody-producing cell, only one heavy-chain and one light-chain allele directs the synthesis of an active polypeptide (*allelic exclusion*). Moreover, in an individual cell, the synthesis of κ and λ light chains is mutually exclusive – only one of the two light-chain loci is active (*isotypic exclusion*). If an antibody-producing cell did in fact express two heavy and two light-chain alleles then, as illustrated in Fig. 3, a total of ten different antibodies would be produced by that cell. If each allele were expressed with equal frequency, then any particular antibody molecule would represent only 10 or 20% of this population and therefore the response to a given antigen by this cell would be diluted by production of the other nine antibodies. Moreover, many of the antibodies in this population have mixed specificities, that is the 'left-hand side' of the antibody is different from the right (see Fig. 3). These antibodies would therefore not be able to participate in multivalent cross-linking reactions, a hallmark of antigen–antibody reactions. Thus, allelic and isotypic exclusion have distinct advantages for the

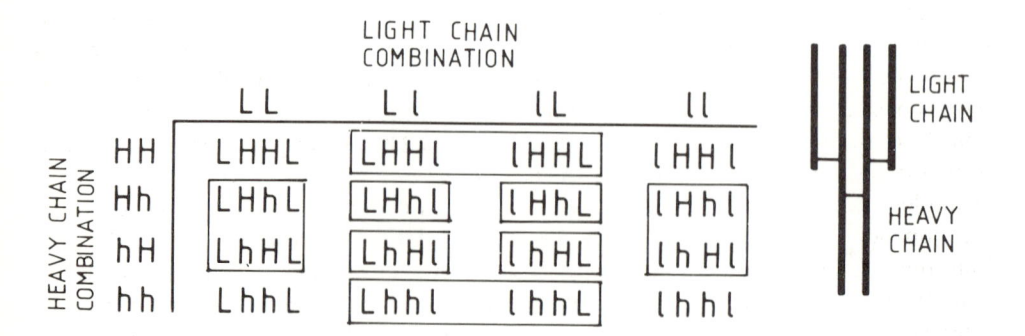

	LIGHT CHAIN COMBINATION			
	L L	L l	l L	l l
HH	LHHL	LHHl	lHHL	lHHl
Hh	LHhL	LHhl	lHhL	lHhl
hH	LhHL	LhHl	lhHL	lhHl
hh	LhhL	Lhhl	lhhL	lhhl

Fig. 3. Combinatorial association of two different light chains (L and l) and two heavy chains (H and h) in the assembly of an antibody molecule (right). Combinations yielding identical molecules are boxed. These antibodies also have mixed specificities (see text).

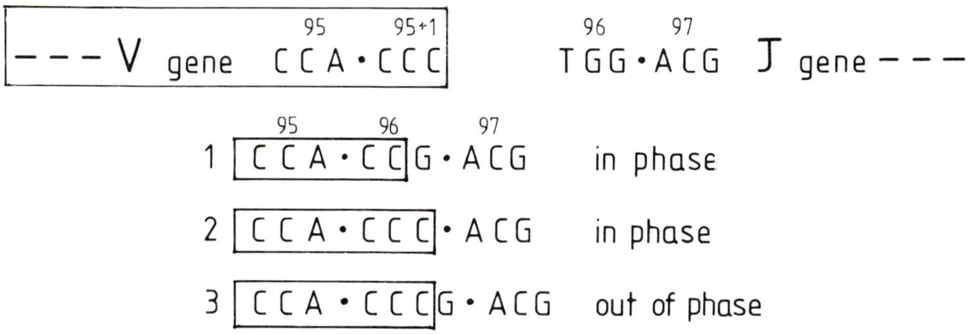

```
                        95    95+1              96     97
---V gene    C C A · C C C      T G G · A C G   J gene---

              95       96      97
1   C C A · C C G · A C G      in phase

2   C C A · C C C · A C G      in phase

3   C C A · C C C G · A C G    out of phase
```

Fig. 4. Generation of an out-of-phase V–J junction.

immune system. What are the mechanisms underlying these phenomena?

Since translocation of a V gene is required for activation of an immunoglobulin locus, an initially attractive explanation for allelic exclusion was that one homologous chromosome is rearranged while the other is not, thus activating only one allele[16]. However, in most plasmacytomas, B-lymphomas and normal B-cells, rearrangement is not in fact confined to just one allele[17–24]. It is now evident that there is imprecision in the process of V gene translocation and that this imprecision generates non-functional (i.e. excluded) genes at a high frequency.

Detailed analyses of V_κ genes in mouse myelomas have revealed several examples of aberrant rearrangements that cause allelic inactivation. V genes may be joined to non-J loci and therefore do not encode complete immunoglobulin polypeptides[20]. In addition, V genes may be joined incorrectly to authentic J segments[18,24]. Fig. 4 shows an example of an incorrect V–J junction. Two variant junctions that could have been formed (as in Fig. 2) are shown in examples 1 and 2. Example 3 is an incorrect junction[24], since 'codon' 96 now has four nucleotides. Thus, the translational reading frame is changed and a nonsense codon is soon encountered, resulting in a truncated, nonfunctional polypeptide. Indeed, unless there is a special mechanism favouring the joining of V and J segments in-phase, 2/3 of V–J fusions will be out of phase.

Allelic exclusion can therefore be viewed probabilistically – in a particular cell it is likely that only one *functional* allele will be generated. One consequence of a *strict* probabilistic model is that not all antibody-producing cells will exhibit allelic exclusion. Thus, if the frequency of *productive* rearrangement on each allele is 30%, on a strict probabilistic model 9% of cells will have two correctly rearranged alleles[24]. In fact, the ratio of single:double-producing lymphocytes in the mouse appears to be more than 100:1 (Ref. 25). To predict such a ratio with a strict probabilistic model requires that the frequency of productive rearrangement on each allele be 2%. As a consequence, some 96% of lymphocytes would have neither allele rearranged[24]. Such a model applied to both the κ and H-chain loci would be particularly costly in cells since the frequencies must be multiplied – a 100:1 discrimination at both the κ and H-chain loci would leave only 0.2% of lymphocytes active[24]. The problem of cell loss is reduced if the mechanism is not strictly probabilistic. Thus, it has been suggested[17] that the generation of a functional κ chain gene on one allele may halt or depress rearrangement of the other allele, perhaps by causing the cell to move out of a period of active rearrangement. A higher frequency of productive rearrangement could be tolerated since it would not cause a corresponding increase in the number of double producers and would reduce the number of cells lost due to lack of rearrangement.

How is the discrimination between κ and λ production achieved? It is striking that, in the mouse and the human, κ genes are almost always rearranged in λ-producing cells, whereas λ genes are rarely rearranged in κ producers. This suggests that genes are either rearranged before λ genes or are rearranged at a much higher frequency. Thus, a cell which had failed to rearrange either κ allele in a functional fashion would then have the opportunity to activate a λ chain allele. It is not known whether the preferential rearrangement of κ genes is predetermined or merely probabilistic, perhaps resulting from the much larger number of V_κ and J_κ genes giving rise to an increased frequency of recombination[21].

Therefore, the flexibility in the V-J joining machinery that allows expansion of the antibody repertoire, also acts to extinguish immunoglobulin alleles.

Acknowledgements

I am grateful to Wendy Moses for preparing this manuscript. The author holds an Australian N.H. and M.R.C. 'C. J. Martin' postdoctoral fellowship.

References

1 Gally, J. A. (1973) in *The Antigens* (Sela, M., ed.) Vol. 1, pp. 162-298, Academic Press, New York
2 Wu, T. and Kabat, E. (1970) *J. Exp. Med.* 132, 211-250
3 Padlan, E. A. (1977) *Q. Rev. Biophys.* 10, 35-65
4 Gough, N. (1981) *Trends Biochem. Sci.* 6, 203-205
5 Bernard, O., Hozumi, N. and Tonegawa, S. (1978) *Cell* 15, 1133-1144
6 Max, E. E., Seidman, J. G. and Leder, P. (1979) *Proc. Natl. Acad. Sci. U.S.A.* 76, 3450-3454
7 Early, P. W., Huang, H., Davis, M., Calame, K. and Hood, L. (1980) *Cell* 19, 981-992
8 Sakano, H., Maki, R., Kurosawa, Y., Roeder, W. and Tonegawa, S. (1980) *Nature (London)* 286, 676-683
9 Gough, N. M. and Bernard, O. (1981) *Proc. Natl. Acad. Sci. U.S.A.* 78, 509-513
10 Sakano, H., Kurosawa, Y., Weigert, M. and Tonegawa, S. (1981) *Nature (London)* 290, 562-565
11 Cory, S., Kemp, D. J., Bernard, O., Gough, N., Gerondakis, S., Tyler, B., Webb, E. and Adams, J. M. (1981) *ICN-UCLA Symposium – Immunoglobulins and Idiotypes* (in press)
12 Kemp, D. J., Cory, S. and Adams, J. M. (1979) *Proc. Natl. Acad. Sci. U.S.A.* 76, 4627-4631
13 Weigert, M., Perry, R., Kelley, D., Hunkapiller, T., Schilling, J. and Hood, L. (1980) *Nature (London)* 283, 497-499
14 Bernard, O. and Gough, N. M. (1980) *Proc. Natl. Acad. Sci. U.S.A.* 77, 3630-3634
15 Pernis, B., Chiappino, G., Kelus, A. S. and Gell, P. G. H. (1965) *J. Exp. Med.* 122, 853-865
16 Brack, C., Hirama, M., Lenhard-Schuller, R. and Tonegawa, S. (1978) *Cell* 15, 1-14
17 Alt, F. W., Enea, V., Bothwell, A. L. M. and Baltimore, D. (1980) *Cell* 21, 1-12
18 Altenburger, W., Steinmetz, M. and Zachau, H. G. (1980) *Nature (London)* 287, 603-607
19 Cory, S., Adams, J. M. and Kemp, D. J. (1980) *Proc. Natl. Acad. Sci. U.S.A.* 77, 4943-4947
20 Seidman, J. G. and Leder, P. (1980) *Nature (London)* 286, 779-783
21 Coleclough, C., Perry, R. P., Karjalainen, K. and Weigert, M. (1981) *Nature (London)* 290, 372-378
22 Heiter, P. A., Korsmeyer, S. J., Waldmann, T. A. and Leder, P. (1981) *Nature (London)* 290, 368-372
23 Nottenburg, C. and Weissman, I. L. (1981) *Proc. Natl. Acad. Sci. U.S.A.* 78, 484-488
24 Bernard, O., Gough, N. M. and Adams, J. M. (1981) *Proc. Natl. Acad. Sci. U.S.A.* (in press)
25 Vitetta, E. and Krolick, K. (1980) *J. Immunol.* 124, 2988-2990

Nicholas Gough is at the European Molecular Biology Laboratory, Postfach 102209, 6900 Heidelberg, F.R.G.

Surprising complexity in the gene locus encoding mouse λ light-chain immunoglobulins
by Nick Gough

The mouse immunoglobulin gene system provided another surprise recently when it became apparent that one of the light-chain gene families has a surprisingly large number of constant region genes. Approximately 95% of immunoglobulin molecules in the mouse contain light chains of the kappa (κ) type and only about 5% bear lambda (λ) light-chains. The κ light-chain gene family contains several hundred different variable region (V) genes, encoding the antigen-binding domain of the antibody, and a unique constant region (C) gene. (See Ref. 1 for a recent review concerning the arrangement and rearrangement of immunoglobulin V and C genes.) In contrast, the λ light-chain gene family contains only a limited number of V genes – probably just two. The paucity of Vλ genes presumably results in a very reduced repertoire of antigenic specificities. Thus, fewer clones of antibody-producing cells expressing λ chains are produced and selected, giving rise to the reduced representation of λ chains in normal mouse serum. It comes as a surprise therefore, to learn that there are at least four different genes encoding λ-chain constant regions.

It has been known for some time that there are two different types of λ light-chains, since the light chain synthesized by the myeloma MOPC 315 (designated λ2) differs from the prototype λ chain (λ1) at 29 positions in the constant region and about 13 in the variable region. Recently, however, several myeloma light-chains that had been identified as λ2 on the basis of their antigenic properties and characteristic C-terminal peptides, were found to differ at five positions in the constant region from the λ2 amino acid sequence and about 30 positions from λ1. These new chains were designated λ3 (Ref. 2).

Fig. 1 (bottom section) shows the organization of the genes encoding these three different Cλ sequences, as well as a previously unsuspected fourth lambda C gene. based on recent work by Blomberg et al.[3] and Miller et al.[4]. The Cλ1 and Cλ3 genes are very close to each other and, at a different (possibly distant) location, the Cλ2 and Cλ4 genes are similarly linked. Neither the relative orientation of these two clusters, nor the distance between them, is known, except that both clusters are on chromosome 16 (Ref. 5). Whereas each C gene has its own J segment, located about 1.3 kilobases to the 5' (left hand) side, there is only one V gene for each pair. Since there are only two Vλ genes[6] then there must be some sharing of these between the four C genes. Indeed, Blomberg et al. have demonstrated that the Vλ1 gene (often expressed with the Cλ1 gene) has been translocated next to the Cλ3 gene in a λ3-producing myeloma. It is unlikely, however, that both V genes are free to associate with all four C genes, but rather that each V gene is associated with a particular pair of C genes (Fig. 1).

Even though the Cλ1 and Cλ3 genes are very closely linked and share the one Vλ1 gene, the frequency of expression of the Cλ1 gene is much higher than that of Cλ3. What is the basis for this discrimination? Blomberg et al. point out that the J segment, located just upstream from each C gene, is an obvious site for regulation. During the somatic development of antibody-producing cells, a V gene is joined to a J segment adjacent to a C gene, thereby mobilizing that V and C gene (see Ref. 1). If the Vλ1 gene joins more frequently or more accurately to the Jλ1 segment than to the Jλ3 segment, then Cλ1 chains will be produced more often. Two short blocks of

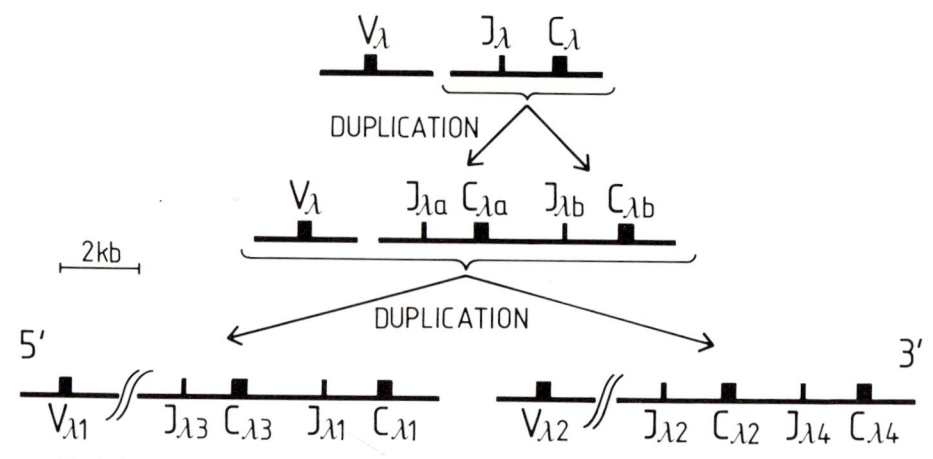

Fig. 1. Arrangement of the mouse λ light-chain genes and a possible pathway for their evolution.

reasonably conserved nucleotides that occur immediately 5' to all J segments, and in the inverse orientation, 3' to V gene segments are attractive candidates for mediating V–J joining, but their mode of action is unclear.

How did this gene complex evolve? Since the general arrangement of the $C_{\lambda 3}/C_{\lambda 1}$ and $C_{\lambda 2}/C_{\lambda 4}$ clusters are very similar, the two clusters probably arose by duplication from a common precursor cluster, rather than by independent pathways (Fig. 1). A precursor λ-chain locus, consisting of V, J and C gene segments, is shown at the top. Duplication of the J–C region of this locus would give rise to the precursor cluster, in which the two J + C genes share one V gene. The sequence of these two J + C genes would drift independently of each other. Duplication of the entire V–JC–JC complex would generate two such clusters. Since, in this scheme, the $C_{\lambda 2}$ and $C_{\lambda 3}$ sequences have a common immediate precursor ($C_{\lambda a}$ in Fig. 1), as do $C_{\lambda 1}$ and $C_{\lambda 4}$, then the primary sequences of $C_{\lambda 2}$ and $C_{\lambda 3}$ should be closer to each other than to $C_{\lambda 1}$ or $C_{\lambda 4}$ and similarly $C_{\lambda 1}$ and $C_{\lambda 4}$ should bear strong homology. Certainly, the amino acid sequence of the $C_{\lambda 3}$ chain is more homologous to $C_{\lambda 2}$ that to $C_{\lambda 1}$ (Ref. 2). It will be interesting to see if the nucleotide sequences of the different C_λ genes and their flanking sequences confirm this scheme.

Finally, it remains to be seen if these four different lambda chains have different biological properties, as do the four different gamma heavy-chains, which display a similar degree of sequence homology.

References

1. Gough, N. (1981) *Trends Biochem. Sci.* 6, 203–205
2. Azuma, T., Steiner, L. A. and Eisen, H. N. (1981) *Proc. Natl Acad. Sci. U.S.A.* 78, 569–573
3. Blomberg, B., Traunecker, A., Eisen, H. and Tonegawa, S. (1981) *Proc. Natl Acad. Sci. U.S.A.* 78, 3765–3769
4. Miller, J., Bothwell, A. and Storb, U. (1981) *Proc. Natl Acad. Sci. U.S.A.* 78, 3829–3833
5. D'Eustachio, P., Bothwell, A. L. M., Tokaro, T. K., Baltimore, D. and Ruddle, F. H. (1981) *J. Exp. Med.* 153, 793–800
6. Brack, C., Hirama, M., Lenhard-Schuller, R. and Tonegawa, S. (1978) *Cell*, 15, 1–14

NICK GOUGH

European Molecular Biology Laboratory, Postfach 10.2209, 6900 Heidelberg, F.R.G.

RNA synthesis and processing

Comparative subunit composition of the eukaryotic nuclear RNA polymerases

Marvin R. Paule

Multiple eukaryotic RNA polymerases have enormously complex subunit compositions. However, comparison studies reveal that these polypeptides fall into several categories which may reflect their functional role in gene transcription.

In eukaryotic cells there are several separate genetic systems possessing enzymes which transcribe information from DNA into an RNA copy. In mitochondria, chloroplasts and bacteria there is a single DNA-dependent RNA polymerase which transcribes all of the genes encoded in the organelle DNA. In contrast, transcriptional duties in eukaryotic nuclei are divided between three separate DNA-dependent RNA polymerase isozymes. Originally identified in the late 1960s, the structural and functional relationships between these polymerases have since been studied intensively because of their involvement in the expression and regulation of genetic information.

Each RNA polymerase isozyme transcribes a different and non-overlapping set of nuclear genes: RNA polymerase I (or A) transcribes only the precursor ribosomal RNA transcription unit; RNA polymerase II (or B) synthesizes heterogeneous nuclear RNA and class III (or C) polymerase transcribes transfer RNA precursors, 5S RNA and a few additional RNAs of unknown function. Each class of polymerase thus catalyzes a reaction which is quite distinct from any other polymerase in that the region of the DNA recognized as a template is different. Yet, in a mechanistic sense, all of these reactions are the same: they involve DNA sequence recognition and binding, initiation of an RNA chain *de novo*, elongation of the nascent RNA in response to the sequence of the DNA template and, finally, termination and release of the RNA product. This diversity, yet overlap of function, is mirrored in the architecture of the RNA polymerases; though there are major structural differences between the polymerase classes, there is evidence to suggest that there are significant structural homologies as well.

Eukaryotic RNA polymerases have complex subunit compositions

Prokaryotic and eukaryotic RNA polymerases are large enzymes having a native mol. wt of 500,000–700,000 and it was initially anticipated that their subunit structures would be similar. Supporting this idea, an overall architectural similarity does exist in that bacterial, plant and animal RNA polymerases consist of two very large subunits in association with an assortment of smaller polypeptides. However, unlike bacterial RNA polymerase which consist of only 4–5 subunits, the eukaryotic nuclear RNA polymerases are baroque examples of multisubunit enzymes: 10–15 distinct polypeptide chains make up each polymerase (see Table I for examples). These subunits are conveniently divided into two groups: (1) *the large subunits*, those above 100,000 mol. wt and (2) the *small subunits*, those below 50,000 mol. wt. Only class III RNA polymerases possess subunits which are intermediate in size.

The large subunits

One of the most striking features of the RNA polymerases is their high mol. wt subunits. These range from about 120,000 to 240,000 mol. wt depending upon the enzyme source. Each is a single polypeptide chain and these subunits therefore represent some of the largest protein molecules found in the cell.

Each RNA polymerase has two different large subunits. Comparative peptide mapping studies[1,2] indicate that each member of the pair of large subunits of a given polymerase is unique and unrelated to the other member of the pair. Further, immunological studies show that the large subunits of one polymerase class are unrelated to the large subunits of any other polymerase class[3]. These findings establish that the enzyme classes are, in part at least, separate gene products and that the three classes of enzyme are distinct on a structural as well as on a functional basis.

The different enzyme classes exhibit differential sensitivities to α-amanitin, a mushroom toxin[4,22]. In animal and plant cells, RNA polymerase II is sensitive to very low toxin concentrations, the class III enzyme is sensitive to high concentrations and polymerase I is resistant. However, in yeast the class I polymerase is sensitive to α-amanitin while the class III polymerase of both yeast and insects is resistant. Amatoxins act by direct interaction with one of the large subunits of RNA polymerase II (Refs 5,6, reviewed in 23) and presumably inhibit the class III and class I enzymes by the same mechanism. Thus, examples can be found in which each of the RNA polymerase classes is inhibited by α-amanitin suggesting, not surprisingly, that there may be an evolutionary relationship between one of the large subunits in all of the enzymes. Further, this finding, coupled with the isolation of amanitin resistant cell lines possessing α-amanitin resistant RNA polymerases[7], demonstrates that this large subunit is a functional subunit of the RNA polymerase *in vivo*.

When based upon the large subunits, simple relationships between members of a polymerase class are somewhat complicated because within each RNA polymerase class there are often heterogeneous structural forms. These forms can sometimes be resolved by electrophoresis or ion exchange chromatography and arise from both size (see below) and charge heterogeneity of the enzyme subunits[4,8,9]. The heterogeneity observed is often a reflection of the heterogeneity in the large subunits. For example, it is extremely common to find two to three forms of the class II RNA polymerase in purified preparations. These are most often detected by the appearance of three or more large subunits with non-integral stoichiometries, which sum to a stoichiometry of two subunits. The available evidence supports the notion that these size isomers of polymerase II arise from proteolytic modification of the largest subunit of the enzyme[1,10]. Some cellular extracts contain proteolytic activities which will convert one polymerase II form to another by digestion of the largest subunit; the ratio of one form to another can often be altered by the inclusion of proteolytic inhibitors in the isolation buffers. However, all polymerase II preparations contain some of the lower molecular weight form and appropriate mixing experiments support the notion that it is present in the original cellular extract as well[1,11]. However, it is not clear whether the modification is of physiological significance. Since all of the heterogeneous forms appear to have nearly equivalent enzymatic activity, it is conceivable that they all function *in vivo* but, there is no evidence at present that these forms have different transcriptional ranges or activities. Until efficient *in vitro* functional tests are developed for the polymerases, the question of the biological role played by these and other multiple forms of the polymerases will be difficult to answer.

The small subunits

The small RNA polymerase subunits are a much more complex group of proteins. Each RNA polymerase may have up to a dozen small polypeptides associated with it. They range in mol. wt from approximately 50,000 to less than 10,000. Both acidic and basic proteins are found in this group whereas the large subunits are all very basic polypeptides.

The small polypeptide composition of RNA polymerases is difficult to determine and universally applicable conclusions about the subunit architecture of the enzymes have been slow to emerge. The sizes of cognate small subunits vary somewhat from species to species making phylogenetic comparisons difficult. Also, since a subunit of 15,000 mol. wt represents only 2% by weight of a native protein of 700,000, the presence and molar ratio of such subunits is extremely difficult to assess in any but the best enzyme preparations. This has resulted in uncertainty about the small subunit composition of the polymerases from many sources. Thus, the phylogenetic approach has been disappointing. However, a comparison of all three RNA polymerases purified from single sources has revealed much more than inter-species comparisons of a single enzyme class. Unfortunately, these studies have been severely hampered by the very low yields of RNA polymerase III obtainable from most eukaryotic cells so that the number of carefully studied enzyme sets is distressingly low. Nevertheless, as more sets of class I, II and III RNA polymerases from single sources have been studied, an order has begun to emerge and a few important generalizations can be made. The small subunits can be divided into three categories: (1) those which appear to be unique to a given enzyme class, (2) those which are common to all nuclear RNA polymerases, and (3) those which are found in two RNA polymerase classes.

Each RNA polymerase contains a number of small subunits which appear to be unique to that class of enzyme. Though it is experimentally supported and generally accepted that the large polypeptides are polymerase subunits, concern has revolved around the subunit status of the small polypeptides, especially the unique ones. The belief that these polypeptides are in fact functional subunits of the polymerase is inferred from the following: (1) These polypeptides co-purify with the polymerase and their presence is coincident with the enzyme activity peak following several types of protein purification procedures; (2) they are present in a constant and often integral stoichiometry from enzyme preparation to enzyme preparation; (3) the native molecular weights of the RNA polymerases are consistent with those calculated by summing the weights of all of the enzyme associated polypeptides. (This argument is only suggestive, however, because measurement of the native molecular weight is not very precise); (4) removal of some of these polypeptides by ion exchange chromatography or by antibody precipitation effects the activity or the catalytic properties of the enzyme[12,13]; (5) finally, the subunit composition of RNA polymerases from a large and diverse range of eukaryotes are remarkably similar in general makeup including the presence of similar sets of unique and common small subunits.

Since each RNA polymerase class is responsible for RNA synthesis from a distinct set of transcription units, it appears reasonable for each enzyme to possess some subunits which are unique to its individual makeup. However, since all attempts to dissociate and reassociate the eukaryotic RNA polymerases (and there have been many attempts!) have been unsuccessful, the subunit status of these unique polypeptides remains an open question.

Some small subunits are common to all RNA polymerase classes

There is strong evidence that there are three subunits common to all RNA polymerase classes (Table I). Support for this comes mainly from studying the enzyme structures of the small free-living amoeba, *Acanthamoeba castellanii*, from yeast, several higher plant sources and from genetic studies in yeast. Thonart et al.[14] have shown that yeast mutants with temperature-sensitive impairment of RNA synthesis fall into three complementation groups. Valenzuela et al.[16] argue, from one and two-dimensional electrophoretic data, that there are three subunits common to the three purified yeast RNA polymerases; Buhler et al.[3] had previously shown these same three polypeptides to be common to the class I and II RNA polymerases of yeast by electrophoretic, isoelectric pH and peptide mapping techniques. D'Alessio et al.[16], using one and two dimensional electrophoretic techniques and isoelectric pH measurements found the three polypeptides to be identical in all of the purified

TABLE I. Subunit comparisons of several eukaryotic nuclear RNA polymerases

Yeast			Acanthamoeba			Wheat germ			Cauliflower		
I	II	III	I	II	III	I	II	III	I	II	III
185[a]	205		185	193		200	220		190	180	
	(170)	160		(178)	169			150	(170)		150
137	145	128	133	152	138	125	140	130	125	140	130
		82			82			94			70
48					52			55			50
44	46		41.5	[40–			[42–			40	
41		41	39	38.5]	39	38	40]	38	38		
			37		37			30			
36					34			28			
	33.5	34	35		30		25	25			
					28.5	24		24.5			
							21	20.5			27
28*	28*	28*	22.5*	22.5*	22.5*	20	20	20	25*	25*	25*
								19.5			24
24	24	24									23
20	18	20	17.5	18	17.5	17.8	17.8	17.8	22	22	22
			15.5	15.5	15.5	17.0	17.0	17.0		19	17.8
14.5†	14.5†	14.5†	13.3†	13.3†	13.3†			16.3	17.5†	17.5†	17.5†
				12.5				16.0		17.0	
12.3	12.5	11		12.0				14		16.2	
			<10	<10	<10					16.0	
										14.0	

[a] numbers represent mol. wts × 10^{-3}
* stoichiometry of 2, basic in charge
† non-integral stoichiometry, acidic in charge
( ) polypeptide arises due to proteolysis
☐ shown to be identical polypeptides by two or more criteria
— underlined polypeptides have been shown to have identical molecular weights

RNA polymerases of *Acanthamoeba*. Jendrisak[11] has found three peptides of identical molecular weight in wheat germ RNA polymerases and Guilfoyle[17] has recently concluded from one and two-dimensional electrophoresis, isoelectric pH studies and immunological cross reactivity studies that the same three subunits are found in all cauliflower nuclear enzymes.

These three common polypeptides can be categorized by isoelectric pH, stoichiometry and relative molecular weight so as to form a cognate set of subunits: (1) the largest is a very basic polypeptide (pI $\simeq$ 9) present in a stoichiometry of two; (2) the smallest subunit is acidic (pI $\simeq$ 4.5–5.0) and has a variable non-integral stoichiometry between one and two. This apparent non-integral stoichiometry may be an artifact arising from the use of dye binding to determine stoichiometries. This method assumes that all proteins bind dye equally, but certain proteins, perhaps this subunit included, bind the dye anomalously – probably because of an unusual amino acid composition. The finding of a non-integral stoichiometry for this subunit may in fact reflect conservation of an unusual amino acid composition. (3) The third common subunit is intermediate in - molecular weight, acidic in charge (pI $\simeq$ 4.5–5.0) and is present in a single copy per native polymerase molecule.

These common polypeptides probably function in the general mechanism of DNA-dependent RNA synthesis. Significantly, Valenzuela et al.[18] have presented evidence that the third peptide listed above is required for activity of yeast RNA polymerase I and it may therefore play some catalytic role in all of the enzymes. In the yeast RNA polymerases, this subunit is also highly phosphorylated both *in vivo* and *in vitro* suggesting a possible regulatory role as well[19]. In the class II polymerase it is the sole phosphorylated subunit detected. One can tentatively conclude that these three common subunits, probably in association with the large subunits by analogy to the bacterial polymerase, form the basic functional core of the eukaryotic nuclear DNA-dependent RNA polymerases.

Some small subunits are shared by only two polymerase classes

In addition to the 'core' polypeptides, there are several subunits which are common to only two of the RNA polymerase classes. In the lower eukaryotes (yeast and *Acanthamoeba*) there is an approximately 40,000 mol. wt subunit which is common to polymerases I and III. In yeast, this subunit is acidic[3] while in *Acanthamoeba* it is basic[17]. Also, in the *Acanthamoeba* enzymes this band has been resolved into two polypeptides (37,000 and 39,000 daltons) by high resolution polyacrylamide gel electrophoresis. Based upon one dimensional gel electrophoresis[11], the wheat germ class I and III polymerases also contain common subunits in the same mol. wt range (41,000 in yeast; 38,000 in wheat germ) leading to the speculation that a subunit of about 40,000 is common to all class I and III, but not class II RNA polymerases. The lower eukaryotic class I and III enzymes also have in common a lower mol. wt, acidic subunit with a stoichiometry of one (20,000 mol. wt in yeast; 17,500 mol. wt in *Acanthamoeba*). The plant RNA polymerases do not appear to have this latter low mol. wt subunit common to polymerases I and III, so the universality of this finding must await studies of other higher eukaryotic systems. It is noteworthy that in yeast this subunit is phosphorylated *in vivo* and *in vitro* by a yeast protein kinase[19]. D'Alessio et al.[16] have detected charge heterogeneity in the cognate subunit of *Acanthamoeba* suggesting that it might be similarly modified.

Subunits common to class I and III polymerases which are modified are perhaps not surprising since these two enzyme classes are both involved in the transcription of stable RNAs which are incorporated into ribosomes and thus these enzymes might participate in some common regulatory networks.

Conclusions and speculations

Clearly the nuclear RNA polymerases of eukaryotic cells are structurally among the most complex enzymes studied to date. The question of 'Why so many subunits?' is thus often asked. Unfortunately, very little is known about the functions of specific subunits and thus we must resort to speculation rather than to hard experimental data to answer this question. The machinery needed for macromolecular biosynthesis is extremely complex. In the case of protein synthesis, the 30 or so proteins needed to catalyze the reaction, subunits of the catalyst so to speak, are assembled into a large macromolecular complex called the ribosome. This unit, in association with a relatively few other components, catalyzes all of the necessary steps to assemble the primary protein sequence. The DNA replication machinery on the other hand is not assembled and it has been an extremely slow and arduous task to identify and isolate all of the 15 or so components needed to replicate a DNA molecule. It is reasonable to assume that the synthesis of an RNA molecule, especially from a template which is highly complexed with proteins as in eukaryotes, also requires the participation of a large number of components. These, as in the case of the ribosome, may be neatly assembled into a macromolecular complex. Supporting this notion is the recent finding that purified RNA polymerases in association, with perhaps only one or two additional proteins, are capable of recognizing promotors and faithfully initiating, elongating and, at least for polymerase III, terminating the synthesis of an RNA transcript from a purified DNA template[20,21]. Much work remains to be done to clarify the functional role of each of the polymerase subunits in this complex process. One conclusion is irrefutable, however, the eukaryotic nuclear RNA polymerases are as complex and enigmatic as the genes which they transcribe.

References

1 Guilfoyle, T. V., Jendrisak, J. J. (1978) *Biochemistry*, 17, 1860–1866
2 D'Alessio, J. M. and Paule, M. R. (unpublished results)
3 Buhler, J.-M., Iborra, F., Sentenac, A. and Fromageot, P. (1976) *J. Biol. Chem.* 251, 1712–1717
4 Roeder, R. G. (1976) in *RNA Polymerase* (Losick, H. and Chamberlin, M., eds), pp. 285–329, Cold Spring Harbor Press, New York
5 Brodner, O. G. and Wieland, T. (1976) *Biochemistry*, 16, 3480–3484
6 Ingles, C. J., Guialis, A., Lam, J. and Siminovitch, L. (1976) *J. Biol. Chem.* 251, 2729–2734
7 Somers, D. G., Pearson, M. L. and Ingles, C. J. (1975) *J. Biol. Chem.* 250, 4825–4831
8 Kedinger, C., Gissinger, F. and Chambon, P. (1974) *Eur. J. Biochem.* 44, 421–436
9 Sklar, V. E. F., Schwartz, L. B. and Roeder, R. G. (1975) *Proc. Natl Acad. Sci. U.S.A.* 72, 348–352
10 Sentenac, A., Dezélée, S., Iborra, F., Buhler, K.-M., Huet, J., Wyers, F., Ruet, A. and Fromageot, P. (1976) in *RNA Polymerase* (Losick, H. and Chamberlin, M., eds), pp. 763–778, Cold Spring Harbor Press, New York
11 Jendrisak, J. (1980) in *Genome Organization and Expression in Plants* (Leaver, C. V., ed.), pp. 77–92, Plenum Press, New York
12 Huet, J., Buhler, J. M., Sentenac, A. and Fromageot, P. (1975) *Proc. Natl Acad. Sci. U.S.A.* 72, 3034–3038
13 Huet, J., Dezélée, S., Iborra, F., Buhler, J. M., Sentenac, A. and Fromageot, P. (1976) *Biochimie*, 58, 71–80
14 Thonart, P., Bechet, J., Hilger, F. and Burney, A. (1976) *J. Bact.* 125, 25–32
15 Valenzuela, P., Bell, G. I., Weinberg, F. and Rutter, W. J. (1976) *Biochem. Biophys. Res. Comm.* 71, 1319–1325
16 D'Alessio, J. M., Perna, P. J. and Paule, M. R. (1979) *J. Biol. Chem.* 254, 11282–11287

17 Guilfoyle, T. J. (1981) *Biochemistry*, 19, 5966–5972
18 Valenzuela, P., Bell, G. I. and Rutter, W. J. (1976) *Biochem. Biophys. Res. Comm.* 71, 26–31
19 Bell, G. I., Valenzuela, P. and Rutter, W. J. (1977) *J. Biol. Chem.* 252, 3082–3091
20 Wu, G.-J. (1978) *Proc. Natl Acad. Sci. U.S.A.* 75, 2175–2179
21 Weil, P. A., Luse, D. S., Segall, J. and Roeder, R. G. (1979) *Cell*, 18, 469–484
22 Chambon, P. (1975) *Ann. Rev. Biochem.* 44, 613–638
23 Chambon, P. (1974) in *The Enzymes* (Boyer, P. D., ed.), Vol. 10 pp. 261–331, Academic Press, New York

Marvin R. Paule is at the Department of Biochemistry, Colorado State University, Fort Collins, Colorado 80523, U.S.A.

Control of adenovirus gene expression

Lennart Philipson and Ulf Pettersson

It is now possible to study the control mechanisms involved in the expression of adenovirus genes. This review examines a regulatory function that appears to operate early after adenovirus infection; this may regulate the maturation of mRNAs from several of the early transcription units.

The adenoviruses have double-stranded DNA genomes which theoretically could code for 30–40 polypeptides. The linear DNA carries inverted terminal repetitions about 100 nucleotides long and a protein (mol. wt. 55,000) is covalently linked to each 5'-terminus. Adenovirus infection of permissive cells involves two distinct phases which are separated by the initiation of viral DNA replication. The early phase lasting 5–6 h, prepares the infected cell for viral DNA replication and late transcription. At the onset of the late phase a drastic change in viral gene expression takes place and large amounts of structural proteins accumulate up to 24 h post infection when the productive cycle is completed by the death of the host cell and release of the progeny.

After infection, the uncoated viral DNA is rapidly transported to the nucleus of the cell, where a nucleosome structure is probably generated with the aid of cellular histones [1,2]; transcription of the viral DNA starts within 1 h. At least five early promoters have been identified and the mature messenger RNAs from these transcription units are complementary to around 35% of the viral genome [3,4]. By electron microscopy 20 distinct early viral mRNA species have been identified [5] but only 7–10 early viral polypeptides have so far been detected [6].

The activation of a promoter at map position 16.3 appears to be the key event which initiates late transcription. This promoter controls at least 18 distinct mRNAs which encode most of the viral structural proteins. In addition, 2 or 3 promoters outside the major late transcription unit are active at intermediate or late times of the infectious cycle [7]. Several, but not all, of the early promoters are also active during the late phase. A total of 22 late mRNAs have been identified by R-loop analysis [8] but only 12 proteins have been detected [9]. Fig. 1 illustrates the location of early and late regions on the adenovirus genome and their corresponding gene products. It is now possible to study control mechanisms that are involved in adenovirus gene expression.

The adenovirus system – a tool for studying gene expression in mammalian cells

During productive infection, adenovirus genes are expressed in an ordered fashion and since the viral genome does not specify its own RNA or DNA polymerases this expression is probably controlled mainly by mechanisms which operate in normal uninfected mammalian cells. Both the cellular DNA polymerases α and γ have been implicated in the replication of the viral DNA [10,11] and RNA polymerases II and III are involved in the transcription of the viral genome. Although the adenoviruses like other prokaryotic and eukaryotic DNA viruses, have a compressed genome in which several genes share

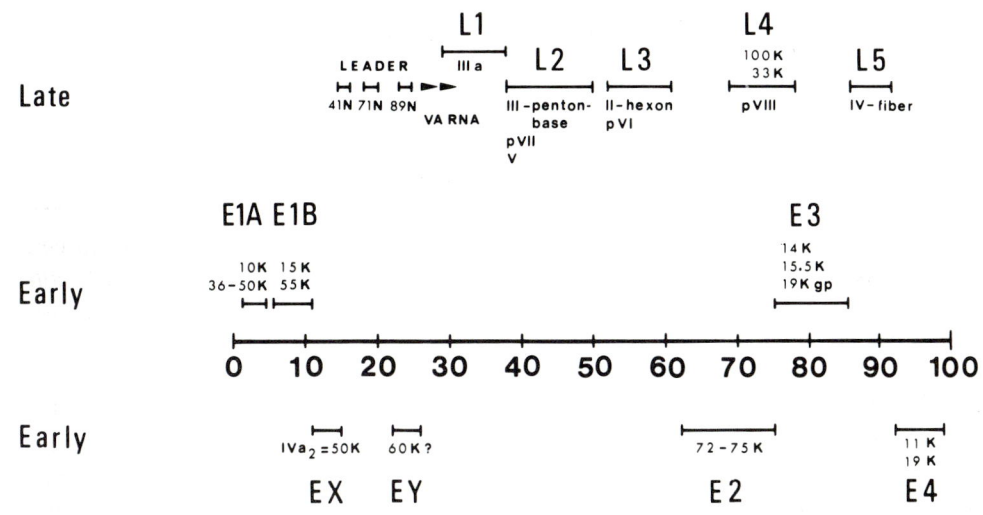

Fig. 1. A map of the early and late regions of the adenovirus genome. The five major early transcription units are denoted E1A, E1B, E2, E3 and E4. Two additional early regions which have recently been identified are referred to as EX and EY respectively. The molecular weights $\times 10^3$ (K) of the polypeptides encoded in the different regions are also indicated. The late regions and their protein products controlled by the major late promoter at 16.3 map units are referred to as L1–L5. The location of the 5'-terminal leader fragment contained in all late mRNAs is also included as well as the genome position of the low molecular weight virus-associated RNAs (VA). The Roman numerals refer to the virus structural proteins [18] which in some cases may be present in precursor forms indicated by the prefix p.

DNA sequences, the adenoviruses may provide a model system for the study of regulatory events which operate when mammalian cellular genes are expressed. In particular, the efficient transcription from the major late adenovirus promoter provides an opportunity to investigate the mechanism by which RNA polymerase II initiates transcription because both the initiation site of the late promoter and the 5'-end of the nascent transcript have been identified by sequence analysis [12,13].

Maturation of adenovirus mRNA involves the removal of intervening sequences in the original transcript followed by ligation of the remaining RNA sequences destined to be part of the final mRNA [8,9,14]; this mechanism, known as RNA splicing, occurs also in uninfected mammalian cells. An analysis of the adenoviral mRNA splicing patterns indicates that at early times splicing primarily removes sequences within the coding regions, whereas at late times splicing usually connects a 5'-terminal leader segment to viral mRNAs. Early in the infectious cycle the machinery for host macromolecular synthesis probably takes care of the incoming genome, and the early gene products then modify the cellular machinery to prepare for the shift to late adenovirus gene expression.

Adenovirus early genes

The regions on the viral genome which are transcribed early in adenovirus infection were first identified by hybridization of mRNA to the separated strands of restriction enzyme fragments of adenovirus DNA. Many different methods, including hybridization of nascent nuclear RNA chains to an ordered series of restriction fragments have been used to reveal the initiation sites for early transcription [15]. Four early regions have been identified which have been designated E1–E4 (Fig. 1). There are, however, at least five independent initiation sites for transcription and the E1 region has been subdivided in two transcription units E1A and E1B [15,16]. The 5'-ends of all five transcription units have been identified by sequence analysis of the capped 5'-terminal T1 oligonucleotides [17] whereas the 3'-termini have been identified in only a few cases. All true early transcripts are spliced to generate the true early mRNAs. In many cases the mRNAs from the five early transcription units contain overlapping sequences and the differences between individual RNAs often consist of subtle alterations in the pattern by which sequences from the primary transcripts are spliced together. Fig. 2 illustrates the mRNAs originating from the early transcription units. In several cases it has been possible to correlate specific polypeptides with particular early regions and a total of 10 polypeptides have unequivocally been assigned to the early transcription units. Because some of these 10 polypeptides appear to be coded by overlapping regions on the viral DNA they may not have unique amino acid sequences [18]. Polypeptides which are derived from the different early regions are indicated in Fig. 1. Since the adenovirus genome specifies about 20 individual early messenger RNAs (Fig. 2) the discovery of many more early proteins can be expected.

Recently two additional regions of the adenovirus genome that are expressed early after infection were identified [19]. The mRNA sequences from both these regions are present in extremely low copy-numbers prior to the onset of DNA replication but increase thereafter. One of the two 'new' early regions is located between map positions 11 and 16 and encodes a polypeptide (mol. wt 50,000) which has been designated IVa$_2$. The IVa$_2$ polypeptide is present in very low amounts in purified preparations of virus particles and it may have a role in viral morphogenesis [20]. The second 'new' early region is located between map positions 19.8 and 23.5. No polypeptide has yet been assigned to this region but a temperature-sensitive mutant, ts36, that maps in this region, is defective for viral DNA replication which corroborates the observation that sequences from this region are transcribed early after infection.

A temporal control of early gene expression

Pulse-labelling cells at different times

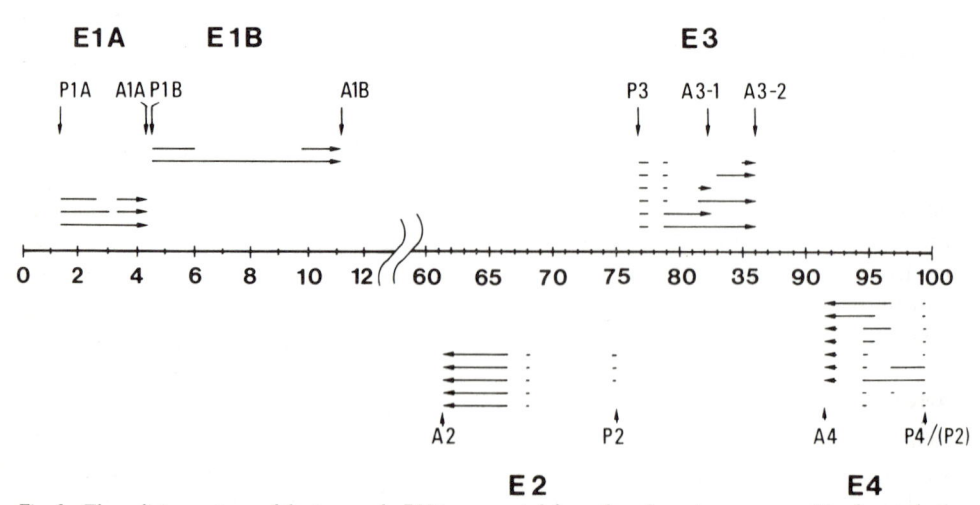

Fig. 2. The splicing pattern of the true early RNAs generated from the adenovirus genome. The five early transcription units are indicated as E1A—E4 as in Fig. 1 and the promoter and polyadenylation sites are indicated with P and A, respectively. The map is based on electron microscopic analysis of R-loops between DNA and RNA as reported previously [5].

after infection revealed that the early proteins are produced in an ordered fashion [18]. The proteins from the E3 and E4 regions are synthesized before the products from the E2 and E1B regions appear. The E1A proteins could not be detected in this initial study. When protein synthesis is inhibited by cycloheximide, 2 h after infection large amounts of viral mRNAs accumulate first from the E3 and E4 regions followed by the mRNAs from the E2 and E1B regions [18]. This suggests that the sequential appearance of the early proteins is not controlled by viral proteins which cannot be synthesized in the presence of the inhibitor. However, when another inhibitor of protein synthesis like emetine is added at the onset of infection, only mRNA from the E1A region accumulates in the cytoplasm; mRNAs from the other four transcription units cannot be detected [21]. This effect is only observed if inhibition of protein synthesis occurs at the time of infection and in the case of cycloheximide the block must already be introduced 2 h before infection. A kinetic analysis of mRNA accumulation from the five early transcription units confirms the ordered appearance of the early mRNAs and suggests that inhibitors of protein synthesis at late times in the early phase prolong the half-life of viral mRNAs, thus leading to an accumulation of viral mRNA [22]. The results taken together suggest that protein products from region E1A may control the transcription from the other four early regions. The E1A region may therefore be of special importance in controlling early adenovirus gene expression.

Early region E1A

This region of the genome extends between map positions 1.5 and 4.5 as indicated in Fig. 2 and specifies three different mRNAs [16]. mRNAs selected by hybridization to restriction fragments from this region, when translated *in vitro*, specify a heterogeneous set of polypeptides with mol. wts of 36,000 M_r–52,000 M_r [16]. These proteins from region E1A are phosphorylated, but how they control the expression from the other early transcription units of the adenovirus genome remains unknown. Recently cDNA copies of mRNAs from this region have been cloned in *E. coli* and subsequently their nucleotide sequences determined; from these the amino acid sequences of two E1A proteins have been deduced (see Fig. 3 and [23]). An additional mRNA from this region has been observed by electron microscopy but neither its nucleotide sequence nor the corresponding gene product have been identified with certainty. One or more of these three polypeptides specified by the E1A region appear to control the expression of other early genes.

Mutations affecting the function of the E1A region

The properties of certain mutant strains of adenoviruses have shed more light on the function of the E1A region of the genome. A deletion mutant of adenovirus 5, called dl 312, lacks part of the viral genome between map units 1.5 and 4.0 which spans the region between the promoter (5'-end) and poly(A) addition site (3'-end) of the E1A mRNAs (Fig. 2). When HeLa cells are infected with this mutant no mRNAs from any of the early regions can be detected in the cytoplasm but low amounts of nuclear RNA transcripts from all the early transcription units are present in the cell nuclei [24]. This indicates that the E1A gene products are required for the processing of the other early gene transcripts.

Studies with other mutants of adenovirus 5 have provided similar but not identical results. These mutants have a different host range from the wild type and they can no longer replicate in HeLa cells but can repli-

                                    450   500
                                     | ↓
GGG TCA AAG TTG GCG TTT TAT TAT AGT CAG CTG ACG TGT AGT GTA TTT ATA CCC GGT GAG TTC CTC AAG AGG CCA CTC TTG AGT GCC

 514   550   560   600
  | | | |
AGC GAG TAG AGT TTT CTC CTC CGA GCC GCT CCG ACA CCG GGA CTG AAA ATG AGA CAT ATT ATC TGC CAC GGA GGT GTT ATT ACC GAA GAA
                                                              Met Arg His Ile Ile Cys His Gly Gly Val Ile Thr Glu Glu

                                              650
                                               |
ATG GCC GCC AGT CTT TTG GAC CAG CTG ATC GAA GAG GTA CTG GCT GAT AAT CTT CCA CCT CCT AGC CAT TTT GAA CCA CCT ACC CTT CAC
Met Ala Ala Ser Leu Leu Asp Gln Leu Ile Glu Glu Val Leu Ala Asp Asn Leu Pro Pro Pro Ser His Phe Glu Pro Pro Thr Leu His

      700   750
       | |
GAA CTG TAT GAT TTA GAC GTG ACG GCC CCC GAA GAT CCC AAC GAG GAG GCG GTT TCG CAG ATT TTT CCC GAC TCT GTA ATG TTG GCG GTG
Glu Leu Tyr Asp Leu Asp Val Thr Ala Pro Glu Asp Pro Asn Glu Glu Ala Val Ser Gln Ile Phe Pro Asp Ser Val Met Leu Ala Val

                    800   850
                     | |
CAG GAA GGG ATT GAC TTA CTC ACT TTT CCG CCG GCG CCC GGT TCT CCG GAG CCG CCT CAC CTT TCC CGG CAG CCC GAG CAG CCG GAG CAG
Gln Glu Gly Ile Asp Leu Leu Thr Phe Pro Pro Ala Pro Gly Ser Pro Glu Pro Pro His Leu Ser Arg Gln Pro Glu Gln Pro Glu Gln

                              900   G   950
                               | |
AGA GCC TTG GGT CCG GTT TCT ATG CCA AAC CTT GTA CCG GAG GTG ATC GAT CTT ACC TGC CAC GAG GCT GGC TTT CCA CCC AGT GAC GAC
Arg Ala Leu Gly Pro Val Ser Met Pro Asn Leu Val Pro Glu Val Ile Asp Leu Thr Cys His Glu Ala Gly Phe Pro Pro Ser Asp Asp

      974   1000   1050
       | | |
GAG GAT GAA GAG GGT GAG GAG TTT GTG TTA GAT TAT GTG GAG CAC CCC GGG CAC GGT TGC AGG TCT TGT CAT TAT CAC CGG AGG AAT ACG
Glu Asp Glu Glu Gly Glu Glu Phe Val Leu Asp Tyr Val Glu His Pro Gly His Gly Cys Arg Ser Cys His Tyr His Arg Arg Asn Thr

                                                                1100   1112
                                                                  | |
GGG GAC CCA GAT ATT ATG TGT TCG CTT TGC TAT ATG AGG ACC TGT GGC ATG TTT GTC TAC AGTAAGTGAAAATTATGGGCAGTGGGTGATAGA
Gly Asp Pro Asp Ile Met Cys Ser Leu Cys Tyr Met Arg Thr Cys Gly Met Phe Val Tyr Ser

                                                                                                      1229
                                                                                                        |
GTGGTGGGTTTGGTGTGGTAATTTTTTTTTTTAATTTTTACAGTTTTGTGGTTTAAAGAATTTTGTATTGTGATTTTTTTAAAAG GT CCT GTG TCT GAA
                                                                                      Pro Val Ser Glu

    1250   G   T T
     | | |
CCT GAG CCT GAG CCC GAG CCA GAA CCG GAG CCT GCA AGA CCT ACC CGC CGT CCT AAA ATG GCG CCT GCT ATC CTG AGA CGC CCG ACA TCA
Pro Glu Pro Glu Pro Glu Pro Glu Pro Ala Arg Pro Thr Arg Arg Pro Lys Met Ala Pro Ala Ile Leu Arg Arg Pro Thr Ser
                                                                                Leu Val

            1350   T   1400
             | |
CCT GTG TCT AGA GAA TGC AAT AGT AGT ACG GAC AGC TGT GAC TCC GGT CCT TCT AAC ACA CCT CCT GAG ATA CAC CCG GTG GTC CCG CTG
Pro Val Ser Arg Glu Cys Asn Ser Ser Thr Asp Ser Cys Asp Ser Gly Pro Ser Asn Thr Pro Pro Glu Ile His Pro Val Val Pro Leu

                    1450   1500 T
                     | |
TGC CCC ATT AAA CCA GTT GCC GTG AGA GTT GGT GGG CGT CGC CAG GCT GTG GAA TGT ATC GAG GAC TTG CTT AAC GAG CCT GGG CAA CCT
Cys Pro Ile Lys Pro Val Ala Val Arg Val Gly Gly Arg Arg Gln Ala Val Glu Cys Ile Glu Asp Leu Leu Asn Glu Pro Gly Gln Pro
                                                                                                                    Ser
                          1543   1550   1600
                            | | |
TTG GAC TTG AGC TGT AAA CGC CCC AGG CCA TAA GGT GTA AAC CTG TGA TTG CGT GTG TGG TTA ACG CCT TTG TTT GCT GAA TGA GTT GAT
Leu Asp Leu Ser Cys Lys Arg Pro Arg Pro ***

                          1632
                            |
GTA AGT TTA ATA AAG GGT GAG ATA ATG TTT AAAAAAAAAAA......

Fig. 3. Nucleotide sequence of region E1A from ad5. The DNA sequence was determined by van Ormondt et al. [30]. Two spliced mRNAs, the 12 S and the 13 S mRNAs, are transcribed from this region. Both mRNAs have a common 5'-end which is located at nucleotide 499 [17] and their common poly(A) tract follows nucleotide 1632. The splice in the 12 S mRNA joins nucleotides 974 and 1229 and the splice in the 13 S mRNA joins nucleotides 1112 and 1229. The amino acid sequence of the polypeptide which is specified by the 13 S mRNA is shown. Sequences which are absent in the 12 S mRNA are underlined. A promoter-like sequence (----) is indicated as well as the hexanucleotide sequence AATAAA which precedes the poly(A) tail of all eukaryotic mRNAs. The figure is modified from [23].

cate in another line of human cells (293 cells) which is transformed and carries in its chromosomes 14% of the left-hand end of the viral DNA including region E1 [25]. When host range mutants with a lesion in the E1A region were used to infect non-permissive HeLa cells, nuclear transcripts from the E1A region could be identified [26]. However, transcripts from all the other early genes could not be detected in the cell nuclei with the methods used. This suggests that functional E1A proteins are required for the initiation of transcription of the other early regions, but more sensitive techniques might be needed to establish unequivocally the absence of transcription. The results obtained with the two types of mutants are at variance in one respect; the dl 312 mutation appears to allow low levels of transcription from all early regions whereas the host range mutants seem to be blocked for initiation of transcription. One conceivable explanation is the fact that dl 312 shows a remarkable leakiness at high multiplicities of infection, suggesting that stocks of mutant dl 312 virus may often contain some revertants with an intact E1A region. Another, perhaps more likely possibility, is that the E1A protein is not absolutely essential for transcription from all early regions. The role of the E1A region may then be to enhance the efficiency of the lytic cycle.

Clearly the results of studies with dl 312 and host range mutants indicate that the polypeptide products of the E1A region are required for the expression of the other early regions, although the precise mechanism of this control remains uncertain. We expect that further work on the E1A gene products will clarify one control mechanism for mRNA biogenesis in this model system.

Adenovirus genes with pleiotropic effects

When adenoviruses infect permissive human cells the viruses replicate and in so doing kill the cells. When, however, adenoviruses infect non-permissive rodent cells the virus fails to replicate and the cells survive. In a minority of the survivors part of the viral genome may become integrated into cellular chromosomes and as a result change or transform the growth properties of the cells. Such transformed cells in many ways resemble the cells of tumours induced by injecting the virus into new-born rodents.

All mutants with lesions in the E1A region are defective not only for replication in permissive hosts but also for transformation. If the E1A region, which is present and expressed in all transformed cells, is required for a post-transcriptional step in the maturation of mRNA it is feasible that its gene products could alter the processing or splicing pathway of cellular as well as of viral transcripts. In non-permissive cells expression of the viral E1A region might activate cellular genes which in turn might lead to the induction of cellular DNA synthesis, which probably is a prerequisite for transformation. Permanent transformation may only occur in the rare case when region E1A DNA sequences become integrated into chromosomal DNA in a manner which allows expression of the viral genes. Alternatively, E1A mutants may fail to transform non-permissive cells because they do not allow the expression of region E1B, which appears also to be required to make cells express the fully transformed phenotype. This latter alternative is very likely because certain host-range mutants that are defective in transformation have an intact E1A region and lesions in the E1B region [27]. The true transforming gene(s) may therefore reside in the E1B region whose expression is dependent on the E1A region.

The host-range mutants of adenovirus 5, with lesions in the E1B region can grow in human embryonic cells. This suggests that these cells provide a gene product which can replace at least one of the region E1B proteins [28]. Host-range mutants of polyoma virus, which are defective in transformation, can likewise be complemented when grown in embryonic cells [29].

These results indicate that products of the E1B region are functionally identical to cellular gene products present in embryonic cells. Transformation may, therefore, be akin to a gene dosage phenomenon in which the virus introduces into the cell a function normally present in embryonic cells. The present results have narrowed down to a handful of gene products the transforming principle of adenoviruses and have brought us close to understanding the control of transcription of the early viral genes during normal productive infections. We expect that further progress will lead not only to an unequivocal identification of the transforming gene product(s) but also to a deeper understanding of how gene expression in mammalian cells is regulated.

References

1 Sergeant, A., Tigges, M. and Raskas, H. J. (1979) *J. Virol.* 29, 888–898
2 Tate, V. and Philipson, L. (1979) *Nucleic Acids Res.* 6, 2769–2785
3 Pettersson, U., Tibbetts, C. and Philipson, L. (1976) *J. Mol. Biol.* 101, 479–501
4 Flint, S. J. (1977) *Cell* 10, 153–166
5 Chow, L. T., Broker, T. R. and Lewis, J. B. (1979) *J. Mol. Biol.* 134, 265–303
6 Harter, M. and Lewis, J. B. (1978) *J. Virol.* 26, 736–749
7 Sehgal, P. B., Fraser, N. W. and Darnell, J. E. (1979) *Virology* 94, 185–191
8 Chow, L. T., Roberts, J. M., Lewis, J. B. and Broker, T. (1977) *Cell* 11, 819–836
9 Lewis, J. B., Anderson, C. W. and Atkins, J. F. (1977) *Cell* 12, 37–44
10 Ito, K., Arens, M. and Green, M. (1975) *J. Virol.* 15, 1507–1510
11 Fraenkel, G. D. (1978) *J. Virol.* 25, 457–463
12 Ziff, E. B. and Evans, M. E. (1978) *Cell* 15, 1463–1475
13 Akusjärvi, G. and Pettersson, U. (1979) *J. Mol. Biol.* 134, 143–158
14 Berget, S., Moore, C. and Sharp, P. A. (1977) *Proc. Natl. Acad. Sci. U.S.A.* 74, 3171–3175
15 Wilson, M. C., Fraser, N. W. and Darnell, J. E. (1979) *Virology* 94, 175–184
16 Halbert, D. W., Spector, D. J. and Raskas, H. J. (1979) *J. Virol.* 31, 621–629
17 Baker, C. and Ziff, E. (1979) *Cold Spring Harbor Symp. Quant. Biol.* Vol. 44, 415–428
18 Philipson (1979) *Adv. Virus Res.* 25, 357–405
19 Galos, R. S., Williams, J., Binger, M. H. and Flint, S. J. (1979) *Cell* 17, 945–956
20 Persson, H., Mathisen, B., Philipson, L. and Pettersson, U. (1979) *Virology* 93, 198–208
21 Berk, A. J., Lee, F., Harrison, T., Williams, J. and Sharp, P. A. (1979) *Cell* 17, 935–944
22 Nevins, J. R., Ginsberg, H. S., Blanchard, J.-P., Wilson, M., Darnell, J. E. (1979) *J. Virol.* 32, 727–733
23 Perricaudet, M., Akusjärvi, G., Virtanen, A. and Pettersson, U. (1979) *Nature (London)* 281, 694–695
24 Jones, N. and Shenk, T. (1979) *Proc. Natl. Acad. Sci. U.S.A.* 76, 3665–3669
25 Harrison, T., Graham, F. and Williams, J. (1977) *Virology* 77, 319–329
26 Berk, A. J., Lee, F., Harrison, T., Williams, J. and Sharp, P. A. (1979) *Cell* 17, 935–944
27 Frost, E. and Williams, J. (1978) *Virology* 91, 39–50
28 Graham, F. L., Harrison, T. and Williams, J. (1978) *Virology* 86, 10–21
29 Benjamin, T. and Goldman, E. (1974) *Cold Spring Harbor Symp. Quant. Biol.* 39, 41–44
30 van Ormondt, H., Maat, J., de Waard, A. and van der Eb, A. J. (1978) *Gene* 4, 309–328

Lennart Philipson and Ulf Pettersson are at the Department of Microbiology, University of Uppsala, The Biomedical Centre, Box 581, S-751 23 Uppsala, Sweden.

Addendum – *L. Philipson and U. Pettersson*

Since this review was written, numerous reports have dealt with the mapping and sequence analyses of the early adenovirus mRNAs and identification of their protein products. It is now established that both the EX and EY regions (Fig. 1) are transcribed at intermediate times in the infectious cycle from the promoter in the E2 region located at 75 map units [31]. This transcription unit is called E2B and the early E2 unit is referred to as E2A. Messenger RNA from the E2B region codes for the terminal protein covalently linked to the ends of viral DNA and in addition for two proteins (mol. wts 75,000 and 105,000) [31] all of which may be involved in DNA replication. The protein products from almost all early viral mRNAs have been further characterized and those from the transforming regions E1A and E1B (Fig. 2) have, for obvious reasons, been investigated in considerable detail [32].

With the aid of site directed mutagenesis and sequence analysis it has been established that the 13S mRNA from the E1A region (Fig. 3) is necessary for the accumulation of mRNA from the other four early regions, since a mutant defective in splicing of the 12S mRNA, without affecting the 13S mRNA expression [33], generates normal amounts of mRNA from all early regions; by contrast a chain termination mutation in the unique portion of 13S mRNA does not [34].

A set of deletions of sequences around the promoter TATA box for the 13S mRNA has shown that in the region from -44 to $+62$ relative to the transcriptional start site, only the TATA homology is required for wild-type levels of an aberrant but functional 13S mRNA. When the TATA homology is deleted, 5- to 10-fold lower levels of 13S mRNA are accumulated, this level, however, appears sufficient to turn on the other four early transcription units [35]. All these deletion mutants, in contrast to those deleted for the entire E1A region, were also capable of transforming rat embryo fibroblasts. This suggests that the function of the 13S mRNA is not completely dependent on the normal 5'-terminus of the mRNA. Results with one mutant suggest that the 14 N-terminal amino acids are not essential. This study appears to have established for the first time that the TATA box is required for correct initiation in a truly homologous system.

Less progress has been made towards elucidating the function of the protein encoded in the 13S mRNA. The original finding that the other early regions respond differently to the E1A product [36] was corroborated with the deletion mutants [35]. The E1B mRNA appears to be the least dependent on the 13S mRNA product followed by E4 and finally E2 and E3, the latter two showing the strongest dependence on the E1A product. How the 13S mRNA product from the E1A region acts is still unresolved. Evidence has been presented in favour of a transcriptional control [37] but others suggest a major effect on a post-transcriptional step [38]. One of the E1A products can bind to double-stranded DNA (Esche, personal communication) and the E1A region is required for induction of a host cell protein, which also is inducible by heat-shock treatment of the cells (Nevins, personal communication). It has not yet however been unequivocally proven that only the 13S mRNA product is involved in these functions. The possibility that the other strand of viral DNA in this region also encodes protein products is still open since the sequence on the l-strand [30] contains open reading frames in this region and the deletion mutants will affect products from the two strands concurrently. Although adenovirus gene regulation shows considerable complexity at this stage, it is generally agreed that the expression of early adenovirus genes is controlled in a precise manner, probably involving a cascade mechanism. Identification of the molecular events involved may provide the tools to understand control of transcription in mammalian cells.

References

31 Stillmann, B. W., Lewis, J. B., Chow, L. T. and Mathews, M. B. (1981) *Cell* 23, 497–508
32 Esche, H., Mathews, M. B. and Lewis, J. B. (1980) *J. Mol. Biol.* 142, 393–417
33 Montell, C., Fisher, E. F., Caruthers, M. H. and Berk, A. J. (1982) *Nature* 295, 380–384
34 Ricciardi, R. P., Jones, R. L., Cepko, C. L., Sharpe, P. A. and Roberts, B. E. (1981) *Proc. Natl Acad. Sci. U.S.A.* 178, 6121–6125
35 Osborne, T. F., Gaynor, R. B. and Berk, A. J. (1982) *Cell* 29, 139–148
36 Persson, H., Katze, M. and Philipson, L. (1981) *J. Virol.* 40, 358–366
37 Nevins, J. R. (1981) *Cell* 26, 213–220
38 Katze, M., Persson, H. and Philipson, L. (1981) *Mol. Cell. Biol.* 1, 807–813

Patterns and consequences of adenoviral RNA splicing
Ex Pluribus Unum – Ex Uno Plura

Thomas R. Broker and Louise T. Chow

Splicing of primary RNA transcripts of adenovirus DNA increases genetic versatility, introduces a new level of regulation of mRNA synthesis, allows coordinated transcription of multiple genes into polycistronic precursor RNAs with common 5' and 3' terminal structures and renders all their internal messages accessible to translation.

Many eukaryotic messenger RNAs are coded by separate portions of the chromosomal DNA and are generated by one or more rounds of intrastrand splicing (deletion and re-ligation) of long primary transcripts. RNA splicing was discovered in the spring of 1977 during studies of the RNA synthesized by human adenovirus serotype 2 (Ad2) in productively infected cells [1–5]. Within months, the discovery of RNA splicing led to a general redefinition of the basic organization of eukaryotic genes and transcription units which set them distinctly apart from those in bacteria and bacterial viruses. Subsequent research has surveyed the incidence of RNA splicing, its patterns, mechanisms, and biological consequences, but important unanswered questions remain to challenge future efforts.

The human adenoviruses grow lytically in human cultured cells and can transform rat and hamster cells or cause tumors in rodents – hence they are *bona fide* tumor viruses. The review by Philipson and Pettersson (*TIBS*, May, 1980, 135–138) describes the organization of the adenoviral genome, the stages of the infectious cycle, and the role and control of early gene products in both lytically infected cells and in transformed cells. We will focus on the processing of primary viral transcripts into mature mRNAs and emphasize the splicing mechanisms inferred from the RNA structures and the biological consequences of splicing deduced from the DNA and RNA sequences and protein compositions.

Transcription of adenovirus DNA

As illustrated in the transcription and translation maps (Fig. 1), one minor (E2B) and five major (E1A, E1B, E2A, E3, and E4) regions of the adenoviral chromosome are transcribed into RNA at early times after infection [6–9]. At intermediate times or in the presence of inhibitors of DNA replication, several other transcription units located between 9.8 and 39.0 map units are active; these are expressed more fully at late times [8,10–12]. Additional, late RNA species are synthesized after the onset of DNA replication from a single promoter at map unit 16.5, from which transcription extends along the r-strand to near the right end of the chromosome [13–16]. Most primary transcripts are polycistronic in one of two respects: they either contain overlapping genes that are sorted out by RNA splicing within coding regions, or consist of serially arranged messages, any one of which can be brought near the 5' leader by deletion of intervening messages. The maturation of the six

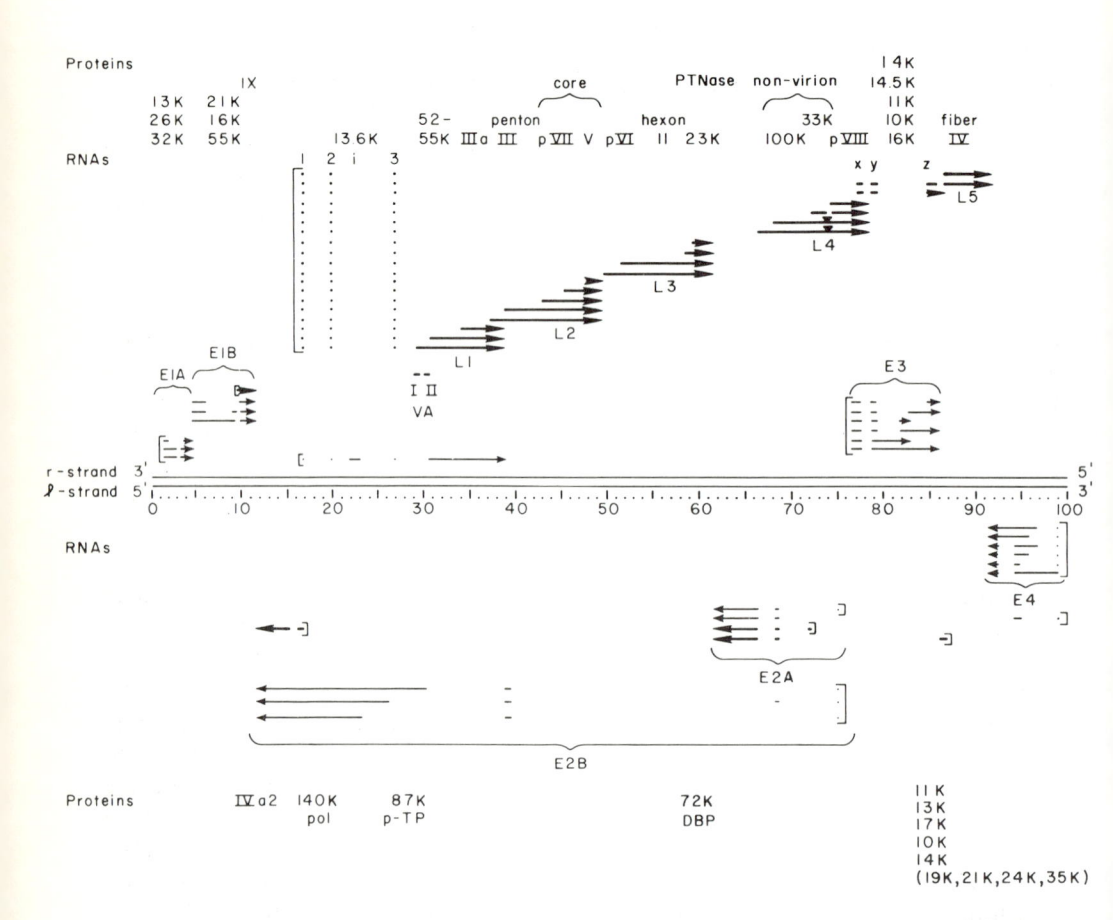

Fig. 1. The map of adenovirus-2 cytoplasmic RNA transcripts, determined by electron microscopy and by nuclease digestion of RNA:DNA heteroduplexes. The 36,500 base-pair chromosome is divided into 100 map units. Arrows show the direction of transcription along the r-strand or l-strand of DNA. The 5' ends of the cytoplasmic RNAs correspond to the locations of promoters [13,14,25] (indicated by vertical brackets). The conserved segments constituting early RNAs [6–8] are depicted by thin arrows and those in late RNAs [2,10,15,16] by thick arrows. Gaps in arrows represent intervening sequences removed from the cytoplasmic RNAs by splicing. Early regions 1A, 1B, 2A, 2B, 3 and 4 are bracketed. The precise coordinates and splicing patterns of region 2B RNA have not yet been determined. At intermediate and late times, region 2 is expressed from several additional promoters and region 3 RNA can be made under the direction of the major r-strand late promoter [8]. The region 2 RNAs may have an additional small deletion near coordinate 66.3 [8]. All derivatives of the late r-strand transcript have the same tripartite leader, the segments of which are labeled 1, 2, 3. At intermediate times, a four-part leader is observed and includes the i segment [31]. It can be in any of the r-strand products, but predominates in the 52K message. The RNAs for protein IV can also contain some combination of ancillary leader segments x, y, z [15]. The correlations of mRNAs with encoded proteins were based on cell-free translations of RNA selected by hybridization to DNA restriction fragments ([5,10,11,12,24,26] Miller, Ricciardi, Roberts, Paterson and Mathews, personal communication). Alternatively spliced RNAs complementary to early regions 1A, 1B, 3, and 4 give rise to multiple proteins; some of them share common peptides. The proteins indicated on the top, middle, and bottom lines of regions 1A and 1B correlate with the three RNA species shown below them. Correlations in regions 3 and 4 have not yet been achieved. Proteins are designated by K (1000 mol. wt) or by Roman numerals (virion components). The late proteins, which are components of the virion capsid, are: II (hexon), III (penton base), IIIa (peripentonal hexon-associated), IV (fiber), V (minor core), pVI (hexon-associated, precursor), pVII (major core, precursor), pVIII (hexon-associated, precursor), IX (hexon-associated).

early and the several intermediate and late primary adenoviral transcripts into seventeen or more early mRNAs and nineteen or more late mRNAs occurs by two processes: 3' end determination, including the addition of polyadenylate (poly(A)), and RNA splicing.

The basic steps in RNA synthesis and processing are summarized in Fig. 2. **RNA transcription is generally initiated about 30 nucleotides downstream from an TAT(A,T)$_{3-6}$ nucleotide sequence in the viral DNA** ([14,17] and D. Hogness, personal communication). Synthesis by RNA polymerase extends continuously for about 1000 to 27,000 nucleotides [13,16]. **The 5' end of all adenoviral transcripts synthesized in the nucleus at both early and late times after infection is quickly modified with a** $^{7\text{-methyl}}$**Gppp 'cap', which is conserved during subsequent RNA processing** [3,18,19]. Some transcriptional units produce RNAs with a common 3' end; others have alternative 3' ends. The 3' poly(A) sequence is usually downstream from an AAUAAA signal in the transcript [20,21].

In addition to methylation of the cap structure at the 5' terminus, several 5' proximal nucleotides and occasional internal adenosine residues are methylated, and these nucleotides are preferentially conserved in fully spliced RNA [22]. However, the roles of methylation, if any, in RNA maturation are not yet determined. The selection and deletion of intervening sequences in the primary transcripts take place in the nucleus prior to, and possibly coincident with, transport to the cytoplasm. The enzyme systems responsible for RNA splicing remain obscure.

Versatility through splicing

Early regions 1A and 1B provide the best examples of the genetic versatility achieved by RNA splicing. These gene blocks are necessary and sufficient for transformation of rodent cells and are implicated in the control of the lytic cycle (Philipson and Pettersson, *TIBS*, May, 1980, 135–38). Region 1A encodes a set of closely related proteins [11,12,23,24]. These are translated from related mRNAs generated by alternative patterns of splicing of a common primary transcript [7,8]. Examination of the DNA sequence has revealed two long open translation regions in different reading frames separated by several protein termination codons near map units 3.1–3.2 [24]. Production of all three mature mRNA species from region 1A involves removal of the sequences from 3.0–3.3; this deletes the termination codons and connects the open frames [21]. The three messages differ in the position of the 5' splice sites within the coding region. Therefore, splicing allows a single genetic region to give rise to multiple proteins that have common amino- and carboxy-termini, a dramatic exception to the classical Beadle and Tatum rule of 'one gene – one polypeptide'. The relative abundances of the different mRNAs change during the course of infection, suggesting that splicing can modulate the production of alternative proteins – a new form of genetic regulation [8,12].

Early region 1B originates from a separate promoter at map unit 4.6 and extends to coordinate 11.2 [6–8,25]. A continuous open reading frame exists between coordinates 4.7 and 9.6 [24]. A short splice is introduced near coordinate 9.8–9.9 [7], downstream from the termination codon (Perricaudet, Le Moullec and Pettersson, personal communication) to generate the mRNA for a protein of mol. wt 55,000 [11,12,23,24]. A more common mRNA from this region, which increases dramatically at late times after infection, has a long deletion between coordinates 6.1 and 9.9 [2,6,8]. This mRNA encodes a mol. wt 15,000 protein entirely before the

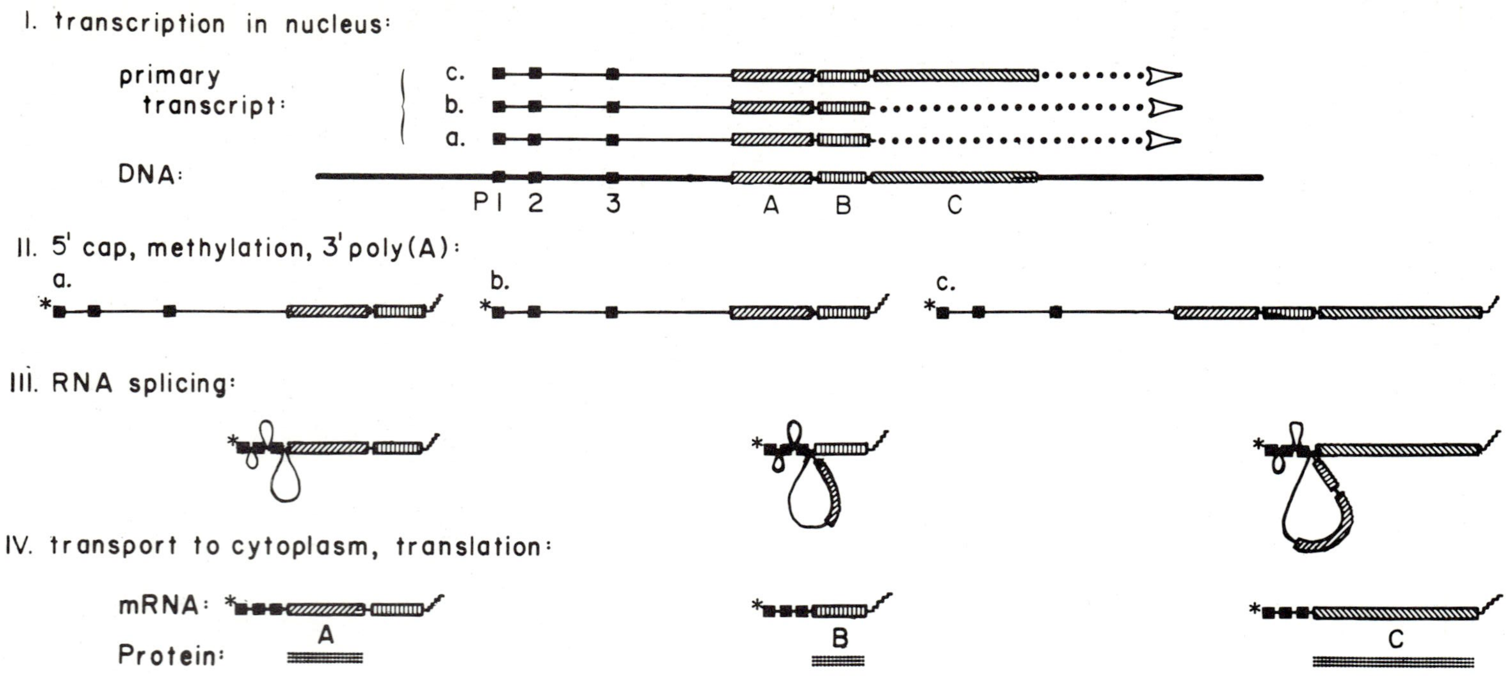

Fig. 2. Transcription and processing of adenoviral late RNA. (I) The DNA is transcribed from the promoter (P) to the right end of the chromosome by RNA polymerase II. The sequences encoding the first, second, and third leader segments are indicated by boxes. Genes A and B are members of the same 'family' and their primary transcript is cut between genes B and C. Gene C is a member of another family and its primary transcript is cut downstream from the gene. The downstream portions of transcripts a, b, and c beyond the cut sites are degraded. (II) A $^{7\text{-methyl}}$Gppp 'cap' is enzymatically coupled to the 5' end of each primary transcript, the cap and some internal nucleosides are methylated, and polyadenylate is added to the 3' end. (III) Sequences destined to be eliminated from mature messages are looped out, and the conserved segments brought together and spliced. The leader segments are joined to the 5' ends of gene transcripts and will recognize alternative members of the same family, as in a and b. But in a long primary transcript that spans two or more families, splicing will eventually join the leaders to a message in the 3' proximal family, as in c. (IV) The spliced RNAs are transported from the nucleus to the cytoplasm and associate with ribosomes. Only the 5' proximal message in a polycistronic message is translated [29], so message a cannot be translated into protein B.

splice in a reading frame which differs from that for the mol. wt 55,000 protein ([11,12,23,24]; Bos et al. (1981) Cell 27, 121–131). In neither RNA is the 3' proximal sequence from map units 9.9 to 11.2 translated. A third RNA from the region is transcribed from a separate promoter near coordinate 9.8 at intermediate to late times [8,10,25]; it coincides with the 3' terminal segment of the early region 1B RNAs, but encodes a protein component of the virus capsid [5]. The splice present in both early region 1B RNAs deletes the promoter and the 18 S rRNA binding sequence of the peptide IX message.

Regions 3 and 4 generate multiple RNAs and proteins [6–8,26], but they have not been correlated with one another and the functions of the proteins have not been defined. Early region 2 encodes a DNA-binding protein (72K) essential for replication at intermediate to late times after infection. The mRNA consists of three or four segments [6–8] and utilizes different promoters at early and late times [8]. The early and late forms of the RNA have different 5' leader segments, but encode the same 72K protein. Therefore splicing allows alternative promoters to be utilized for the expression of the same gene. The transition from early to late promoters for region 2 does not occur in the presence of cycloheximide, an inhibitor of protein synthesis, showing that normal development of the viral infectious cycle depends on positive effectors [8]. The sequences present in the major late 72K mRNA leader at map unit 72 are included in the primary transcript of the early 72K RNA (map units 75.0 to 61.5) but are deleted from the mature RNA. This reveals a subtle control over splicing and, with other evidence, suggests that the presence of a capped 5' end may help potentiate nearby sequences as splice junctions [8].

Splicing of late RNAs

At late times, a major r-strand promoter at coordinate 16.5 is activated, and it gives rise to a long primary transcript [13,14,16]. Polyadenylate can be added at any one of five major and several minor sites [15,16,20]. The decision as to where to nick the nascent RNA and to polyadenylate is made during transcription shortly after the polymerase has moved past the chosen site [16].

All the different late mRNAs processed from this primary transcript have the same tripartite leader sequence of 203 nucleotides derived by splicing from transcripts of the genome at map units 16.6, 19.6, and 26.6 [1,2,15,27,28]. Therefore, only one mRNA is generated from each primary product. The common leader is spliced to one of a dozen different sites, which lie within 1000–4500 nucleotides of the poly(A) at the 3' end. Consequently, families of mRNAs are generated which share the same 3' end and 5' leader, but they have different coding sequences adjacent to the leader. The longer members of a family carry coding regions for more than one protein and therefore are physically polycistronic. But they are functionally monocistronic, because only the 5' proximal message is translated in eukaryotes [29]. The tripartite leader sequence is not translated; however, it contains a sequence which is complementary to the 3' end of 18 S ribosomal RNA and may be involved in binding to ribosomes [14]. The mechanisms that allocate an equitable distribution of the leader to one of several alternative coding regions within each family are unknown. This capability is one of the most important features of RNA splicing because it ensures that each message in the primary transcript will be 5' proximal in some of the mature mRNAs and therefore available for translation.

Splicing at intermediate times

At intermediate times after infection many of the cytoplasmic mRNAs copied from the r-strand, and particularly the one with its main body specified by DNA between map units 30.5 and 39.0, have a four-part, rather than a three-part, leader sequence [8]. The extra segment 'i' is derived from sequences located between the normal second and third leader components (Fig. 1 and 3). These molecules may either be splicing intermediates, not yet fully processed, or mature molecules, in which case splicing is different at intermediate and late times after infection. Furthermore, in the nucleus [30] and, occasionally in the cytoplasm [31], RNA molecules are found in which the third segment of the leader is longer and extends toward, or merges with the i leader. The various leader structures imply that splicing is not processive in the 5' to 3' direction, that certain downstream splices can precede other upstream deletions, and that the splices between ultimately conserved segments do not take place in a single step, but instead occur by two or more partial deletions. Furthermore, the different combinations of leader structures would demand alternative pathways for the RNA processing.

Variations in the message for the fiber protein (IV), the distal RNA produced from the late transcription unit, reinforce these concepts. At least a quarter of the fiber RNAs have extra short leader segments, designated x, y, and z, between the tripartite leader and the main body of the message [15] (Fig. 3). These coincide closely with segments present in certain early RNAs specified by early region 3 (Fig. 1), indicating that early RNA splice signals continue to be recognized at late times [8]. The 'y' leader segment does not affect the translation of the mRNAs [32]. The 1-2-3-y leader is spliced directly to the AUG initiation codon for the fiber protein, which is located 387 nucleotides from the 5' end of the RNA [28]. Different combinations of the x, y, and z leaders are found on the fiber transcripts, again demonstrating that RNA splicing is not tightly coupled along a unique pathway [15].

When human adenoviruses infect monkey CV-1 cells, the infection is abortive. The block has been traced to the failure to produce fiber protein. More specifically, there is a reduction in the percentage of transcripts containing the fiber message [33] and, in about half the RNAs that are made, long nucleotide sequences between the tripartite leader and the main body of the message are not removed [34] (Fig.3). Because these extra coding sequences are adjacent to the 5' leader, the transcripts probably would not function as fiber messages during translation. Adenoviral mut-

A three-part leader joined to the chosen messenger.

```
1 2 i 3    100 K           x y z  Fiber
```

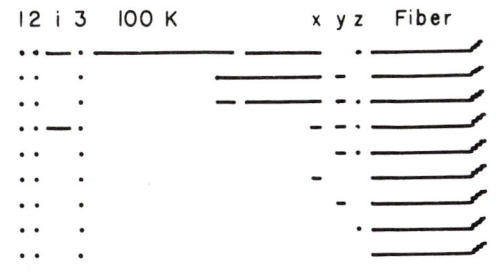

Fig. 3. Some of the alternative structures of fiber gene transcripts [15,34]. All share the tripartite leader segments (1, 2, 3) found on each product of the r-strand transcription unit. The i leader can be present along with any other combination of leader segments and is shown on two forms. The x, y, and z segments coincide with sequences present in early region 3 RNAs and can be present in any combination. The long sequences from the 100K gene region are present in about half the RNAs isolated from monkey cells, but only in a few percent of those from human cells.

ants have been isolated which can grow in monkey cells [35]. The mutations restore the full production of fiber RNA and allow the normal elimination of intervening sequences [34]. They map in the genetic region encoding the DNA-binding protein [35]. This is the first specific eukaryotic protein implicated in mRNA splicing.

Summary

From the various structures of the spliced adenoviral transcripts and correlations with DNA, RNA, and protein sequences, many biological consequences have been inferred. Splicing greatly increases the genetic versatility and economy of the chromosome by allowing DNA sequences to specify multiple mRNAs, introduces a new level of regulation over mRNA production, allows coordinated transcription of multiple genes into polycistronic RNA and provides translational accessibility to all messages, while accommodating the need to preserve both the 5' cap and 3' poly(A) of the RNA to confer stability. The alternative structures of the mRNAs derived from the adenoviral transcription units reveal some of the rules by which splicing proceeds. Further correlations will soon allow many additional insights and will extend the value of adenoviruses as probes of eukaryotic cell processes.

Acknowledgements

Work in the authors' laboratory was supported by the United States National Cancer Institute through a grant to Cold Spring Harbor Laboratory. We thank our colleagues for communication of results prior to publication and express regret that limitation on space did not permit more thorough citations of the extensive literature.

References

1 Berget, S. M., Moore, C. and Sharp, P. A. (1977) Proc. Natl. Acad. Sci. U.S.A. 74, 3171–3175
2 Chow, L. T., Gelinas, R. E., Broker, T. R. and Roberts, R. J. (1977) Cell 12, 1–8
3 Klessig, D. F. (1977) Cell 12, 9–21
4 Dunn, A. R. and Hassell, J. A. (1977) Cell 12, 23–36
5 Lewis, J. B., Anderson, C. W. and Atkins, J. F. (1977) Cell 12, 37–44
6 Kitchingman, G. R., Lai, S-P. and Westphal, H. (1977) Proc. Natl. Acad. Sci. U.S.A. 74, 4392–4395
7 Berk, A. J. and Sharp, P. A. (1978) Cell 14, 695–711
8 Chow, L. T., Broker, T. R. and Lewis, J. B. (1979) J. Mol. Biol. 134, 265–304
9 Galos, R. S., Williams, J., Binger, M-H. and Flint, S. J. (1979) Cell 17, 945–956
10 Persson, H., Pettersson, U. and Mathews, M. B. (1978) Virology 90, 67–79
11 Lewis, J. B., Esche, H., Smart, J. E., Stillman, B., Harter, M. L., and Mathews, M. B. (1979) Cold Spring Harbor Symp. Quant. Biol. 44, 493–504
12 Spector, D. J., Halbert, D. N., Crossland, L. D. and Raskas, H. J. (1979) Cold Spring Harbor Symp. Quant Biol. 44, 437–445
13 Evans, R. M., Fraser, N., Ziff, E., Weber, J., Wilson, M. and Darnell, J. E. (1977) Cell 12, 733–740
14 Ziff, E. B. and Evans, R. M. (1978) Cell 15, 1463–1475
15 Chow, L. T. and Broker, T. R. (1978) Cell 15,

16 Nevins, J. R. and Darnell, J. E., Jr (1978) Cell 15, 1477–1493
17 Baker, C. C. and Ziff, E. B. (1979) Cold Spring Harbor Symp. Quant. Biol. 44, 415–428
18 Gelinas, R. E. and Roberts, R. J. (1977) Cell 11, 533–544
19 Hashimoto, S. and Green, M. (1979) Virology 94, 254–272
20 Fraser, N. and Ziff, E. (1978) J. Mol. Biol. 124, 27–51
21 Perricaudet, M., Akusjärvi, G., Virtanen, A., and Pettersson, U. (1979) Nature (London) 281, 694–696
22 Chen-Kiang, S., Nevins, J. R. and Darnell, J. E., Jr. (1979) J. Mol. Biol. 135, 733–752
23 Green, M., Wold, W. S. M., Brackmann, K. H. and Cartas, M. A. (1979) Virology 97, 275–286
24 van der Eb, A. J., van Ormondt, H., Schrier, P. I., Lupker, J. H., Jochemsen, H., van den Elsen, P. J. DeLeys, R. J., Maat, J., van Beveren, C. P., Dijkema, R. and deWaard, A. (1979) Cold Spring Harbor Symp. Quant. Biol. 44, 383–399
25 Wilson, M. C., Fraser, N. W. and Darnell, J. E. (1979) Virology 94, 175–184
26 Harter, M. L. and Lewis, J. B. (1978) J. Virol. 26, 736–749
27 Akusjärvi, G. and Pettersson, U. (1979) Cell 16, 841–850
28 Zain, S., Gingeras, T. R., Bullock, P., Wong, G. and Gelinas, R. E. (1979) J. Mol. Biol. 135, 413–433
29 Kozak, M. (1978) Cell 15, 1109–1123
30 Berget, S. M. and Sharp, P. A. (1979) J. Mol. Biol. 129, 547–565
31 Chow, L. T., Jewis, J. B. and Broker, T. R. (1979) Cold Spring Harbor Symp. Quant. Biol. 44, 401–414
32 Dunn, A. R., Mathews, M. B., Chow, L. T., Sambrook, J. and Keller, W. (1978) Cell 15, 511–526
33 Klessig, D. F. and Anderson, C. W. (1975) J. Virol. 16, 1650–1668
34 Klessig, D. F. and Chow, L. T. (1980) J. Mol. Biol. (in press)
35 Klessig, D. F. and Grodzicker, T. (1979) Cell 17, 957–966

Thomas R. Broker and Louise T. Chow are Senior Staff Scientists at the Cold Spring Harbor Laboratory, P.O. Box 100, Cold Spring Harbor, New York 11724, U.S.A.

Addendum – T. R. Broker and L. T. Chow

Early region 2B has been identified and characterized within the past 2 years and shown to encode proteins essential for viral DNA replication. Very small amounts of 1-strand transcripts were first observed by filter hybridization [9]. Their structures were determined by electron microscopical heteroduplex analysis [36]. Surprisingly, the primary transcripts originate from the early region 2 promoter at coordinate 75 and extend to the poly(A) site used by the message for the IVa$_2$ protein (which is involved in virion morphogenesis and, possibly, DNA packaging [37]). Alternative splicing creates several overlapping messages, each with two or three leaders (Fig. 1). Hybridization-selection of these RNAs followed by cell-free translation resulted in the identification of at least two major products, a mol. wt 87,000 protein (87 KD) and a mol. wt 140,000 protein (140 KD, originally designated 105 KD) [36,38]. The 87 KD protein is the primer for initiation of DNA replication and ultimately is cleaved to the 55 KD 'terminal protein' which is covalently linked to the 5' end of each viral DNA strand [36–39]. The 140 KD protein has recently been identified as a viral DNA polymerase that copurified with the 87 KD protein [40–42]. Together, the genes for the 72 KD DNA-binding protein, the 87 KD protein and the 140 KD protein constitute a coordinated transcription unit essential for viral replication. Much more DNA-binding protein is needed than either of the other products, and this seems to be achieved by the predominance of polyadenylation at the 3' end of early region 2A, with relatively few RNAs extending through early region 2B. Hence, selection of a poly(A) site and use of alternative acceptors for a common donor splice site together dictate the distribution of the primary transcript to the family of mRNAs, as also seen with the r-strand transcripts.

A substantial portion of the chromosome from several adenovirus serotypes has been sequenced and numerous RNA splice junctions determined [43–46]. In the cases of the 53–55 KD protein [47] and the fiber protein [48], the third leader is spliced to the main body immediately preceding the AUG initiation codon, and it replaces a C at the third base preceding the AUG with an A from the leader. This substitution places the AUG in a more favorable environment for efficient translation [49]. The 'i' leader [8,50] is present in some r-strand transcripts, most notably at early and intermediate times, and turns out to specify a 13.6 KD protein [11,50]. Translation is initiated within the 'i' segment, but it terminates in the third leader. The transient and limited inclusion of the 'i' leader in r-strand products is suggestive of a role in the transition from early to late gene expression.

In all the adenoviral RNA splice junctions sequenced, as in all eukaryotic mRNAs [51], the basic [GU...AG] pattern is present at the boundaries of the intervening sequences. Surprisingly, the alternative partners of common splice donors or acceptors do not share similar sequences around the splice sites. This implies there is no simple scheme by which RNA splice sites could be predicted from the DNA sequence.

Present splicing models invoke secondary, tertiary and quaternary interactions of a largely undefined nature, and include the possible involvement of small ribonuclear protein complexes [52]. RNA splicing has been observed in isolated nuclei [53] and, at low efficiency, in cell-free systems [54,55]. Nevertheless, the participating enzymes and cofactors remain essentially unknown.

Acknowledgements

Recent studies in the authors' laboratory have been supported by an American Cancer Society Research Grant to L.T.C. as well as by a National Cancer Institute Program Project Grant.

References

36 Stillman, B. W., Lewis, J. B., Chow, L. T., Mathews, M. B. and Smart, J. E. (1981) *Cell* 23, 497–508
37 Persson, H., Mathisen, B., Philipson, L. and Petterson, U. (1979) *Virology* 93, 198–208
38 Binger, M.-H., Flint, S. J. and Rekosh, D. M. (1982) *J. Virol.* 42, 488–501
39 Challberg, M. D., Desiderio, S. V. and Kelly, T. J., Jr. (1980) *Proc. Natl Acad. Sci. U.S.A.* 77, 5105–5109
40 Enomoto, T., Lichy, J. H., Ikeda, J. E. and Hurwitz, J. (1981) *Proc. Natl Acad. Sci. U.S.A.* 78, 6779–6783
41 Lichy, J. H., Nagata, K., Friefeld, B. R., Enomoto, T., Field, J., Guggenheimer, R. A., Ikeda, J. E., Horwitz, M. S. and Hurwitz, J. *Cold Spring Harbor Symp. Quant. Biol.* 47 (in press)
42 Stillman, B. W. and Tamanoi, F. *Cold Spring Harbor Symp. Quant. Biol.* 47 (in press)
43 Aleström, P., Akusjärvi, G., Petterson, M. and Petterson, U. *J. Biol. Chem.* (in press)
44 Engler, J. A., Hoppe, M. S. and van Bree, M. P. *Gene* (in press)
45 Gingeras, T. R., Sciaky, D., Gelinas, R. E., Jiang, B. D., Yen, C. E., Kelly, M. M., Bullock, P. A., Parsons, B. L., O'Neill, K. E. and Roberts, R. J. *J. Biol. Chem.* (in press)
46 Broker, T. R. (1981) in *DNA Tumor Viruses* (Revised Second Edition) (Tooze, J., ed.), Cold Spring Harbor Lab, Cold Spring Harbor, N.Y. pp. 937–1060
47 Akusjärvi, G. and Persson, H. (1981) *Nature* 292, 420–426
48 Zain, S., Gingeras, T. R., Bullock, P., Wong, G. and Gelinas, R. E. (1979) *J. Mol. Biol.* 135, 413–433
49 Kozak, M. (1981) *Nucl. Acids Res.* 9, 5233–5252
50 Virtanen, A., Petterson, U., Le Moullec, J. M., Tiollais, P. and Perricaudet, M. (1982) *Nature* 295, 705–707
51 Breathnach, R., Benoist, C., OHare, K., Gannon, F. and Chambon, P. (1978) *Proc. Natl Acad. Sci. U.S.A.* 75, 4852–4857
52 Yang, V. W., Lerner, M. R., Steitz, J. A. and Flint, S. J. (1981) *Proc. Natl Acad. Sci. U.S.A.* 78, 1371–1375
53 Blanchard, J.-M., Weber, J., Jelinek, W. and Darnell, J. E. (1978) *Proc. Natl Acad. Sci. U.S.A.* 75, 5344–5348
54 Weingärtner, B. and Keller, W. (1981) *Proc. Natl Acad. Sci. U.S.A.* 78, 4092–4096

55 Goldenberg, C. J. and Raskas, H. J. (1981) *Proc. Natl Acad. Sci. U.S.A.* 78, 5430–5434

Personal communications now published
Perricaudet, M., Le Moullec, J. M. and Pettersson, U. (1980) *Proc. Natl Acad. Sci. U.S.A.* 77, 3778–3782

Miller, J. S., Ricciardi, R. P., Roberts, B. E., Patterson, B. M. and Matthews, M. B. (1980) *J. Mol. Biol* 142, 455–488

Transcription enhancer sequences: a novel regulatory element by W. Dynan and R. Tjian

Simian virus 40 (SV40) has a circular genome with promoters that direct initiation of transcription in opposite directions from a region near the origin of viral DNA replication. One of these, the early promoter, has been extensively mapped using deletion mutants. A region of the DNA 70-155 base pairs (bp) upstream from the RNA start is necessary and sufficient to direct transcription in a cell-free extract[1]. However, experiments *in vivo* have shown that there is a second regulatory element, the enhancer sequence, that lies immediately upstream from and partially overlaps the early promoter (Refs 2, 3 and P. Berg and M. Fromm, personal communication). The enhancer has the startling property of greatly increasing gene expression *in vivo*, even when this sequence is separated from the promoter and reinserted thousands of base pairs away.

The effects of enhancer sequences can be most clearly demonstrated by measuring the transient expression of mRNA after transfection of structural genes into tissue culture cells. The rabbit β-globin gene directs little β-globin mRNA synthesis when transfected into HeLa cells, despite the presence of globin 5' flanking sequences sufficient to function as a promoter *in vitro*. However, when enhancer sequences from SV40 are inserted into plasmids containing the β-globin transcription unit mRNA production increases 200-fold, and the amount of β-globin synthesized is sufficient to be detected by immunofluorescence[4].

A 366 bp region of the SV40 genome containing the replication origin and the early and late promoters is sufficient to produce the enhancement, and deletion analysis suggests that the important element within this region is a 72 bp direct repeat located approximately midway between the start points for the early and late viral mRNAs. Precise excision of one 72 bp repeat does not destroy enhancer activity, but larger deletions do. The enhancer sequence functions when inserted in either direction at any of several sites upstream or downstream from the β-globin mRNA start point, over distances as great as 3300 bp. These findings, together with the observation that β-globin mRNA produced under enhancer control starts at or near the normal *in vivo* site, argue against the enhancement being simply a consequence of readthrough from viral promoters.

Although the enhancer and the viral DNA replication origin are near each other in the SV40 genome, they are two separate genetic elements. The enhancer can function without an intact origin of replication. However, like the origin, the enhancer is a *cis*-acting element. Little β-globin expression is seen, for example, when β-globin and SV40 sequences on separate plasmids are co-transfected. There is also some evidence for *cis*-acting suppression of the enhancer effect by linked plasmid sequences.

Further insight into the nature of the enhancer comes from a large number of recombinants where the SV40 enhancer sequence was inserted in various ways into plasmids containing the gene for SV40 large T antigen, and expression of T antigen protein was measured by immunofluorescence[5]. In all experiments, expression was strongly dependent on the presence of the enhancer. In most cases, expression was also dependent on the presence of sequences previously thought to be important because they are required for transcription *in vitro* – for example, the 'TATA box' region of the conalbumin or adenovirus major late promoters, or the 5' flanking region of the SV40 early transcription unit.

The mechanism by which the enhancer acts remains completely unknown. The enhancer apparently does not function in cell-free transcription systems. Transcription of SV40 in vitro is not diminished when sequences necessary for enhancer activity are deleted. Moreover, promoters that lack an enhancer, such as the adenovirus major late promoter, give high levels of expression in vitro. The enhancer does not appear to change the number of transfected DNA molecules in the cell. Amplification via DNA replication is not required and there is even some suggestion that replication might diminish the enhancer effect[6]. Enhancement may involve an effect on chromatin structure, an effect on DNA supercoiling, a change in the attachment of the DNA to the nuclear matrix, an increase in the frequency of integration into the cell's chromosomes, or the provision of a bidirectional 'entry site' for RNA polymerases.

Enhancer sequences are not restricted to SV40. A segment of DNA with similar properties is found near the replication origin of polyoma virus[7], in bovine papilloma virus[8] and apparently also in the long terminal repeats of certain retroviruses. It is not yet clear whether enhancers from different viruses are freely interchangeable. The polyoma enhancer functions in cells from several different species, including mouse, rat, mink and humans, but this may reflect the broad host range of polyoma, rather than being a general property of enhancers. There is no striking sequence homology between enhancers from different viral sources.

Are there counterparts to viral enhancers in cellular DNA sequences? Human cells have sequences that are homologous to the SV40 72 bp repeats. Like the repeats, the cellular element is able to increase the frequency with which cells are stably transformed to a new phenotype by exogenous genes, such as the gene for thymidine kinase, but it is not yet known if the cellular sequences increase transcription in a transient assay (S. Conrad and M. Botchan, personal communication).

The idea that there might be cellular genetic elements that are able to activate whole regions of DNA provides an attractive model for co-ordinate expression of clustered multi-gene families during development. However, the instances of enhancement so far observed do not involve normal control of cellular genes but rather viral sequences acting outside the normal control system: for example, the activation of β-globin gene transcription in cells that do not ordinarily express the globin genes. Thus, it remains to be seen what role, if any, enhancers have in the regulation of cellular gene expression.

References

1 Myers, R. M., Rio, D. C., Robbins, A. K. and Tjian, R. (1981) *Cell* 25, 373–384
2 Gruss, P., Dhar, R. and Khoury, G. (1981) *Proc. Natl Acad. Sci. U.S.A.* 78, 943–947
3 Benoist, C. and Chambon, P. (1981) *Nature (London)* 290, 304–310
4 Banerji, J., Rusconi, S. and Schaffner, W. (1981) *Cell* 27, 299–308
5 Moreau, P., Hen, R., Wasylyk, B., Everett, R., Gaub, M. P. and Chambon, P. (1981) *Nucl. Acids. Res.* 9, 6047–6067
6 Treisman, R. and Maniatis, T. (1982) in *Eukaryotic Viral Vectors* Cold Spring Harbor Laboratory (in press)
7 deVilliers, J. and Schaffner, W. (1981) *Nucl. Acids. Res.* 9, 6251–6254
8 Lusky, M., Berg, L. and Botchan, M. (1982) in *Eukaryotic Viral Vectors,* Cold Spring Harbor Laboratory (in press)

W. DYNAN and R. TJIAN

Department of Biochemistry, University of California, Berkeley, CA 94720, U.S.A.

Immunoglobulin RNA processing

Randolph Wall

Recent studies indicate that, in addition to generating messenger RNA, RNA processing plays an important role in controlling the expression of immunoglobulin genes in the development of the immune response.

The immense diversity of antibody specificities and the complex developmental switches in the expression of antibody genes have long intrigued molecular biologists. Molecular cloning has now provided many insights into the structure, multiplicity and origins of diversity in immunoglobulin genes. Parallel studies on immunoglobulin RNA have complemented gene cloning approaches in defining the structure of active immunoglobulin genes (i.e. transcription units) and in resolving the events in messenger RNA (mRNA) processing in eukaryotic cells. This brief review will concentrate on recent studies which indicate that an early developmental transition in the expression of genes for immunoglobulin heavy chains is regulated through RNA processing. These studies provide striking support for the long standing proposal that different types of post-transcriptional processing of RNA may be used to control the expression of eukaryotic genes.

Immunoglobulin mRNA processing in fully differentiated cells

The immunoglobulin protein molecule consists of two identical light chains and two identical heavy chains. Immunoglobulin light chains contain two functional domains: the variable region (V_L) at the NH_2-terminus, which is involved in antibody specificity, and the constant region (C_L) at the COOH-terminus. Heavy chains have a variable region (V_H) which also contributes to antibody specificity, along with either three or four constant region domains (C_H1, C_H2, etc.) homologous to the constant region in the light chains. Coding sequences for variable and constant regions are widely separated in DNA from germline and somatic cells but are rearranged into closer proximity in immunoglobulin-producing cells[1]. However, even in the rearranged active immunoglobulin gene, sequences coding for variable and constant regions (called exons) are separated by non-coding intervening sequences (called introns). The separated variable and constant regions are transcribed directly into large nuclear RNA molecules and then spliced together in a complex series of RNA processing events.

Considerable evidence now indicates that these large nuclear RNA molecules are the precursors to immunoglobulin mRNA; this evidence includes formal proof from pulse-chase experiments in which immunoglobulin mRNA sequences, exclusively in large nuclear RNA molecules, were quantitatively processed into cytoplasmic mRNA after all further transcription was abolished[2-4]. The post-transcriptional processing events which generate immunoglobulin mRNA from large RNA precursors apparently occur in a specific order. One of the earliest events after transcription is the addition of poly(A) at the 3' end of the primary transcript[2-4]. Then the RNA is spliced to remove introns and join the separated exons[2-8]. The fully processed immunoglobulin mRNA molecules appear to be immediatedly transported to the cytoplasm.

Myeloma cells were used in these studies on the processing of immunoglobulin RNA. These tumor cells correspond to fully differentiated antibody-secreting cells, where a substantial fraction (up to 20%) of the total cell protein synthesized is immunoglobulin. The transcriptional and post-transcriptional processes which generate immunoglobulin mRNA in these fully differentiated cells are extremely efficient[3,4]. Reasonable estimates for the transcriptional activity of immunoglobulin genes in myeloma cells approximate to those for ribosomal RNA genes in animal cells[3]. Virtually all of the immunoglobulin mRNA sequences in large RNA precursors are normally processed into cytoplasmic mRNA. However, defects in RNA splicing in variant myeloma cells can cause the loss of immunoglobulin production or can generate variant immunoglobulin chains with internally deleted domains[8]. The high rate of transcription of immunoglobulin genes together with efficient RNA processing and relatively long lifetimes of cytoplasmic mRNA all combine to produce 20–50,000 immunoglobulin mRNA molecules per cell generation in these cells[3,4]. While these features proved to be extremely useful for studying the molecular events in the production of immunoglobulin mRNA, fully differentiated myeloma cells are not promising systems for examing the role of RNA processing in controlling the expression of immunoglobulin genes.

Immunoglobulin RNA processing in B lymphocyte development

Antibody-producing cells undergo transitions or switches in the expression of immunoglobulin genes during the development of the immune response; this makes them promising candidates for studying regulatory mechanisms. The earliest cell type carrying complete immunoglobulin molecules is called the B lymphocyte, and the first class of immunoglobulin on these cells is monomeric membrane IgM (composed of μ heavy chains and light chains). The membrane IgM molecules apparently function as receptors for antigen on the outer membranes of lymphocytes. When these receptors encounter the appropriate antigen, the interaction stimulates the B cell to divide and produce clonal

μ MEMBRANE MRNA:
C_H4 ——— C-TERMINUS$_M$ (41 AA) ——— 3'-UT$_M$ (270 NT)
555   560   595
SerThrGluGlyGluValAsn   PheLysValLysStop
5'———TCCACTGAGGGGGAGGTGAAT———TTCAAGGTGAAATGA———AATAAA———TGA$_{200}$ 3'

μ SECRETED MRNA:
C_H4 ——— C-TERMINUS$_S$ (20 AA) ——— 3'-UT$_S$ (128 NT)
555   560
SerThrGlyLysProThrLeu   CysTyrStop
5'———TCCACTGGTAAACCCACACTG———TGCTATTGA———AATAAA———TGA$_{200}$ 3'

EXON ——— INTRON
UPSTREAM CONSENSUS SPLICE SITE   5'———AGGTAAGTA———3'

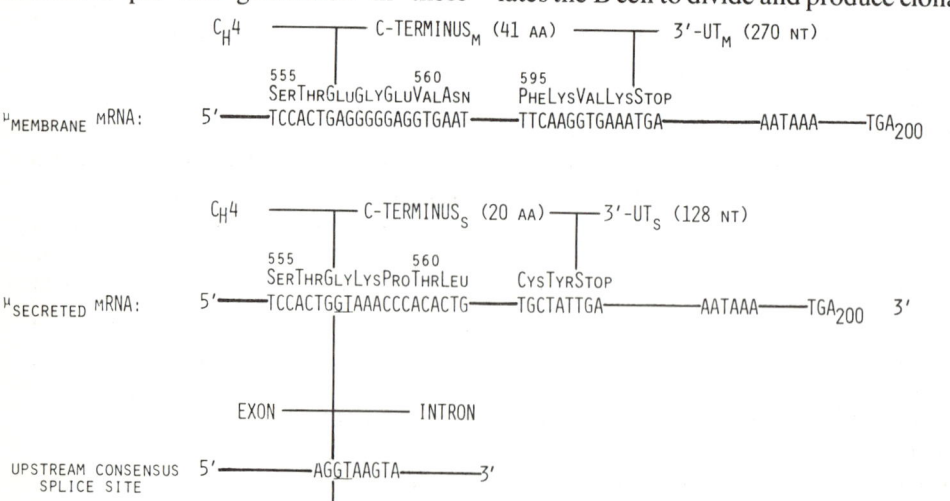

Fig. 1. Selected sequences of membrane and secreted μ mRNAs showing the divergence of these species at the end of the μ CH4 domain. Otherwise, the sequences of both membrane and secreted μ mRNAs encoding the V_H region and $CH1 \to CH4$ regions are identical. The μ_S mRNA sequence at codons 557–558 where M exon splicing occurs closely resembles the consensus upstream splicing site[16].

descendants which secrete pentameric IgM. IgM-secreting cells then undergo switching to produce other classes of immunoglobulins with different heavy chains (IgG, IgA) which retain the antibody specificity of the variable region of the original receptor on the IgM-bearing B lymphocyte.

Considerable indirect evidence from studies of protein structure and of the translation of mRNAs suggested that the μ heavy chains in membrane-bound and secreted IgM were probably identical except for their COOH-terminal sequences[9-12]. Several laboratories demonstrated the presence of multiple μ mRNA species, with two prominent μ mRNA species of 2.7 and 2.4 kilobases (kb) from a variety of B lymphoma tumor cells representative of early stages in B lymphocyte development[10-13]. Because the 2.4 kb μ mRNA was the only species in cells that were secreting pentameric IgM, it was proposed that this species codes for μ_s heavy chains[11,12]. The 2.7 kb μ species was always found in cells making membrane IgM and accordingly was presumed to code for μ_m heavy chains.

Nucleotide sequencing of μ cDNA clones established that the 2.4 kb species is the μ_s mRNA and that the 2.7 kb species is the μ_m mRNA[12,14]. The 2.7 kb and the 2.4 kb μ mRNA are identical up to the end of the μ C$_H$4 domain; thereafter they contain different C-terminal coding regions and 3'-untranslated regions (Fig. 1)[11,12]. The μ_s chain has a hydrophilic C-terminal segment of 20 residues[9,12] which is required for the secretion of μ_s, whereas the μ_m chain has a C-terminal segment of 41 residues which has hydrophobic properties consistent with its being a transmembrane peptide[12].

The origins of the sequences comprising the C-termini of μ_m and μ_s mRNAs were established by mapping μ gene clones which showed that the μ_s C-terminal sequence is continuously encoded with the μ C$_H$4 domain and the μ_m C-terminal sequence is encoded in two separate exons (the M exons) approximately 1.8 kb from the end of the μ_s C-terminal sequence in DNA[14]. The alternate forms of μ mRNA do not seem to be generated by DNA rearrangements because no rearrangement of the M exons of the Cμ genes in μ_s-producing myeloma cells can be detected[14,15]. Instead, the μ_m and μ_s mRNAs appear to be generated from a single μ heavy chain transcription unit by alternate patterns of RNA splicing[12,14]. There is an RNA splicing site at codons 557–558 at the 3' end of the μ C$_H$4 domain, where the μ_m and μ_s cDNA sequences diverge (Fig. 1). It clearly resembles the consensus sequence for an upstream RNA splicing site, AG/GTAAGTA, where the stroke divides the exon from the intron[16]. The μ_m mRNA is generated by RNA splicing between this site and the first M exon, and also between the two M exons (Fig. 2). While μ_m is generated by two additional splices relative to μ_s mRNA, we have proposed[12,14] that the developmentally regulated addition of poly(A), either at the μ_s site or at the second M exon site, determines which μ mRNA will be produced (Fig. 2). This proposal is based on the widespread finding that the addition of poly(A) to the 3' ends of nuclear RNA precursors is a rapid process which precedes RNA splicing[12,14]. The molecular mechanisms that determine the choice of poly(A) sites remain to be resolved. Our current hypothesis for the switch from μ_m to μ_s synthesis is that a factor is induced in an antigen-stimulated B lymphocyte which enforces poly(A) addition at the first poly(A) site, leading to the production of μ_s mRNA only.

Closing remarks

These studies establish that RNA processing, in addition to generating mRNA, plays an important role in controlling a key early transition in IgM expression. Differ-

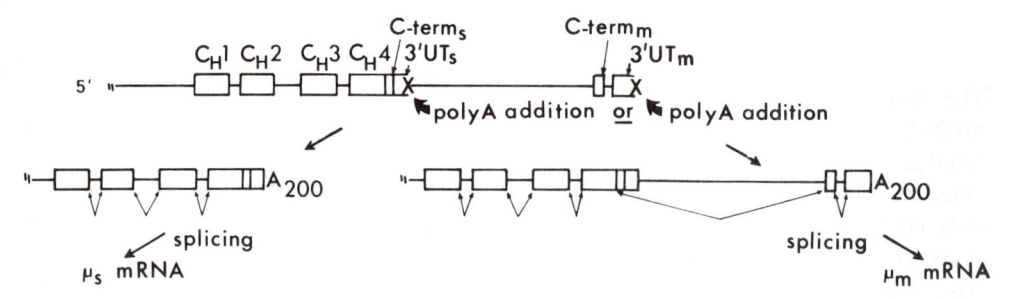

Fig. 2. A model showing the production of either μ_m or μ_s mRNA from a complex transcription unit through the choice of alternate poly(A) addition sites. For simplicity, only the C_H regions and M exons are shown in the μ primary transcript. Alternate poly(A) sites in the μ primary transcript are marked by X. The RNA splicing patterns generating μ_s and μ_m mRNAs are shown below the μ primary transcript.

ent types of post-transcriptional processing of RNA may be used to generate membrane or secreted forms for the other immunoglobulin heavy chain classes. In this regard, it is striking that all immunoglobulin heavy chains now sequenced at either the protein or DNA level have glycine–lysine residues or a GGTAAA nucleotide sequence identical to that at the splicing site at the end of the μ C_H4 domain[12]. RNA processing mechanisms may also be involved in simultaneous production of μ and δ heavy chains in membrane IgM and IgD in B lymphocyte development[12,14]. Experiments are now in progress to test these predictions. Certainly, these examples of the expression of immunoglobulin genes raise the possibility that RNA processing may be involved in the regulation of other eukaryotic genes.

Acknowledgements

My special thanks to Maureen Gilmore-Hebert, Edmund Choi and John Rogers for their outstanding contributions to this Laboratory's studies on RNA processing. Our research program is supported by NIH Grant AIT 3410, CA 12800 and NSF Grant PCM 7924876.

References

1 Flavell, R. A., Glover, D. M. and Jeffreys, A. J. (1978) *Trends Biochem. Sci.* 3, 241–244
2 Gilmore-Hebert, M. and Wall, R. (1978) *Proc. Natl. Acad. Sci. U.S.A.* 75, 342–345
3 Schibler, U., Marcu, K. B. and Perry, R. P. (1978) *Cell* 15, 1495–1509
4 Gilmore-Hebert, M. and Wall, R. (1979) *J. Mol. Biol.* 135, 879–891
5 Rabbitts, T. H. (1978) *Nature* 275, 291–296
6 Gilmore-Hebert, M., Hercules, K., Komaromy, M. and Wall, R. (1978) *Proc. Natl. Acad. Sci. U.S.A.* 75, 6044–6048
7 Perry, R. P., Kelley, D. E. and Schibler, U. (1979) *Proc. Natl. Acad. Sci. U.S.A.* 76, 3678–3682
8 Choi, E., Kuehl, M. and Wall, R. (1980) *Nature* (in press)
9 Kehry, M., Schibley, C., Fuhrman, J., Schilling, J. and Hood, L. E. (1979) *Proc. Natl. Acad. Sci. U.S.A.* 76, 2932–2936
10 Singer, P. A., Singer, H. H. and Williamson, A. R. (1980) *Nature* 285, 294–299
11 Alt, F. W., Bothwell, A. L. M., Knapp, M., Mather, E., Koshland, M. and Baltimore, D. (1980) *Cell* 20, 293–301
12 Rogers, J., Early, P., Carter, C., Calame, K., Bond, M., Hood, L. and Wall, R. (1980) *Cell* 20, 303–312
13 Perry, R. P. and Kelley, D. E. (1979) *Cell* 18, 1333–1339
14 Early, P., Rogers, J., Davis, M., Calame, L., Bond, M., Wall, R. and Hood, L. (1980) *Cell* 20, 313–319
15 Calame, K., Rogers, J., Early, P., Davis, M., Livant, D., Wall, R. and Hood, L. (1980) *Nature* 284, 452–455
16 Rogers, J. and Wall, R. (1980) *Proc. Natl. Acad. Sci. U.S.A.* 77, 1877–1879

Randolph Wall is at the Molecular Biology Institute and Department of Microbiology and Immunology, UCLA School of Medicine, Los Angeles, CA 90024, U.S.A.

Addendum – R. Wall

The discovery of the two different μ mRNAs coding for membrane and secreted μ immunoglobulin heavy chains provided the first instance of chromosomal gene control by post-transcriptional RNA processing mechanisms. We predicted, and have now confirmed, that all immunoglobulin heavy chain genes have a similar gene organization and use alternative RNA processing pathways for the production of membrane and secreted mRNA species. This complex gene organization and RNA processing model has also been expanded to account for the co-expression of μ and δ heavy chains in IgM and IgD on the surface of B lymphocytes. These findings establish that RNA processing mechanisms play a central role in regulating immunoglobulin gene expression in B cell differentiation.

Two recent studies suggest that post-transcriptional control mechanisms are likely to be a widespread feature in eukaryotic gene expression. The chromosomal genes for yeast invertase and rat calcitonin also use alternative RNA processing pathways to generate functionally distinct forms of the yeast enzyme and tissue specific forms of the mammalian hormone, respectively. Clearly, RNA processing mechanisms extend the versatility of gene expression and provide a new dimension in the regulation of genes in eukaryotic cells.

The mechanism of tRNA splicing

Richard C. Ogden, Gayle Knapp, Craig L. Peebles,
Jerry Johnson and John Abelson

Transfer RNA precursors that contain intervening sequences have been isolated from a yeast mutant. These precursors are substrates in vitro for activities in yeast which excise the intervening sequence and ligate the intermediate tRNA halves. The characteristics of this splicing reaction are discussed.

The phenomenon of non-colinearity between a eukaryotic gene and its mature product has been shown to be general. This finding poses a new question regarding the pathway of expression of these genes: how are the noncontiguous sequences encoding the mature RNA brought together to form the functional molecule? Accumulating evidence indicates that transcription of such genes gives rise to RNA precursors containing both expressed and intervening sequences. The intervening sequences are subsequently removed in a process called splicing[1]. The discovery that precursors to certain transfer RNAs accumulate in a temperature-sensitive mutant of the yeast *Saccharomyces cerevisiae*[2] and that these precursors contain intervening sequences[3] has considerably facilitated the study of the tRNA splicing reaction in the context of yeast tRNA biosynthesis.

tRNA splicing in yeast

The tRNA precursors accumulating in the yeast mutant have been isolated and in some cases the primary sequences are known. Without exception, they are matured at the 5' and 3' termini and thus are intermediates in the biosynthetic pathway from primary transcript to mature tRNA. In addition, several of the modified bases observed in the mature tRNA are found in the precursors. The modified bases of the dihydro-U and TΨC loops are usually present in high partial molar yields.

On the other hand, modifications of bases in the anticodon loop are generally not observed in the precursors. For example, 6-isopentenyl adenosine in tRNA^{Tyr}, the Y base in tRNA^{Phe} and 2'-O-methyl modifications are not found. The precursors are all substrates for an activity found in yeast which converts them to mature-sized tRNA.

In the precursors of different tRNAs the intervening sequences have been shown to be located at the same position with respect to the mature tRNA, but no sequence homologies have been noticed and the sizes of the intervening sequences vary from 14 to 60 nucleotides. Fig. 1 shows the nucleotide sequences of the five identified pre-tRNAs. In addition, the similarities observed among these sequences are summarized in this figure. The sequences of pre-tRNA^{Tyr} and pre-tRNA^{Phe} are drawn in secondary structures which have maximized base pairing as determined by the calculation of the most favorable free energy. The structures of pre-tRNA^{Tyr} and pre-tRNA$^{Ser}_{UCG}$ are consistent with studies of the nuclease sensitivity of the precursors[4,5]. In each case, the intervening sequence is located at the 3' side of the base immediately adjacent to the anticodon. The acceptor stem and dihydro-U and TΨC stems and loops appear as they are found in the tRNA cloverleaf structure. The anticodon stem is intact, but is augmented by a second helical region. This second helix always includes the anticodon which base

See p. 190 for legend.

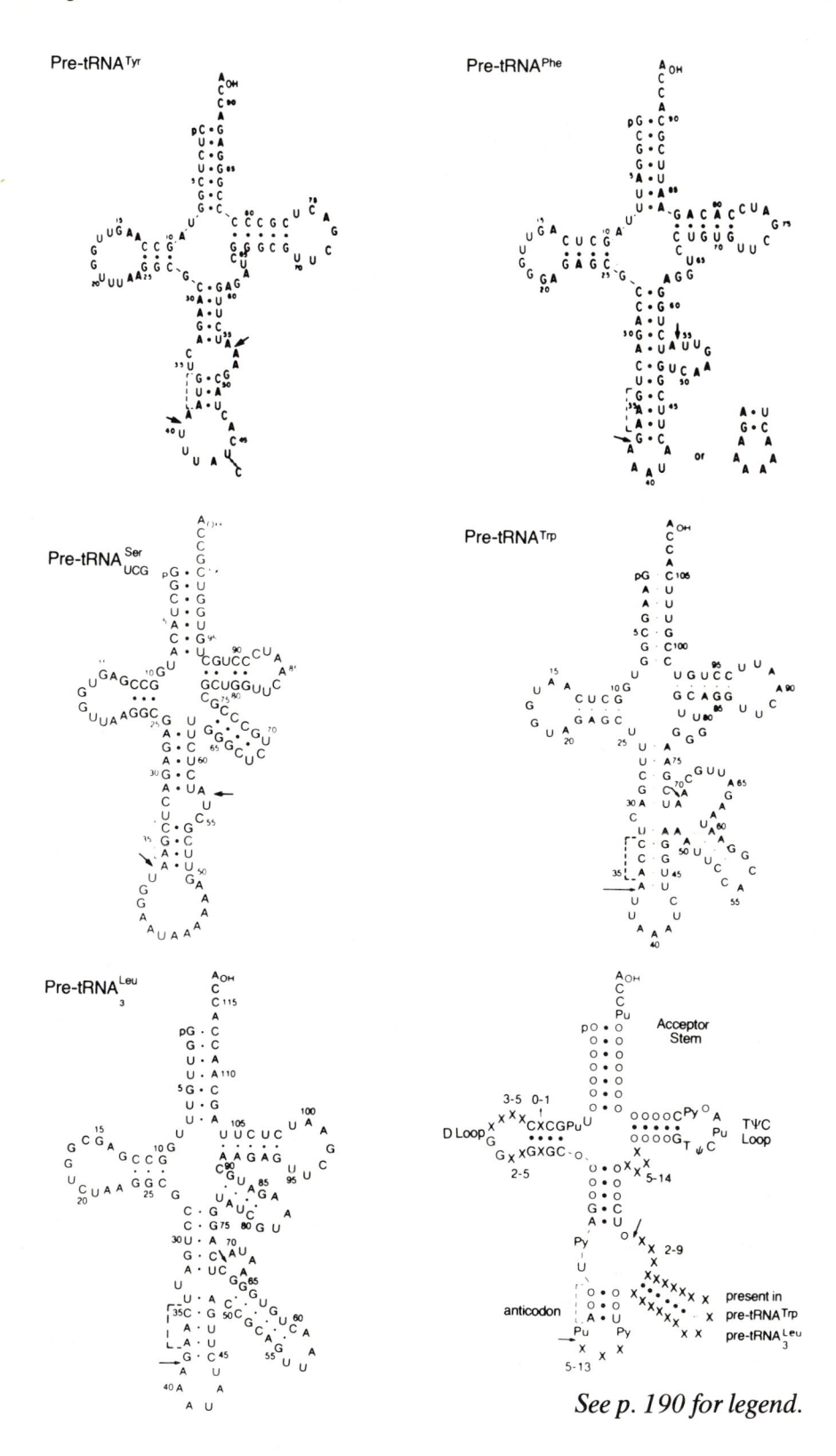

pairs with a complementary region in the intervening sequence. In two of the precursors, pre-tRNA^{Trp} and pre-tRNA$_3^{Leu}$ which contain large intervening sequences, a majority of the 'additional' nucleotides can be contained in a hairpin loop structure which has a substantial base-paired stem. Unifying features of these structures which may be important in recognition of the precursor substrate by the RNA splicing system will be discussed below.

Two simplified models for the mechanism of tRNA splicing can be considered. In the first, an exchange of phosphodiester linkages between the mature tRNA and intervening sequences could generate the mature tRNA sequence and excise the intervening sequence as a circular RNA molecule. This postulated one-step 'recombination' reaction would ensure the joining of cognate portions of the tRNA. Furthermore, no free termini would be exposed to nuclease or phosphatase inactivation. An alternative model would propose a two-stage process: endonucleolytic excision of the IVS as a linear molecule followed by ligation of the resultant intermediates. Such a pathway might involve a single multifunctional enzyme or two or more separable activities which could have different requirements.

Accumulating evidence suggests that the *in vitro* splicing of precursor tRNAs, as observed in a crude yeast extract, occurs through two separate enzymatic steps[6]. First, half-sized tRNA molecules are observed as kinetic intermediates in the tRNA splicing reaction. Second, the same intermediates accumulate in the *in vitro* reaction under two different conditions: in the absence of ATP and in the presence of mature tRNA. Third, these intermediates can be isolated and are substrates for a ligase activity in the same extract which in the presence of ATP produces the mature tRNA sequence. Thus, the splicing reaction can be divided into two reactions: an endonucleolytic step in which the precursor is cleaved twice, in an ATP-independent reaction, to yield the intervening sequence and half-tRNA molecules and a ligation step which requires ATP and produces the mature-sized tRNA product. However, it is important to remember that these two reactions may actually occur via a concerted mechanism when they are not experimentally uncoupled *in vitro*.

Characterization of the half-tRNA molecules and intervening sequences has proved to be equally important to an understanding of the mechanism of the splicing reaction. There are several possible ways in which the excision of the intervening sequence could occur. The structure of the excised intervening sequence would reflect the way in which it had been removed. These possible structures are enumerated for pre-tRNA^{Tyr} in Fig. 2. First, the reaction might proceed via a concerted mechanism so that the intervening sequence is removed as a circle. This could happen if the splicing occurred via concerted reaction with a reciprocal exchange at the two points of cleavage. Alternatively, the product could be a unique linear molecule. There are two possible classes of linear products: those with 5'-phosphate and 3'-hydroxyl termini and those with

Fig. 1. (p. 189) Nucleotide sequences of five yeast tRNA precursors. The nucleotide sequences of the precursors have all been arranged in secondary structures similar to those derived for pre-tRNA^{Tyr} and pre-tRNA^{Phe}. In the cases of pre-tRNA^{Trp} and pre-tRNA$_3^{Leu}$ additional hairpin helices in the intervening sequence contribute to the favorable free energy of the secondary structures. The arrows indicate the splice points. A composite secondary structure is also shown which summarizes the constant and variable positions of nucleotides. Variable nucleotides in the mature portion of the precursor are denoted by (○). Variable positions in the intervening sequence are denoted by (—X—).

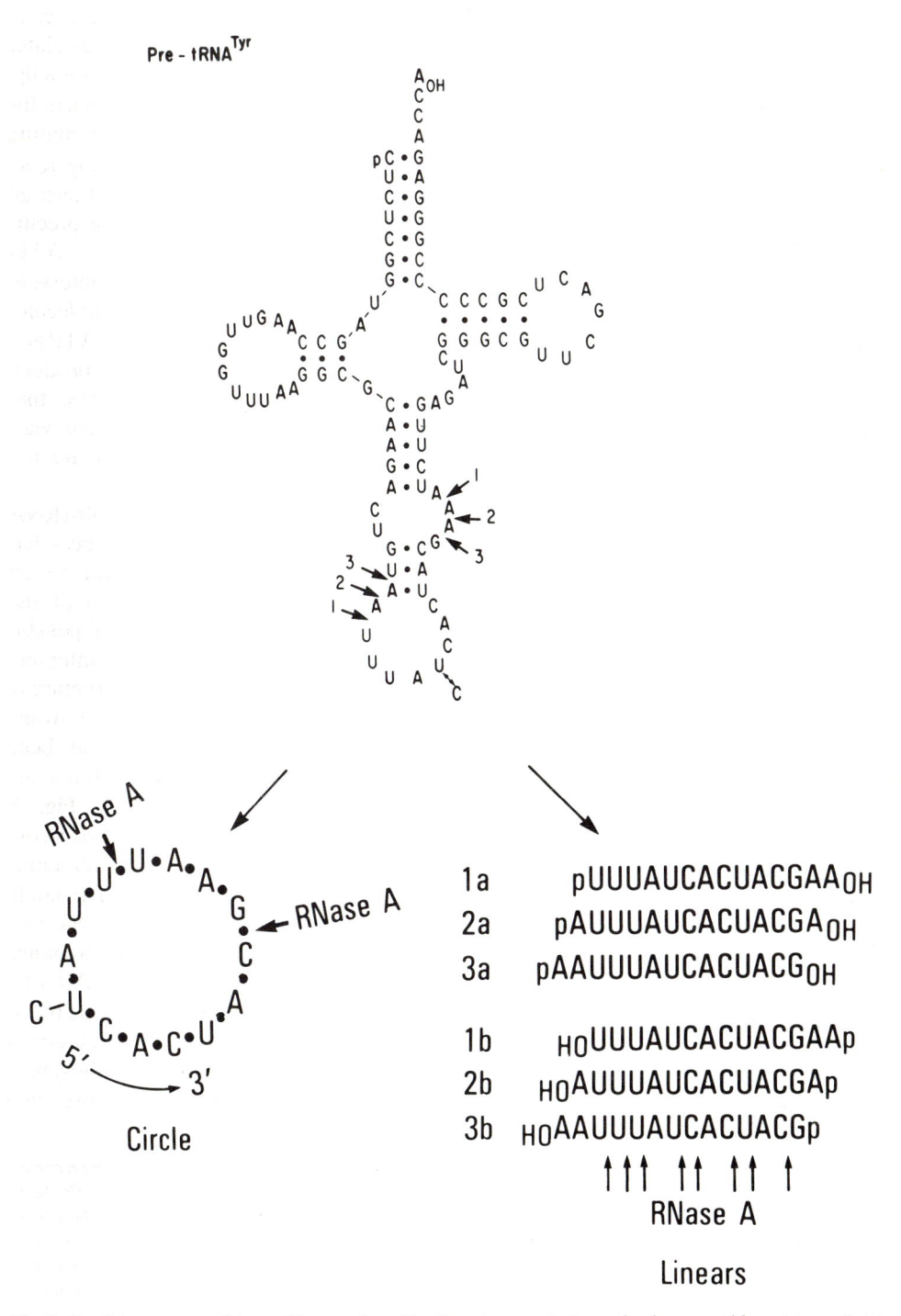

Fig. 2. Possible structure of the excision product. Numbered arrows indicate the three possible excision points to which the linear sequence permutations correspond. Unnumbered arrows indicate RNase A cleavage sites.

5'-hydroxyl and 3'-phosphate termini. In both of these classes there are three possible sequence permutations in the case of pre-tRNA[Tyr] because of the repeated sequence at the cleavage sites. Another possibility is that a mixture of all three linear sequences would be produced in the splicing reaction.

Digestion of the intervening sequence from pre-tRNA[Tyr] with RNase A can distinguish among the various possibilities shown in Fig. 2. The results of the analysis indicate that, in the case of pre-tRNA[Tyr] and all other precursors, the intervening sequence is excised as an unique linear molecule with 5'-hydroxyl and 3'-phosphate termini[7].

If the cleavage reaction is a simple scission of two phosphodiester bonds, the previous data imply that the half-tRNA molecules produced in the ATP-independent reaction should also have 3'-phosphate and 5'-hydroxyl termini. Fingerprint analysis of half-tRNA molecules from the five precursors whose sequences are given in Fig. 1 has shown this to be the case[7].

Mechanism of the tRNA splicing reaction

These results have strong implications concerning the mechanism of precursor cleavage and subsequent ligation. Several features of one mechanism that may be constructed are novel when compared with other processing enzymes. First, in all endonucleolytic biosynthetic ribonucleases previously described, the scission of the phosphodiester chain produces 5'-phosphate and 3'-hydroxyl ends. This of course, is not a chemical necessity as there are many endonucleases (e.g., RNase A and RNase T1) which produce 3'-phosphate termini. However, these enzymes are generally considered to be degradative enzymes. The position of the terminal phosphate left by the splicing endonuclease is also surprising with respect to the ligation step which must then occur. To splice the ends the ligase must join a 3'-phosphate to a 5'-hydroxyl. This requirement is not a feature of the T4 RNA ligase or of the known DNA ligases which specifically join 5'-phosphate to 3'-hydroxyl termini. We have shown that the 3'-phosphate is required for the ligase reaction[7] and that 5'-half-tRNA molecules from which the 3'-phosphate has been selectively removed are no longer substrates in the ATP-dependent joining reaction.

It is interesting to speculate whether or not yeast pre-tRNA splicing might share common features with the splicing of mRNA precursors in other eukaryotes. It is probable that with regard to the initial recognition and excision of the intervening sequence the pre-tRNA splicing system will be unique (see discussion below). On the other hand the ligation step of pre-tRNA splicing may be a common feature of all splicing reactions. If the 3'-phosphate requirement proves to be a feature of splicing reactions in general it may provide a useful means of identification and purification of this class of enzyme.

How do the pre-tRNA splicing enzymes recognize the precursor, cleave it correctly to release the intervening sequence, and precisely ligate the half-tRNA molecules? This is perhaps a premature question since we do not know that there is only one enzyme for each activity. However, preliminary enzyme purification has not revealed activities which distinguish among the nine precursors. Furthermore, mature tRNA selectively inhibits the ligation reaction[6], but the inhibition is not tRNA species specific. For example, pure yeast tRNA[Phe] inhibits equally the processing of pre-tRNA[Phe] and pre-tRNA[Tyr]. Because of this and because RNA processing enzymes in general tend to have multiple roles (cf. RNase P, RNase III) we shall assume for the purposes of discussion that a single enzyme system recognizes all of the tRNA precursors shown in Fig. 1.

It is now clear that the recognition cannot be sequence specific as the intervening sequences themselves are quite different among the tRNA species. However, there are, some interesting structural similarities which may be of paramount importance in recognition by the splicing endonuclease. The position of the intervening sequence, as determined by analysis of the excised intervening sequences, is the same in all five precursors – immediately adjacent to the nucleotide that usually is hypermodified in the mature tRNA. This means that the cleavage sites must occur in all precursors in exactly the same position relative to the mature portion of the molecule. It may also be significant that in all five cases the anticodon can always form base pairs with a complementary region in the intervening sequence. In some cases adjoining nucleotides are also base paired so that the anticodon helix may be extended to the helix of the anticodon stem.

tRNA splicing in other eukaryotes

The biosynthesis of tRNA in other eukaryotes is known to involve splicing. tRNA genes from a variety of eukaryotes including *Xenopus*[8] have been shown to contain intervening sequences. In the case of *Xenopus laevis*, transcripts of *Saccharomyces cerevisiae* tRNA genes have been used to investigate the splicing reaction. It has been established that nuclear extracts of *Xenopus laevis* oocytes can support RNA polymerase III directed transcription of yeast tRNA genes which contain intervening sequences[9]. Yeast tRNA precursors are also produced when tRNA genes are injected into the nucleus of intact oocytes[10] and in both cases, the mature-sized tRNAs, arising from recognition of the yeast precursor transcripts by a *Xenopus* splicing system, are found. The features of this heterologous splicing reaction have been investigated in a series of elegant experiments employing the *in vivo* oocyte injection technology[11]. Furthermore, it has been demonstrated in this heterologous system that the removal of the intervening sequence of a yeast tRNA^{Tyr} precursor occurs at a defined time (and cellular location) in the maturation process. Splicing occurs subsequent to 5' and 3' terminal maturation and certain base modifications and prior to transportation of the precursor from the oocyte nucleus to the cytoplasm.

In addition to purification of the activities involved in splicing, we are attempting to probe the mechanism of the splicing reaction by the use of substrate analogs. As well as investigating the role of the phosphate and hydroxyl termini in the ligation step[7], we have constructed altered yeast tRNA genes and investigated the ability of *in vitro* transcripts of these mutant genes to participate in the heterologous (*Xenopus*) splicing reaction and in the yeast splicing system. In the initial experiments[12] we have successfully transcribed a yeast tRNA$_3^{Leu}$ gene containing (in addition to the intervening sequence of 32 bp) an insert of 21 bp in the middle of the intervening sequence. The mature-ended *in vitro* transcript is a substrate for the activities in *Xenopus* and in yeast responsible for excision of the intervening sequence and ligation of the halves. The reaction appears to proceed analogously to the splicing of *in vitro* substrates. It is interesting that the inserted DNA can extend a putative base-paired region in the central portion of the intervening sequence (Fig. 3) and causes no perturbations at the sites of excision. A study of the structurally significant regions of the precursor with respect to recognition and cutting by the endonuclease awaits purified enzyme and a refined approach to defined mutagenesis.

Conclusion

The splicing of yeast tRNA precursors

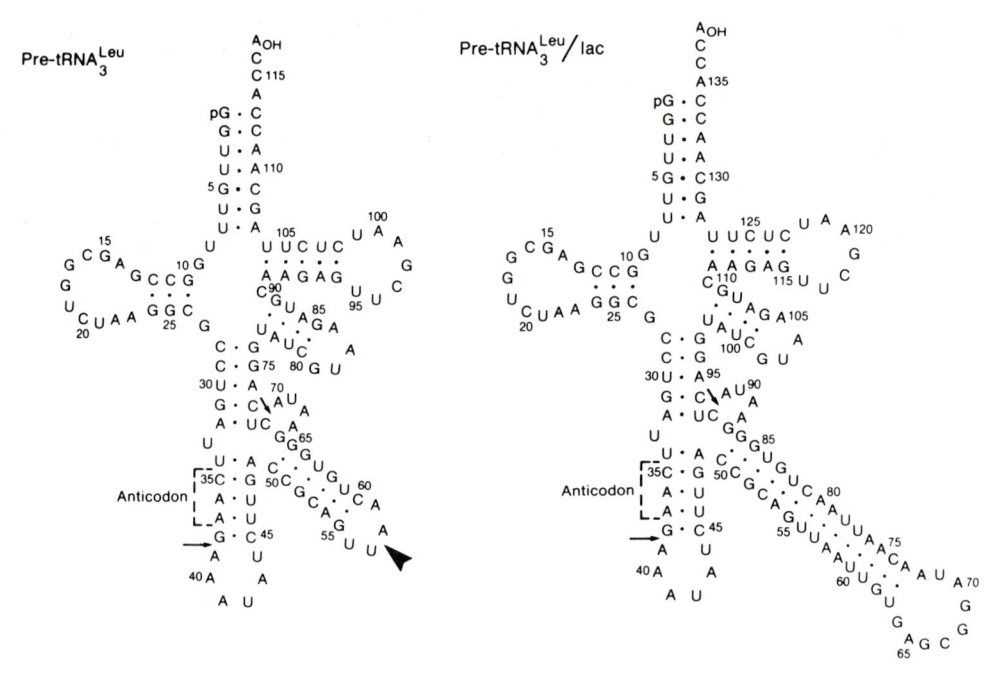

Fig. 3. Effect of lacDNA insertion on the structure of pre-tRNA₃^Leu. Pre-tRNA₃^Leu: Proposed secondary structure of the in vivo *precursor from the yeast mutant. Small arrows denote splice points and the large arrowhead indicates the HpaI site. Pre-tRNA₃^Leu/lac: Proposed structure of the mature* in vitro *transcript of the mutant tRNA₃^Leu gene.*

has been operationally separated into two distinct steps: an endonucleoytic cleavage of the pre-tRNA which produces the intervening sequence and two half-tRNA sized molecules and a ligation step in which the halves are joined to produce the mature-sized tRNA.

This separation of activities has allowed us to characterize the products and intermediates of the splicing reaction and thereby to gain an insight into the mechanism of splicing. The cleavage of the precursor occurs by a simple scission of two phosphodiester bonds leaving 5'-hydroxyl and 3'-phosphate termini on the intervening sequence and the halves. The 3'-phosphate is required for the subsequent ligation of the 5'-half to its cognate 3'-half. The probable participation of this 3'-phosphate may be a common feature of all eukaryotic splicing reactions.

In addition, the characterization of the products and intermediates of the splicing reaction and comparison of the known pre-tRNA sequences have shown common structural features among the precursors which may have significance for recognition by the activities involved.

References

1 Abelson, J. (1979) *Ann. Rev. Biochem.* 48, 1035–1069
2 Hopper, A. K., Banks, F. and Evangelidis, V. (1978) *Cell*, 14, 211–219
3 Knapp, G., Beckmann, J. S., Johnson, P. F., Fuhrman, S. A. and Abelson, J. (1978) *Cell*, 14, 221–236
4 O'Farrell, P. Z., Cordell, B., Valenzuela, P., Rutter, W. J. and Goodman, H. M. (1978) *Nature (London)*, 274, 438–445
5 Etcheverry, T., Colby, D. and Guthrie, C. (1979) *Cell*, 18, 11–26
6 Peebles, C. L., Ogden, R. C., Knapp, G. and Abelson, J. (1979) *Cell*, 18, 27–35
7 Knapp, G., Ogden, R. C., Peebles, C. L. and Abelson, J. (1979) *Cell*, 18, 37–45

8 Muller, F. and Clarkson, S. G. (1980) *Cell*, 19, 345-353
9 Ogden, R. C., Beckmann, J. S., Abelson, J., Kang, H. S., Söll, D and Schmidt, O. (1979) *Cell*, 17, 399-406
10 De Robertis, E. M. and Olson, M. V. (1979) *Nature (London)*, 278, 137-143
11 Melton, D. A., De Robertis, E. M. and Cortese, R. (1980) *Nature (London)*, 284, 143-148
12 Johnson, J. D., Ogden, R. C., Johnson, P. F., Abelson, J., Dembeck, P. Itakura, K. (1980) *Proc. Natl Acad. Sci. U.S.A.* 77, 2564-2568

Richard Ogden, Gayle Knapp, Graig Peebles and John Abelson are at the Department of Chemistry, B-O17, University of California San Diego, La Jolla, CA 92093, U.S.A.; and Jerry Johnson is at the Department of Biochemistry, University of Wyoming, Laramie, WY 82070, U.S.A.

The 5' ends of influenza viral messenger RNAs are donated by capped cellular RNAs

Robert M. Krug, Michele Bouloy and Stephen J. Plotch

Eukaryotic messenger RNAs and other RNAs containing a fully methylated cap structure (m^7GpppX^m) stimulate the transcription of influenza viral RNA in vitro. RNA fragments containing the cap and the ensuing 10–14 nucleotides are cleaved from the mRNA and are used as primers to initiate viral RNA transcription. In the infected cell a similar priming by capped host cell RNAs is found.

Background

Influenza virus contains eight RNA segments, ranging from about 2500 to 900 nucleotides in length (reviewed in Ref. 1). The seven larger segments each code for a different structural protein of the virus (or virion), while the smallest segment codes for two different non-structural proteins via overlapping nucleotide sequences[1-3]. The genome is 'negative-stranded', the viral messenger RNA (mRNA) being complementary to the virion RNA (vRNA) and the virion containing the enzyme system which transcribes the vRNA into mRNA. This transcriptase complex is associated with the internal ribonucleoproteins of the virus, which contain four of the virion structural proteins[4]. Assays for transcription can be carried out using either detergent-treated virions or purified viral ribonucleoproteins.

One feature which distinguishes influenza virus from other non-oncogenic RNA viruses is that the host nuclear DNA-dependent RNA polymerase II must function if the virus is to replicate. Actinomycin D and α-amanitin (a specific inhibitor of RNA polymerase II) inhibit virus replication[5,6]. In mutant cells containing an α-amanitin-resistant RNA polymerase II, the replication of the virus is also resistant to this drug[7,8] showing that the host RNA polymerase II is required for virus replication. This host function is required for all viral RNA transcription. When added at the beginning of infection, actinomycin D or α-amanitin inhibits primary transcription but, neither inhibit viral RNA transcription *in vitro* catalysed by the virion-associated transcriptase[9,10]. Thus, in an apparent paradox, these drugs inhibit the functioning of the virion transcriptase when it is introduced into the cell to carry out primary transcription, but not when it is assayed *in vitro*.

We have now resolved this apparent paradox and can explain the need for the host cell RNA polymerase II. In fact, we had proposed the correct explanation several years ago, from two sets of data[16-18]. First, viral RNA transcription *in vitro*, catalysed by the virion-associated transcriptase is greatly enhanced by the addition of a primer dinucleotide, ApG or GpG[11,12,14]. We therefore proposed that the transcription of viral RNA *in vivo* also requires a primer and that this primer is an RNA synthesized by RNA polymerase II[11,12]. This explains why α-amanitin inhibits viral RNA transcription *in vivo* but not *in vitro*[9]. Second, because we did not detect capping and methylating enzymes associated with purified virions[13], we proposed that the 5' terminal methylated cap structure found on viral mRNAs *in vivo*[15] is derived from these primer RNAs[13].

Capped RNAs serve as primers for influenza viral RNA transcription *in vitro*

We first identified primer RNAs in rabbit reticulocyte extracts, where they were shown to be globin mRNAs[16]. Pure β-globin mRNA stimulated viral RNA transcription about 80-fold and, on a molar basis, was about 1000 times more effective as a primer than ApG. Other capped eukaryotic mRNAs were also found to be extremely effective as primers[16-18]. The viral RNA transcripts primed by these eukaryotic mRNAs were effectively translated into virus-specific proteins in cell-free systems[16].

To show that the cap of the primer mRNA was transferred to the viral mRNA during transcription *in vitro*, we used as a primer globin mRNA containing ^{32}P in the cap (prepared by enzymatically recapping β-eliminated globin mRNA in the presence of [α-^{32}P]GTP)[17]. After transcription in the presence of unlabeled nucleoside triphosphates, the resulting viral mRNA segments were shown to contain ^{32}P-labeled cap structures. To determine whether sequences in addition to the cap were transferred, we compared the size of the viral mRNA segments primed by globin mRNA with that of the segments primed by ApG[17]. The latter segments initiate exactly at the 3' end of the virion RNA (vRNA) templates[19,20]. Gel electrophoretic analysis indicated that the segments primed by globin mRNA were 10–15 nucleotides larger than those primed by ApG[17]. The segments primed by other mRNAs were also about 10–15 nucleotides larger[18]. Thus, about 10–15 nucleotides together with the cap, were transferred from globin and other mRNA primers to the viral mRNA.

These studies did not identify the bases which were transferred from the primer mRNA. We therefore labeled globin mRNA *in vitro* with ^{125}I to high specific activity, thereby labeling the C residues, and used this globin mRNA as a primer in the presence of unlabeled nucleoside triphosphates[21]. The ^{125}I-labeled region transferred to each of the viral mRNA segments had the same T1 ribonuclease and pancreatic ribonuclease fingerprints, indicating that the same region of the globin mRNA primer was transferred to each of the segments. The predominant sequence transferred, comprising 75% of the total, was identical to the first 13 nucleotides (plus the cap) at the 5' terminus of β-globin mRNA, which has the sequence shown in Fig. 1[22]:

Because only the C residues were labeled with ^{125}I, we could not conclude with certainty that all 13 5' terminal nucleotides were transferred from β-globin mRNA. The results, however, indicated that the β-globin mRNA donated at least the first eight, but no more than the first 14, 5' terminal bases to the viral mRNAs. It was almost certain that the ...UUUU... sequence (residues 9–12) originated from β-globin mRNA since there are no A residues in the 12-base 3' terminal sequence of the vRNA templates (3' UCGUUUUCGUCC...; Ref. 19,20). Thus, the predominant sequence transferred from β-globin mRNA includes the first 12, 13 or 14 5' terminal nucleotides.

Minor ^{125}I-labeled oligonucleotides resistant to RNase T1 were also found in the viral mRNAs. Fourteen per cent of the

m⁷Gpppm⁶AmpC(m)pApCpUpGpCpUpUpUpGpApCp...
1   8   13 15

Fig. 1.

RNase T1 products were the sequences CUUUGp, CUUGp, CUGp and CGp, which can be presumed to derive their C- and U-residues from β-globin mRNA (residues 8–11) and which must obtain their G-residue by transcription. Similarly, the m⁷Gpppm⁶AmC(m)AGp, containing 11% of the ¹²⁵I label, must clearly derive its m⁷Gpppm⁶AmC(m)p from β-globin mRNA, could derive its next A-residue either from the β-globin mRNA or as a result of transcription, and must obtain its G-residue as a result of transcription. Thus, 14 and 11% of the time 5' terminal, capped pieces of β-globin mRNA 8–11 bases, and 2–3 bases in length, respectively, were transferred. These pieces, as well as the predominant 12–14 nucleotide piece, were most likely linked to G as the first base transcribed.

The predominant sequence transferred from β-globin mRNA is not complementary at its 3' end to the 3' terminal common sequence of the vRNA templates. This and other data eliminate the possibility that hydrogen-bonding is needed for the stimulation of transcription. For example, capped, 5' terminal fragments of mRNAs as short as 14–23 nucleotides in length are effective primers. The active 5' fragments of this size derived from β-globin mRNA and alfalfa mosaic virus (A1MV) RNA 4 do not contain a sequence complementary to the 3' end of the vRNA segments. In addition, capped ribopolymers, such as capped poly A and capped poly AU, which lack a sequence complementary to the 3' end of vRNA, are effective primers[23].

In the absence of hydrogen-bonding between the primer mRNA and the template vRNA, the most likely mechanism for priming is that shown in Fig. 2. A 5'

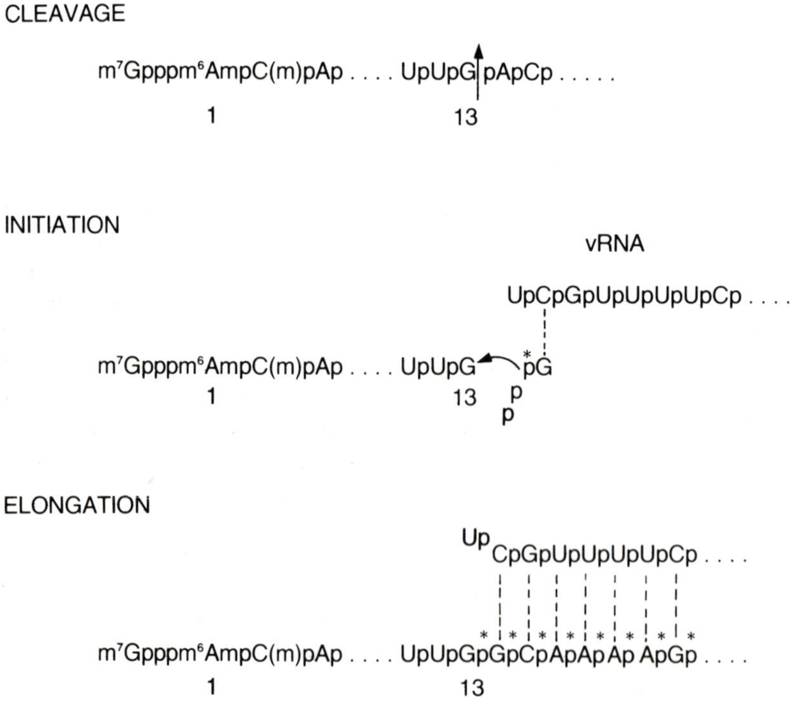

CLEAVAGE

m⁷Gpppm⁶AmpC(m)pAp .... UpUpG│pApCp .....
1   13

INITIATION

                                    vRNA
                                    UpCpGpUpUpUpUpCp ....
m⁷Gpppm⁶AmpC(m)pAp .... UpUpG⬋ ˙pG
1   13   p
                                  p

ELONGATION

                            Up
                              CpGpUpUpUpUpCp ....
                              │ │ │ │ │ │ │ │
                              * │ * │ * │ * │ * │ * │ * │ *
m⁷Gpppm⁶AmpC(m)pAp .... UpUpGpGpCpApApAp ApGp ....
1   13

Fig. 2. Mechanism for the priming of influenza viral RNA transcription by β-globin mRNA and other capped RNAs.

terminal fragment of a mRNA (β-globin mRNA in the example shown) is cleaved by a virion-associated nuclease, and this fragment is the actual primer that initiates transcription. The stimulation of the initiation of transcription would require a specific interaction of the capped RNA and/or its 5' terminal fragments with one or more proteins in the transcriptase complex. Since G, rather than A, is the first base linked to the primer RNA by the transcriptase, the viral RNA transcripts would not necessarily contain an A complementary to the 3' terminal U of the template vRNA. It is most likely that this G would be directed by the 3' penultimate C of the vRNA, though it could also be directed by the 3' terminal U since the dinucleotide GpG, like ApG, has been shown to initiate transcription exactly at the 3' terminal UC of the vRNA[24]. For an A to be found opposite the 3' terminal U of the vRNA template, it would be necessary for the 5' terminal fragment cleaved from the primer mRNA and used to initiate transcription to have a 3' terminal A.

This mechanism predicts the existence of 5' terminal fragment(s) cleaved from the mRNA primer. In the presence of GTP only it should be possible to isolate cleaved 5' terminal fragment(s) to which one or more G's were added. We have identified both these species. In the presence of [α-^{32}P]GTP and globin mRNA as primer, two terminally labeled species predominated: the fragment resulting from the cleavage of β-globin mRNA at G_{13} having one or two G's added at its 3' end, consistent with endonucleolytic cleavage occurring mainly at this position (as shown in Fig. 2). To identify the cleavage product(s) directly, without addition of G, mRNA primers containing ^{32}P only in their caps were used as substrates. With A1MV RNA 4 (with the 5' terminal sequence cap-GUUUUUAUUUUUAAUU..[25]), the predominant cleavage occurred at A_{13}.

This cleavage product contained a 3' terminal hydroxyl group, as expected for a primer molecule. In the presence of GTP, one, two or three G's were added to the 3' end of this cleavage product. By analysing several different capped primers, we have found that the cleavage is predominantly at a purine (A or G) at a position 10–13 nucleotides from the 5' end. Thus, with capped poly AU, cleavage occurred essentially only after A residues, and with brome mosaic virus (BMV) RNA 4 (with the 5' terminal sequence cap-GUAUUAAUAAUGUC...[26], cleavage occurred at A_{10} and G_{12}, but not at U_{11}.

The specific interaction between the primer mRNA and one or more transcriptase proteins, which leads to the stimulation of the initiation of transcription, probably involves the 5' terminal methylated cap. Only capped mRNAs are active as primers[16-18]. Removal of the m[7]G of the cap of a mRNA chemically or enzymatically eliminates all priming activity, and most of this activity can be restored by enzymatically recapping the mRNA[17]. The cap must contain methyl groups, since reovirus mRNAs with 5' GpppG ends are not active as primers[18]. Indeed, we have recently shown that each of the two methyl groups in the cap, the 7-methyl on the terminal G and the 2'-O-methyl on the penultimate base, strongly influences the priming activity of a mRNA[27]. Of particular interest is the effect of the 2'-O-methyl group. To demonstrate the importance of this group, we used several plant viral RNAs containing the monomethylated cap 0 structure, m[7]GpppG. BMV RNA 4 stimulated the transcription of influenza viral RNA only about 10–15% as effectively as globin mRNA. After enzymatic methylation of the 2'-O-group of the penultimate base (G) of the cap of BMV RNA 4, its priming activity was increased about 14-fold. Qualitatively similar results were obtained with other plant viral

RNAs: priming activity increased 3- to 20-fold following 2'-O-methylation. This is the first instance in which the 2'-O-methyl group of the cap has been shown to have a strong and clear-cut effect on a specific function of a mRNA. Consequently, the fully methylated cap 1 structure, m^{7}GpppX^m, which is found in all mammalian cellular mRNAs and most animal viral mRNAs, is more stringently required for priming influenza viral RNA transcription than for translation in cell-free systems.

The cap 1 structure is presumably recognized at the initiation step of viral RNA transcription (see Fig. 2), thereby mediating the stimulation of initiation. Recent experiments indicate that the cap is actually recognized at the initial step of the reaction by the virion-associated endonuclease: the specific cleavage of a mRNA described above was found to occur only when the mRNA contained a fully methylated cap structure. Clearly, it will be of great interest to identify the virus-associated protein(s) which recognize the cap 1 structure and to compare it to the cap-recognizing protein associated with eukaryotic ribosomes[28].

Capped RNAs serve as primers for transcription of the influenza viral RNA *in vivo*

The priming of viral RNA transcription by capped host RNAs also occurs in the infected cell[29]. The viral mRNA synthesized in the infected cell should contain 10–15 nucleotides at its 5' end, including the cap, which are not viral-coded and this is exactly what was found. First, gel electrophoretic analysis indicated that the segments of viral mRNA synthesized *in vivo* were 10–15 nucleotides longer at their 5' end than the ApG-primed segments synthesized *in vitro*. In addition, when viral mRNA was tritiated in the methyl groups *in vivo* and then hybridized to vRNA, the 5' terminal cap structure of the mRNA was not protected against pancreatic or T1 RNase digestion[29]. The only [^{3}H]methyl-labeled residues remaining in the double-strands were m^{6}A's although approximately one-third of these residues were also released. As each molecule of viral mRNA synthesized *in vivo* contains an average of three m^{6}A residues[15], these results indicate that one of these m^{6}A's is in the 5' terminal sequence that is not viral coded. Thus, our results strongly suggest that host cell mRNAs and/or their precursors serve as primers for viral RNA transcription in the infected cell, and that they donate their cap and 10–15 nucleotides, one of which is m^{6}A, to the resulting viral mRNA molecules[29]. Because our results from experiments *in vitro* show that most capped RNAs are effective primers, it is extremely likely that many different host capped RNAs serve as primers for transcription *in vivo*. Consequently, the initial sequence at the 5' end of viral mRNA *in vivo* should be heterogeneous. Consistent with such heterogeneity, two different bases, Am or Gm, have been found at the 5' penultimate position[15]. This heterogeneity has recently been confirmed by direct sequence analysis[30].

The synthesis of these host cell mRNA primers can be presumed to constitute the α-amanitin-sensitive step (RNA polymerase II function) required for viral RNA transcription. A critical, unanswered question is why new and continuous synthesis of these host mRNA primers is required. It may be that pre-existing capped RNAs are in ribonucleoprotein structures (including polyribosomes) and cannot be used by the virion transcriptase. Also, the requirement for new synthesis may be due at least in part to the site of viral RNA transcription in the infected cell. The presence of m^{6}A residues in the viral mRNA[15] suggests that viral RNA transcription may have a nuclear phase, and other experiments measuring the steady-state level of viral mRNA[9,10] suggest that the nucleus is the site of at least primary transcription. Finally, recent evidence suggests that the two mRNAs specified by the smallest genome segment are related to each other by a splicing event[31], a result consistent with nuclear transcription. If viral RNA is transcribed in the nucleus, then the need for continued host cell mRNA synthesis may reflect the fact that the amount of available cellular mRNA and/or its precursors in the nucleus is limited and rapidly depleted. Clearly, it will be of great interest to establish where both primary and amplified transcription occurs in the infected cell.

Possible implications for cellular mRNA synthesis

Do these results with influenza viral RNA transcription shed any light on cellular functions? For example, is the recognition of methylated cap (cap 1) structures unique to the transcription of influenza viral RNA, or does it also occur during cellular processes involving transcription? One possibility is that the methylated cap structure is required for the proper processing of cellular heterogeneous nuclear RNA. An initial step in this processing, cleavage of heterogeneous nuclear RNA near its 5' end[32] is comparable to the cleavage of the primer RNA near its 5' end which occurs in the transcriptase reaction for the influenza virus (see Fig. 2). The cellular enzyme(s), like the virion enzyme, may require the presence of the cap 1 structure. A later step in the processing of heterogeneous nuclear RNA, the ligation of the 5' terminal region on to distal regions of the RNA[32], could also involve cap recognition. This hypothesis can be tested when the appropriate cellular enzyme(s) are identified and characterized. At present, the cap-recognizing protein(s) and cleavage enzyme(s) in influenza virions may serve as a useful model system.

Acknowledgements

The research of the authors was supported by U.S. Public Health Service Grants CA 08748 and AI 11772; and by U.S. Public Health Service International Fellowship TW 02590-01 to M.B.

References

1 Palese, P. (1977) *Cell* 10, 1–10
2 Lamb, R. A. and Choppin, P. W. (1979) *Proc. Natl. Acad. Sci. U.S.A.* 76, 4908–4912
3 Inglis, S. C., Barrett, J., Brown, C. M. and Almond, J. W. (1979) *Proc. Natl. Acad. Sci. U.S.A.* 76, 3790–3794
4 Rochovansky, O. (1976) *Virology* 73, 327–338
5 Barry, R. D., Ives, D. R. and Cruickshank, J. G. (1962) *Nature (London)* 194, 1139–1140
6 Rott, R. and Sholtissek, C. (1970) *Nature* 228, 56
7 Lamb, R. A. and Choppin, P. W. (1977) *J. Virol.* 23, 816–819
8 Spooner, L. L. R. and Barry, R. D. (1977) *Nature (London)* 268, 650–652
9 Mark, G. E., Taylor, J. M., Broni, B. and Krug, R. M. (1979) *J. Virol.* 29, 744–752
10 Barrett, T., Wolstenholme, A. J. and Mahy, B. W. J. (1979) *Virology* 98, 211–225
11 Plotch, S. J. and Krug, R. M. (1977) *J. Virol.* 21, 24–34
12 Plotch, S. J. and Krug, R. M. (1978) *J. Virol.* 25, 579–586
13 Plotch, S. J., Tomasz, J. and Krug, R. M. (1978) *J. Virol.* 28, 75–83
14 McGeoch, D. and Kitron, N. (1975) *J. Virol.* 15, 686–695
15 Krug, R. M., Morgan, M. M. and Shatkin, A. J. (1976) *J. Virol.* 20, 45–53
16 Bouloy, M., Plotch, S. J. and Krug, R. M. (1978) *Proc. Natl. Acad. Sci. U.S.A.* 75, 4886–4890
17 Plotch, S. J., Bouloy, M. and Krug, R. M. (1979) *Proc. Natl. Acad. Sci. U.S.A.* 76, 1618–1622
18 Bouloy, M., Morgan, M. A., Shatkin, A. J. and Krug, R. M. (1979) *J. Virol.* 32, 895–904
19 Skehel, J. J. and Hay, A. J. (1978) *Nucl. Acids Res.* 4, 1207–1219
20 Robertson, J. S. (1979) *Nucl. Acids Res.* 6, 3745–3757
21 Robertson, H. D., Dickson, E., Plotch, S. J. and Krug, R. M. (1980) *Nucl. Acids Res.* 8, 925–942
22 Lockard, R. E. and RajBhandary, U. L. (1976) *Cell* 9, 747–760
23 Krug, R. M., Broni, B. A., LaFiandra, A. J., Morgan, M. A. and Shatkin, A. J. (1980) *Proc. Natl. Acad. Sci. U.S.A.* (in press)

24 Hay, A. J. and Skehel, J. J. (1979) *J. Gen. Virol.* 44, 599–608
25 Koper-Zwarthoff, E. C., Lockard, R. E., Alzner-DeWeerd, B., RajBhandary, U. L. and Bol, J. T. (1977) *Proc. Natl. Acad. Sci. U.S.A.* 74, 5504–5508
26 Dasgupta, R., Shih, D. S., Saris, C. and Kaesberg, P. (1975) *Nature (London)* 256, 624–628
27 Bouloy, M., Plotch, S. J. and Krug, R. M. (1980) *Proc. Natl. Acad. Sci. U.S.A.* 77, 3952–3956
28 Sonenberg, N., Rupprecht, K., Hecht, S. and Shatkin, A. J. (1979) *Proc. Natl. Acad. Sci. U.S.A.* 76, 4345–4349
29 Krug, R. M., Broni, B. A. and Bouloy, M. (1979) *Cell.* 18, 329–334
30 Dhar, R., Channock, R. M. and Lai, C.-J. (1980) *Cell.* (in press)
31 Lamb, R. A. and Lai, C.-J. (1980) *Cell.* (in press)
32 Darnell, J. E. Jr (1978) *Science* 202, 1257–1260

R. M. Krug, M. Bouloy and S. J. Plotch are at the Molecular Biology and Genetics Unit of the Graduate School, Memorial Sloan-Kettering Cancer Center, New York, NY 10021, U.S.A.

M. Bouloy – present address: Institut Pasteur, 75015, Paris, France.

Protein synthesis

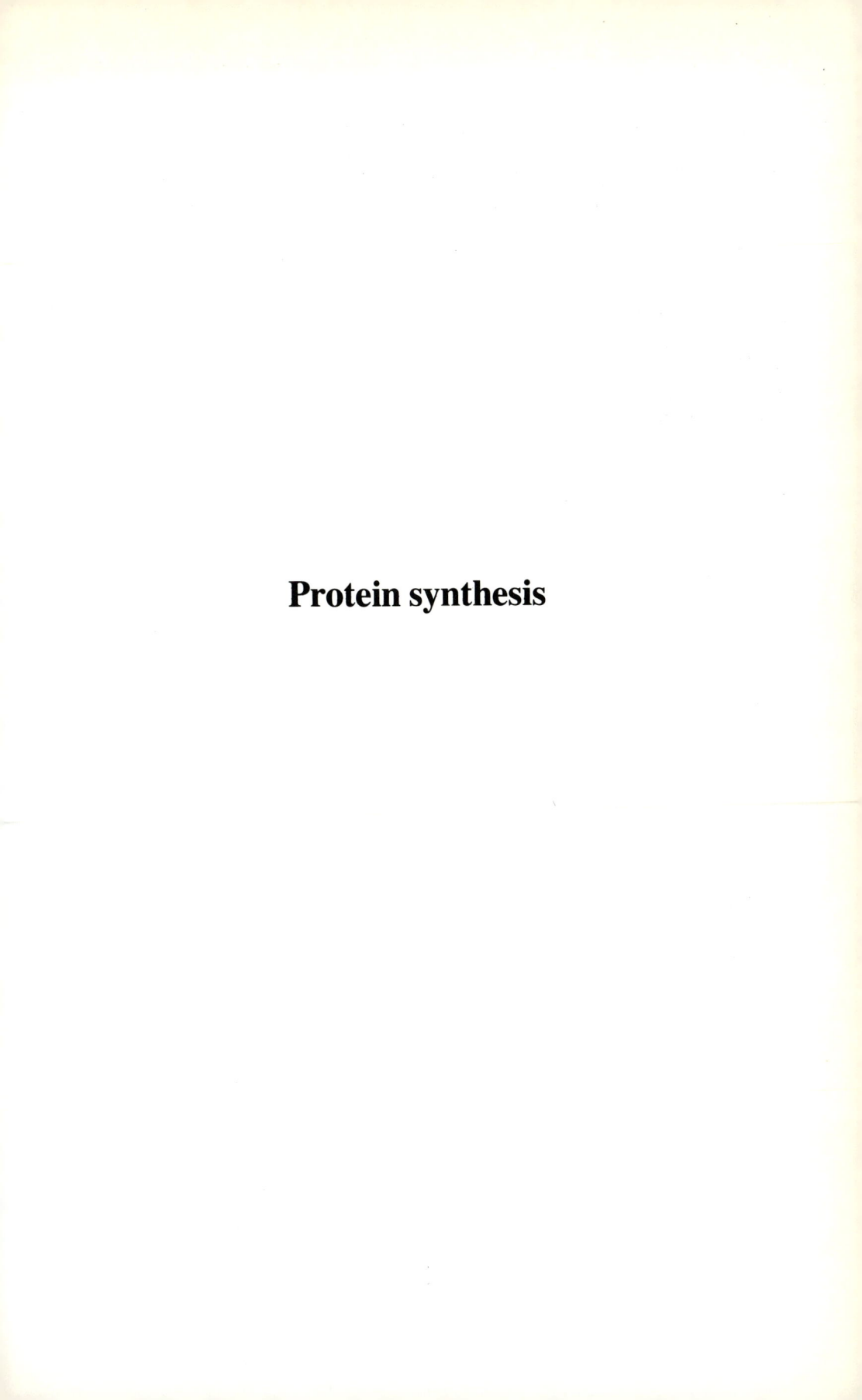

The initiation of protein synthesis
Tim Hunt

This article, the first of a three-part series on protein synthesis, describes some of the current ideas about the mechanism and control of the initiation of protein synthesis. The author's interests are in eukaryotic systems, but where there seem to be differences between the process in pro- and eukaryotes, or particularly striking points in prokaryotes these are mentioned.

The problem

Proteins are assembled on ribosomes in an ordered stepwise manner according to the sequence of bases in messenger RNA. Understanding this process requires answers to many questions which fall into various categories. What are the natures of the steps in the overall process? What is the molecular mechanism of each step? What catalyses these steps? Above all, what features of the process give it its high accuracy, which is estimated as less than one mistake per 10,000 amino acids [1]? What controls the rate of protein synthesis? And so on. Even after more than 20 years of detailed studies, many of these questions can only be answered partially, and it is the purpose of this article and its two successors to give an up-to-date account of where we stand now.

The overall process of protein synthesis can be divided into three stages (bearing in mind the cyclical nature of the process): initiation, elongation and termination of the polypeptide chain. Throughout the process a common recurring theme emerges; at every step there exists a high degree of specificity, implying a tight association of two, or usually many more components. Yet the very next step usually requires that this tight binding be loosened as a new set of equally specific associations occurs, and so on hundreds of times for each protein made. This requires energy supplied in the form of ATP and GTP; all told, it takes the equivalent of four high energy phosphate groups split to make one peptide bond, half of this total being required to charge tRNA with amino acids, the other half being used for reactions occurring on the ribosome.

The main question about the process of initiation is how ribosomes find the correct starting place on mRNA, and how the process is controlled, for alterations in the rate or nature of the proteins synthesized seem mainly to occur at initiation.

There are many possible approaches to these questions, each with particular advantages and disadvantages. In yeast and bacteria, genetic analysis may in principle be used to determine how many different components are involved and, together with biochemical analysis, what each individual component does. In higher eukaryotes, only biochemistry is possible at present, but it is an odd sort of biochemistry which relies heavily on the use of specific inhibitors of protein synthesis to reveal the existence of intermediates, as well as the more conventional purification and reconstruction approach. Particularly in the case of such a complex process as protein synthesis, simple biochemistry can (and has) misled the unwary, for, until the nature of all the reactions is understood, appropriate assays cannot be devised; yet the purpose of the studies is to determine what these reactions are. It is quite possible to identify a protein synthesis factor, and to purify it to homogeneity, without knowing what it actually does in real life, and without having any very obvious way of finding out except by luck. We are just beginning to rectify this situation by raising monoclonal antibodies against each factor, and studying the effect of these reagents in crude cell-free systems, a process formally analogous to studies of mutants in those organisms where it is possible. I think we may be in for some surprises.

Beginning at the end

In both bacterial and eukaryotic protein synthesis, ribosomes can usually only bind to mRNA as separate subunits at different times, the small subunit joining the mRNA before the large one. When and how the subunits are separated after termination is not clear; most likely they separate as they get off the mRNA (it may even be essential for them to dissociate in order to disengage the message), and they are prevented from joining together by a protein which acts as an 'anti-association factor'; in bacteria it is called IF-3, in eukaryotes eIF-3. This factor binds to the smaller ribosomal subunit and prevents the larger subunit from binding; under physiological conditions, the two subunits bind together with high affinity to form inactive 70 S or 80 S ribosomes if this factor is absent. Actually, eukaryotic cells usually contain a sizeable pool of vacant 80 S ribosomes which do not participate in protein synthesis unless they can be dissociated into subunits. This dissociation occurs very slowly compared with the rate at which the active ribosomes present cycle from subunits through polysomes and back to subunits again, though there are conditions under which almost all the ribosomes enter this inactive pool – for instance, under various stressful circumstances – and equally rapidly leave it, as when cells are stimulated from a resting state to grow very rapidly. It is not clear what regulates these fluxes. The concentration of native subunits is, by contrast, rather constant, and is probably set by the concentration of eIF-3 in the cell, except when inhibitors have been added (both aurine tricarboxylate and poly I cause massive accumulation of subunits in reticulocyte cell-free systems). The question arises as to whether eIF-3 participates in removing ribosomes from mRNA after termination, or whether some termination factors bring about dissociation of ribosomes (and mRNA); most of the assays for termination measure the release of completed proteins rather than release of ribosomes from mRNA or their dissociation into subunits.

The role of initiator tRNA

In eukaryotes, the first step in the initiation pathway (Fig. 1) after the formation of native subunits is the binding of initiator tRNA [Met-tRNA$_f$] to the 40 S subunit, in a reaction that requires the initiation factor eIF-2 and GTP (but which also proceeds in the presence of non-hydrolysable analogues of GTP). This is somewhat surprising, given that tRNA normally only binds to ribosomes at the direction of codons in mRNA, and indeed there was some resistance to the idea at first. The evidence for the non-codon-directed binding of Met-tRNA$_f$ is simple and compelling. First, 40 S ribosomes bind Met-tRNA$_f$ in the absence of mRNA, and second, they will not bind mRNA in the absence of bound Met-tRNA$_f$. The situation in bacteria is not as clear cut, because the 30 S/fMet-tRNA$_f$ complexes are not so stable and mRNA seems to bind to 30 S subunits to some extent in the absence of fMet-tRNA$_f$; nevertheless, in bacteria the presence of fMet-tRNA$_f$ greatly stabilizes the binding of mRNA, and some people have argued that bacterial ribosomes follow the same obligatory ordered binding as

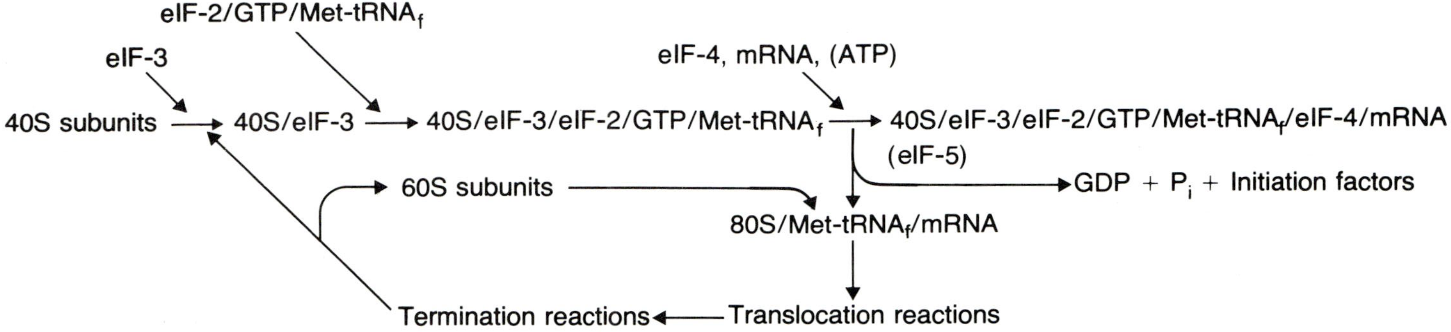

```
40S subunits ──→ 40S/eIF-3 ──→ 40S/eIF-3/eIF-2/GTP/Met-tRNAf ──→ 40S/eIF-3/eIF-2/GTP/Met-tRNAf/eIF-4/mRNA
                ↑              ↑                                ↑                                          │
              eIF-3      eIF-2/GTP/Met-tRNAf              eIF-4, mRNA, (ATP)                              │ (eIF-5)
                                                                                                           ↓
                                                                                      GDP + Pi + Initiation factors
                                          60S subunits ──→                                  │
                                                          80S/Met-tRNAf/mRNA  ←─────────────┘
                                                                    │
                              Termination reactions ←── Translocation reactions
```

Fig. 1. The sequence of reactions in initiation. 40 S ribosomes produced by the termination step bind eIF-3 to prevent them binding directly to 60 S subunits. Subsequently Met-tRNAf, mRNA, and 60 S subunits join the complex in that order. Note that mRNA would normally have several ribosomes already engaged in its translation, so that where the diagram specifies 'mRNA', this would normally be polysome structures; the resulting 40 S/polysome structures are readily detectable among the more conventional polysomes containing integral numbers of ribosomes. For simplicity's sake, the factors that catalyse the attachment of mRNA are referred to as eIF-4; in fact, at least three factors, 4A, 4B, and 4C have been resolved. Factors that play a catalytic role, rather than forming stoichiometric complexes are shown in parentheses.

in eukaryotes. In eukaryotes, this step has received particular attention as a possible site of regulation, since there are many circumstances in which initiation appears to be reversibly blocked at this step. Whereas it is usually possible to detect 40 S/Met-tRNAf complexes in cells or cell-free systems, they sometimes disappear, and when they do, protein synthesis stops. For example, this happens in reticulocyte lysates when they are either short of haemin, or when low concentrations of double-stranded RNA are added. The mechanism of this inhibition is the subject of intense debate; whereas it has been shown that these conditions lead to the phosphorylation of eIF-2, and that protein kinases specific for this initiation factor are activated under these circumstances, there are inconsistencies in the story. There are also suspicions that other factors may play a part, such as oxidation of particular sulphydryl groups on the protein. It is also possible that the 'energy charge' of the cell may regulate the rate of initiation, for GDP has a 100-fold higher affinity for eIF-2 than GTP, but does not allow the binding of Met-tRNAf.

Binding the mRNA

Once Met-tRNAf is bound (together with eIF-3, eIF-2 and GTP) the 40 S subunit is competent to bind mRNA. This requires additional factors, collectively termed eIF-4, and ATP; non-hydrolysable analogues of ATP, or ADP will not substitute. It is not known how much ATP is required, an interesting point, particularly since its function is unknown. Is one ATP hydrolysed per mRNA binding, or are many? Equally, the precise roles of eIF-4a and eIF-4b are obscure, though both are necessary for mRNA binding in fractionated systems. Early reports that they had ATPase activity have not been confirmed.

So much for factors; the question of most interest in this story is of how ribosomes find the initiation site on mRNA. The mechanism seems to be different in bacteria and mammals. In bacteria there is very good evidence that the 16 S ribosomal RNA plays a central role, by forming base pairs with a site on mRNA just to the left of the initiation codon, as first suggested by Shine and Dalgarno [2] purely on the basis of sequence analysis. Their conjecture is supported by several experimental findings. First, there is a good correlation between the extent of potential base pairing between particular mRNAs and ribosomal RNA and the frequency of initiation on that mRNA. More compellingly, an interaction between mRNA and ribosomal RNA was demonstrated directly in a classic experiment by Joan Steitz and Karen Jakes [3]. They formed initiation complexes between E. coli ribosomes and a ^{32}P-labelled fragment of phage R17 RNA which contained the initiation site of the phage A protein. They next digested these complexes with colicin E3, an enzyme that specifically cuts 16 S RNA about 50 bases from its 3' end, where the putative interaction with mRNA was expected. Finally, they added SDS to dissociate the proteins from the RNA and ran the mixture on an acrylamide gel. The labelled mRNA ran as if it were hydrogen bonded to the rRNA fragment. Suitable controls ruled out various trivial explanations for the interaction.

Analysis of mRNA sequences and rRNA sequences in eukaryotes has not revealed any such corresponding complementarities, or any other regular features which might direct ribosomes to particular places on the mRNA. However, the ribosomes always seem to start at the first AUG sequence in the mRNA, in marked contrast to bacterial ribosomes which can select internal AUG codons to make a beginning. Indeed, bacterial ribosomes can initiate synthesis on circular (i.e. endless) messages, whereas eukaryotic ribosomes cannot, as Marilyn Kozak has elegantly shown [4]. The current orthodoxy, spelled out in detail in a review by Kozak [5], suggests that eukaryotic ribosomes bind at or near the 5' end of mRNA, using the 'cap', m7G5'pppX . . . as a guide. This accounts for the observation that improperly capped mRNAs are translated at a much lower frequency than their correctly capped counterparts, and that the translation of capped mRNAs is strongly inhibited by competitors such as m7GTP. However, there are messages lacking caps which serve as excellent templates for protein synthesis, such as EMC RNA which begins pU . . . and in which the initiation codon seems to be at least 400 nucleotides from the 5' end; how these messages work is unclear. Also unclear is which, if any, of the initiation factors recognizes the cap. A protein of M_r 24,000 has been isolated from ribosomes by affinity chromatography on m7GDP Sepharose [6], and this same protein can be crosslinked to mRNA when it is bound in initiation complexes, yet the protein does not belong to the canonical set of initiation factors, and was not recognized as being necessary for initiation when the factors were being isolated, purified and characterized, though it was probably present as a contaminant of eIF-3 and eIF-4B. It is suggestive that extracts of poliovirus-infected cells are bad at translating capped mRNA [7] (polio RNA is uncapped), and that the capacity of such a system to translate capped mRNA can be restored by adding a protein fraction containing this cap-binding protein (and not eIF-4B as first reported). It seems however that other factors may be involved, and it is not known what polio infection does to the 24,000 cap binding protein to inactivate it; indeed, it has not been shown directly that this protein is inactivated in these extracts. Work on this point is presently very active.

The idea that eukaryotic ribosomes bind at the very start of the mRNA and work their way towards the first AUG codon, possibly using ATP as an energy source for such movement is supported by the observation that certain messages (tobacco mosaic virus RNA for example) which have a long stretch of nucleotides between the cap and the initiator AUG codon can bind more than one ribosome, one at the AUG and one to the left where no AUG exists [8]. This is made very obvious when edeine is added. This oligopeptide antibiotic is a highly specific inhibitor of initiation which allows 40 S subunits to bind mRNA while completely preventing the attachment of 60 S ribosomes to the 40 S/mRNA complex. In reticulocyte lysates, the presence of edeine leads to the formation of dimers and trimers of 40 S subunits bound to globin mRNA, and I believe that they stop at the initiation codon; but Kozak found larger numbers of 40 S subunits bound to mRNA in a rather more highly fractionated system and believes that the 40 S subunits can wander all the way down the message in the presence of this inhibitor. Whatever the case, in the absence of edeine but in the presence of inhibitors which arrest ribosomes at the initiation codon after they have synthesized one peptide bond (e.g. sparsomycin or diphtheria toxin), one finds additional ribosomes to the *left* of the one bound at the true start, but *not* to the right, reinforcing the idea that they must enter at the 5' end of the message, and cannot bind internally without having 'walked' there. It is not known why or how bacterial ribosomes differ from mammalian or plant ribosomes to account for these differences in their behaviour.

Joining the 60 S subunit

The 60 S subunit probably joins the 40 S/mRNA/Met-tRNA$_f$ complex when the Met-tRNA$_f$ anticodon has engaged the initiator AUG codon, but this is not certain. The joining reaction requires an addi-

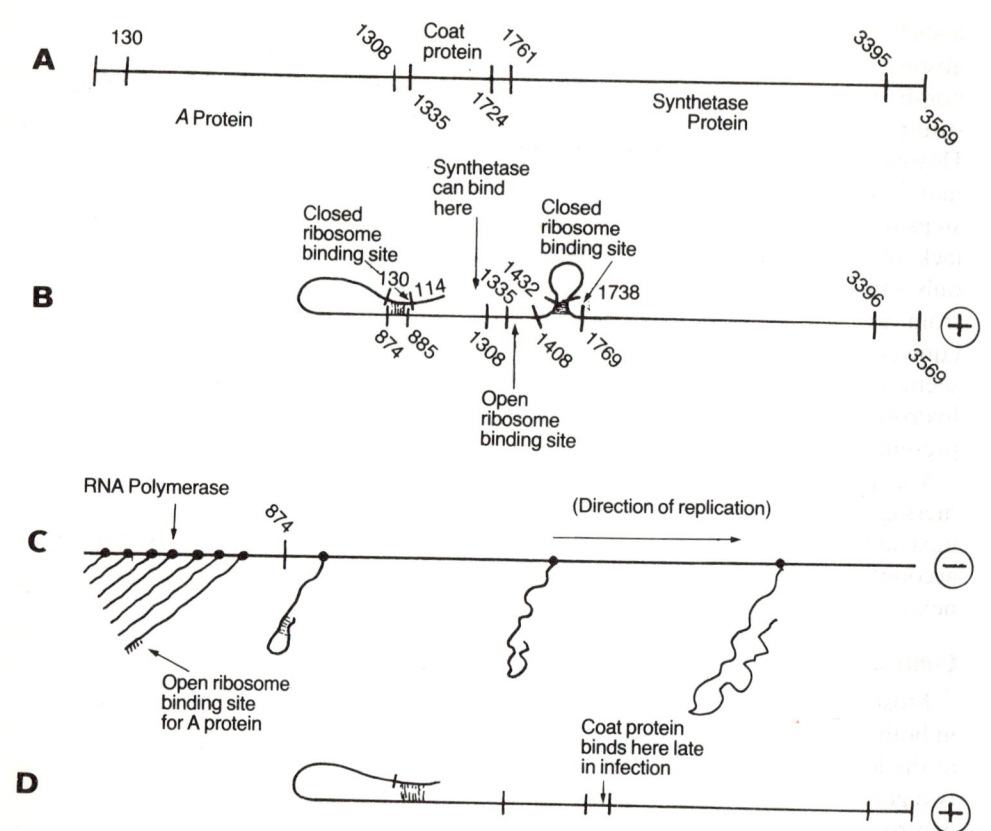

Fig. 2. A simplified diagram of the E. coli RNA phages. (A) shows a map of the three major genes, 'A' protein, coat protein and synthetase (the recently identified lysis protein is omitted for clarity; it overlaps the coat and synthetase genes in a different phase). (B) shows the RNA–RNA interactions that seem to be significant for the control of translation, and ignores the very extensive base-pairing that occurs throughout the genome[12]. It shows the postulated interaction between the start of the 'A' gene and residues about 870–880 downstream, and the interactions between the start of the synthetase gene and residues 1408–1432 lying upstream in the middle of the coat gene. These interactions prevent ribosomes binding to the initiation sites of these genes, leaving only the coat protein open for translation in the early phase of infection. The synthetase gene is opened up by ribosomes engaged in the translation of the coat gene, and when enough synthetase has been made, it binds to a site close to the start of the coat gene, indicated in (B), thereby stopping any more ribosomes binding to the RNA and clearing the RNA for transcription into the (−) strand shown in (C). This (−) strand serves as a template for the synthesis of progeny (+) strands, which can be translated in their nascent state; 'A' protein can be made from these replicative complexes until the polymerase passes residue 875 or thereabouts, when the ribosome binding site is shut down by the base-pairing indicated in (B). Finally, as indicated in (D), the 'A' gene is closed by this secondary structure, and although the synthetase gene is opened up by ribosomes making coat protein, the coat protein itself binds to the RNA at or near the initiation site of the synthetase gene, which is thereby effectively closed late in infection when sufficient coat protein has accumulated. This map is drawn to scale according to the data of Fiers et al., for MS2[12].

tional initiation factor called eIF-5, and entails a concerted set of reactions in which (1) the Met-tRNAf becomes bound in the P-site of the 60 S subunit (i.e. it can form a peptide bond with the incoming amino acid or with puromycin), (2) the GTP which bound the eIF-2 and Met-tRNAf to the 40 S subunit is hydrolysed to GDP and Pi, and (3) all the initiation factors leave the ribosome [9]. Presumably the 60 S ribosome cannot join until after the eIF-3 has left, since this factor inhibits subunit association. Release of all initiation factors absolutely requires GTP hydrolysis, and if non-hydrolysable GTP analogues are substituted, the 60 S subunit cannot bind. However, the details of this 'step', which may in reality be composed of a series of steps which we fail to distinguish owing to lack of specific inhibitors (edeine is the only inhibitor of this step) are shrouded from us. Interestingly, bacteria lack an equivalent of eIF-5, and subunit joining seems to proceed as soon as GTP has been hydrolysed, very likely by the GTPase present in the 50 S subunit.

We are thus poised at the start of the message proper, awaiting the arrival of the next aminoacyl tRNA and next month's account by Brian Clarke of what happens next.

Control of initiation

Most of the control of protein synthesis in both pro- and eukaryotes appears to act at the level of initiation. Such control is of two general kinds which I shall call quality control and quantity control. By quality control I mean the following: a situation in which ribosomes, faced with a variety of different mRNAs, 'choose' to translate a restricted set of those messages in certain circumstances, whereas under other conditions they translate a different set. Quantity control refers to circumstances in which the frequency of initiation on all messages is raised or lowered across the board. Both types of control have been found, and in both cases one can point to specific examples where plausible mechanisms have been proposed to account for them. The best examples of quality control are found in bacteria, particularly in the regulation of translation of viral mRNA. For instance, although the genes of RNA viruses like f2, R17 and Qβ are present in equimolar concentration in infected *E. coli*, the proteins specified by these genes are produced in widely different amounts [10]. Studies of mutant viruses show that amber mutations near the start of the coat protein gene lead to a failure to synthesize anything except for the prematurely terminated N-terminal fragment of coat protein, whereas an amber mutation further along the gene leads to *overproduction* of polymerase and normal synthesis of the *A* protein. In cell-free systems it has been found that coat protein shuts off synthesis of polymerase, and that polymerase shuts off synthesis of coat protein. *A* protein is apparently only synthesised using nascent RNA molecules in the replication complex as templates. These findings are accounted for by a model in which secondary structure in the RNA prevents ribosomes from binding to either the *A* cistron or the polymerase cistron, until a ribosome in the process of making coat protein unravels the start of the polymerase gene; but polymerase binds at or near the initiation site of the coat cistron, and coat protein at or near the start of the polymerase gene (Fig. 2). In this way the RNA can be cleared of ribosomes so that it can serve as a template for RNA synthesis, and the synthesis of *A* protein and polymerase be kept well below the synthesis of coat protein which is required in far larger amounts than the others.

Examples of such control in eukaryotes are virtually non-existent. While it is clear that different mRNAs differ in their intrinsic initiation frequency, as in the case of mRNA for α and β globin, which initiates at frequencies of 20 s and 12 s respectively, there are few well-documented cases of frequencies being regulated in response to specific signals. One or two examples seem to occur in early embryonic development, when a large excess of largely inactive ribosomes is present along with a large population of mRNAs laid down during oogenesis. There is beginning to be quite good evidence that after activation of the oocytes of frogs and clams, which is accompanied by a general increase in protein synthesis, changes in the pattern of protein synthesis without corresponding changes in the mRNA population occur [11]. The mechanism is not known, though it is widely believed that proteins associated with mRNAs may specifically repress or enhance their translation. Modification of such proteins either by allosteric or covalent mechanisms could be specified in such a way as to account for alterations in their activity.

The likely role of modifications of eIF-2 activity in the general control of initiation has been mentioned earlier, but the field is currently in such a state of confusion as to warrant a separate review of its own. As for possible alterations in other initiation factors, very little is known apart from the case of poliovirus infection described above. My impression is that most people believe that eIF-2 is the main site of regulation, but that the evidence for this belief is virtually non-existent except in the case of reticulocytes. Since most cells yield poorly active cell-free systems (but nobody really knows why), it is difficult to analyse rate-limiting steps, and analysis in intact cells is extremely difficult if not impossible; all one can say is that polysomes alter in size and number in ways that clearly indicate control at the level of initiation.

References

1 Yarus, M. (1980) *Prog. Nucleic Acid Res. Mol. Biol.* 23, 195–223
2 Shine, J. and Dalgarno, L. (1974) *Proc. Natl. Acad. Sci. U.S.A.* 71, 1342–1346
3 Steitz, J. A. and Jakes, K. (1975) *Proc. Natl. Acad. Sci. U.S.A.* 72, 4734–4738
4 Kozak, M. (1979) *Nature (London)* 280, 82–85
5 Kozak, M. (1978) *Cell* 15, 1109–1123
6 Sonenberg, N., Rupprecht, K. M., Hecht, S. M. and Shatkin, A. (1979) *Proc. Natl. Acad. Sci. U.S.A.* 76, 4345–4349
7 Rose, J. K., Trachsel, H., Leong, K. and Baltimore, D. (1978) *Proc. Natl. Acad. Sci. U.S.A.* 75, 2732–2736
8 Filipowicz, W. and Haennis, A.-L. (1979) *Proc. Natl. Acad. Sci. U.S.A.* 76, 3111–3115
9 Peterson, D. T., Merrick, W. C. and Safer, B. (1979) *J. Biol. Chem.* 254, 2509–2516
10 Lodish, H. F. (1975) in *RNA Phages* (Zinder, N. D. ed.), pp. 301–318, Cold Spring Harbor
11 Rosenthal, E. T., Hunt, T. and Ruderman, J. V. (1980) *Cell* 20 (in press)
12 Fiers, W. *et al.* (1976) *Nature (London)* 260, 500–507

General reviews

Revel, M. and Groner, Y. (1978) *Annu. Rev. Biochem.* 47, 1079–1126
Grunberg-Manago, M. and Gros, F. (1977) *Prog. Nucleic Acid Res. Mol. Biol.* 20, 209–276
Jagus, R., Anderson, W. F. and Safer, B. (1980) *Prog. Nucleic Acid Res. Mol. Biol.* (in press)

Tim Hunt is at the Department of Biochemistry, Tennis Court Road, Cambridge CB2 1QW, U.K.

The elongation step of protein biosynthesis

Brian Clark

Polypeptide chain elongation is conveniently described in three separate steps. Much is now known about the functioning and structures of molecular components involved in each step but little is known about the kinetics of the process and its motive force. This is the second article in our three-part series on protein synthesis.

General features

During protein synthesis polypeptide chain initiation[1] is followed by a series of peptide chain elongation events prior to the process of chain termination[2] releasing the completed polypeptide. This polypeptide may be further trimmed or modified before spontaneously folding into a functional protein.

Elongation is defined as the addition of amino acids one at a time to a growing polypeptide in a sequence dictated by mRNA. The generally accepted scheme for peptide chain elongation which involves the relative movement of the mRNA and ribosome is shown as a simplified cyclic scheme in Fig. 1 which, in essence, dates to J. D. Watson's proposal of 1964[3]. At present there is not enough evidence to suggest that we drop the simple static view of Fig. 1 and think in terms of a more dynamic situation involving activated binding states.

Most of our information about protein biosynthesis has come from *in vitro* studies, involving reassembled components. So far it has not been possible to reconstitute cell free systems able to join amino acids at the *in vivo* rate of about 10 per second at 30°C. Since cell-free systems may work only 1–10% as efficiently as those *in vivo*, the results from such systems need to be confirmed by *in vivo* molecular genetic studies of, for example, *Xenopus* oocytes injected with various components of the protein biosynthetic machinery.

The process of polypeptide chain elongation is conveniently represented as occurring in three steps involving two ribosomal sites. Step I (of Fig. 1), which occurs after the 70 S initiation complex is formed, or during growth of the polypeptide chain, involves the codon-directed *binding of aminoacyl-tRNA* to a vacant ribosomal A-site adjacent to an occupied P-site. Step II is the *peptide bond-forming step* (Fig. 1) during which there is peptidyl transfer from fMet-tRNA$_f^{Met}$ or the peptidyl-tRNA in the P-site, to aminoacyl-tRNA in the A-site. Step III is concerned with a *translocation of the peptidyl-tRNA* back into the P-site to permit another incoming aminoacyl-tRNA to enter the A-site, with concomitant ejection of the discharged tRNA from the P-site. This translocation (Step III in Fig. 1) involves relative movement of the ribosome and mRNA.

Step I – involvement of EF-Tu

Two protein factors called T and G were shown by Lipmann's group, in the mid-1960s, to be required in *E. coli* for the polymerization of amino acids into polypeptides. Subsequently, transfer factors T and G were renamed elongation factors, EF-T and EF-G. EF-T is concerned with Step I and EF-G with Step III. The functioning of both these protein factors requires the hydrolysis of GTP (see Refs 6–8 for reviews). The elongation factor, EF-T, as originally isolated from *E.*

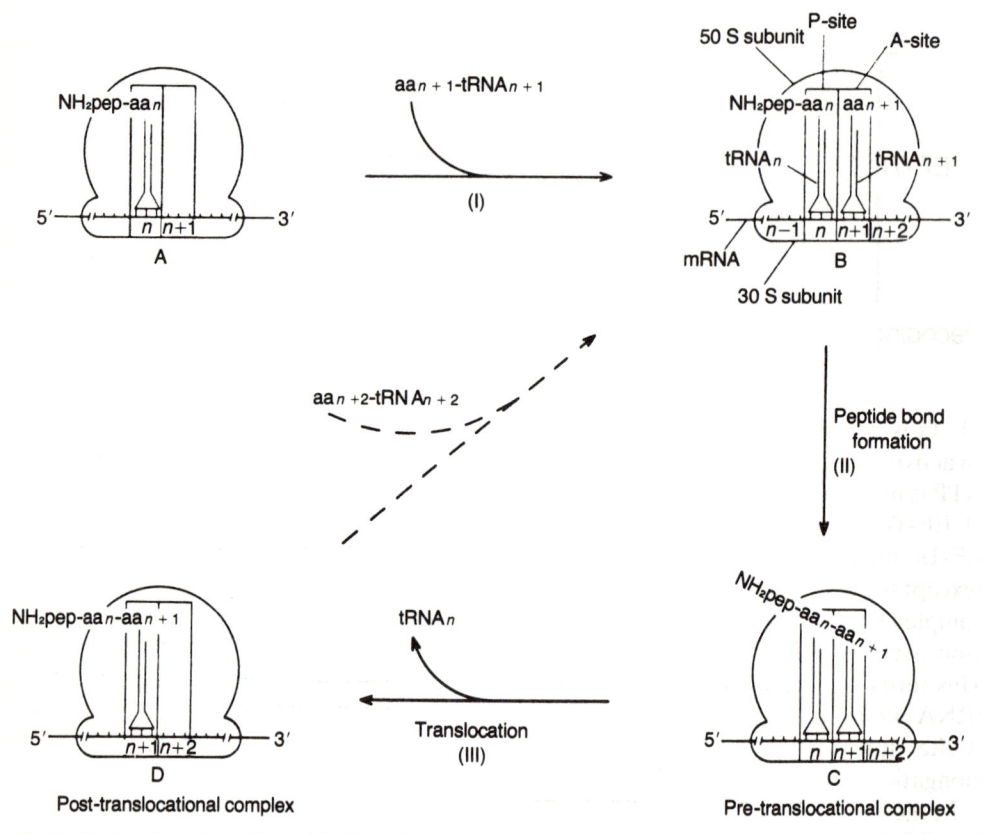

Fig. 1. Cyclic scheme for polypeptide chain elongation to show relative motion between the monoribosome and mRNA. The cycle starts in state A (equivalent to D in the next round) with a growing peptide attached to tRNAn in the peptidyl-tRNA binding site (P-site) of a bacterial 70 S (mono)ribosome[4] decoding codon n of the mRNA (eukaryotic cytoplasmic ribosomes[5] are called 80 S ribosomes). aa-tRNA = aminoacyl-tRNA.

coli, was shown to be composed of two components, EF-Tu (mol. wt 43,000) and EF-Ts (mol. wt 35,000), in equimolar amounts. In contrast to the early findings, it has recently been shown that EF-Tu exists in the cell in a large excess over the other elongation factors and ribosomal proteins[9]. Indeed, EF-Tu is probably the most abundant protein in E. coli constituting 5.5% of the total protein[10]. The ratio of the amounts of elongation factors, however, varies somewhat with growth conditions, for example when the cells are rapidly multiplying, the ratio of EF-Tu:EF-Ts: EF-G:ribosomes is 6:1:1:1 EF-Tu is unique (as far as we know) in that it is specified by two genes, *tufA* and *tufB*, at 72' and 88', respectively, on the E. coli genetic map. Moreover, the *tufA* and *tufB* genes are expressed at different levels so that the *tufA* gene product is in about threefold excess over the *tufB* protein. Whether these two gene products have different primary structures is still not known. How the differential expression of the EF-Tu genes is controlled and how *tufA* is expressed differently from the gene for EF-G, which is probably in the same operon, are intriguing current questions.

EF-Tu and EF-Ts have different functions in polypeptide chain elongation. Only EF-Tu appears to be involved in Step I of elongation in which aminoacyl-tRNA is brought to the 70 S ribosomal A-site;

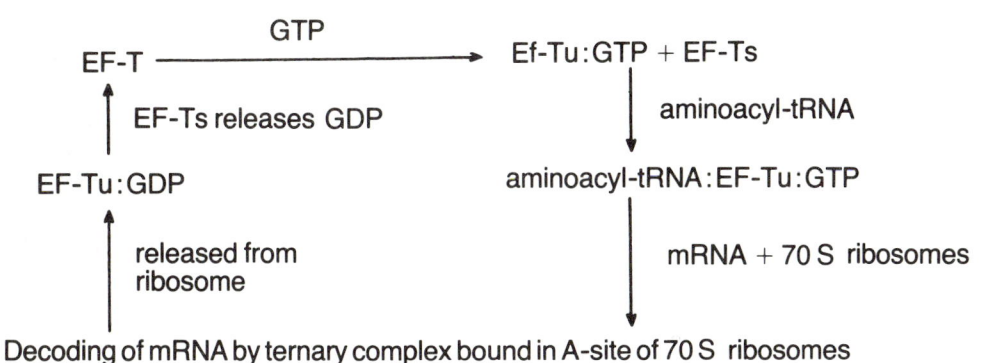

```
                      GTP
    EF-T  ─────────────────────────▶  Ef-Tu:GTP + EF-Ts
     ▲
     │  EF-Ts releases GDP                │  aminoacyl-tRNA
     │                                    ▼
 EF-Tu:GDP                          aminoacyl-tRNA:EF-Tu:GTP

     ▲  released from                    │  mRNA + 70 S ribosomes
     │  ribosome                         ▼
     │
Decoding of mRNA by ternary complex bound in A-site of 70 S ribosomes
```

Fig. 2. Cyclic scheme to show how EF-Tu recycles in the elongation step.

EF-Ts is needed only to recycle EF-Tu in an active form as illustrated in Fig. 2. When GTP is mixed with EF-T, a binary complex of EF-Tu:GTP is formed, displacing the EF-Ts moiety. If any aminoacyl-tRNA (except the initiator) is added to the binary complex EF-Tu:GTP, the ternary complex aminoacyl-tRNA:GTP:EF-Tu results. This ternary complex, not the aminoacyl-tRNA alone, binds to the 70 S ribosomal A-site during Step I of the polypeptide elongation process (see Fig. 1).

Large crystals of a modified EF-Tu:GDP complex suitable for X-ray analysis are being studied by at least three research groups, our own included. The modification, a single cleavage in the EF-Tu polypeptide chain produced spontaneously by an unknown enzyme activity, or by trypsin, is not thought to interfere with main structural integrity of EF-Tu. The primary structure of elongation factor EF-Tu has also recently been established (M. D. Jones, personal communication); interestingly the N-terminus of *E. Coli* EF-Tu is blocked by acetylserine. It should, therefore, be possible to relate in molecular terms the effect of mild tryptic digestion[8,10] to the structure and function of EF-Tu.

Mild tryptic digestion is controllable and can therefore be used as a structural probe. For example, the different complexes of EF-Tu with GTP, GDP or aminoacyl-tRNA are digested at different rates by trypsin. Such experiments comparing the rate of digestion of EF-Tu:GDP and EF-Tu:GTP have suggested different conformations for the two binary complexes. This proposal has support from a number of physical chemical studies using fluorescence and spin label probes, or surface labelling with reactive chemical reagents[8,11]. Possibly, therefore, when GTP forms a binary complex with EF-Tu, the latter's conformation is altered to the correct shape to interact with aminoacyl-tRNA.

We have no clear data about the way in which EF-Tu:GTP binds aminoacyl-tRNA. tRNA is, however, a largely rigid molecule[12,13] and in solution has most flexibility at the single stranded aminoacyl-end. Possibly, all the EF-Tu:GTP needs to do is to make rigid the aminoacyl-end in a recognitory interaction of that region. The rest of the rigid tRNA structure could be unchanged.

We do not know how the amino acid's position on the tRNA is recognized by EF-Tu. The amino acid may be attached to the 2' or 3' hydroxy group of the 3'-terminal adenosine of tRNA according to the particular enzyme that charged the tRNA. Perhaps, EF-Tu selects the 2'-aminoacyl-esters[14], but this is not certain. The charging or loading of tRNA with an amino acid precedes formation of the ternary complex and is catalysed by a set of enzymes[15], called aminoacyl-tRNA synthetases, whose function is also important in ensuring accuracy during the translation process.

Because of the large amount of EF-Tu in bacterial cells, biological roles other than in protein biosynthesis have been investigated. Possible roles for EF-Tu found in plasma membranes and in RNA polymerase are undecided, but EF-Tu does appear to be an essential component of the replicase of RNA bacterophage $Q\beta$[11]. Actually, the large amount of EF-Tu may just be needed to sequester aminoacyl-tRNA in a form which protects the labile aminoacyl ester linkage against chemical hydrolysis at the slightly alkaline cellular pH. This possibility becomes apparent when it is realized that the intracellular concentration of aminoacyl-tRNA is about 0.4 mM, which is of the same order as that of EF-Tu.

At present it is not known how EF-Tu:GDP leaves the ribosome, what the GTP hydrolysis is used for and whether these events are linked to the next step of peptide bond formation.

Step II involves peptidyl transferase

Peptide bond formation constitutes Step II of elongation (see Fig. 1). Since both the peptidyl-tRNA and aminoacyl-tRNA are highly reactive species they should react spontaneously to form the peptide bond. Hence, the ribosome might play a passive part in providing the appropriate sites to permit the reactive species to approach each other. Nevertheless, the formation of the peptide bond on the ribosome is catalysed by an enzyme, peptidyl transferase, a component of the 50 S subunit; probably ribosomal protein L16 contributes at least part of the activity.

What happens to the excess of energy released during peptide bond formation and during GTP hydrolysis is unclear. The energy is not needed apparently for bond formation but may be needed to actuate possible conformational changes of the interacting macromolecules during chain elongation, and for the subsequent relative movement of the ribosome and mRNA in Step III.

Model systems have been devised to obtain information about the enzymically catalysed peptide bond formation and especially about the substrate specificities at the ribosome catalytic site. The use of the aminoacyl-tRNA analogue puromycin to define the P- and A-sites on the ribosomes has been described before in *TIBS*[16]. The initiator tRNA (fMet-tRNA), because it blocked amino acid is removed from the ribosome by puromycin, appears to be located in the P-site. When other aminoacyl-tRNA fragments, blocked and unblocked, and of various lengths, were tested in a model system, it was clear that the enzyme requires an acylaminoacyl-tRNA or acylaminoacyl-tRNA fragment as a substrate in the P-site and recognizes at least the terminal CpCpA.

The substrate requirements of the peptidyl transferase at the A-site have been studied using chemically synthesized analogues of the aminoacyl-adenosine end of aminoacyl-tRNA. There is a strict requirement for a 3'-aminoacyl ester of adenosine with a neighbouring (*cis* vicinal) hydroxy group.

Antibiotic inhibitors and drugs useful in delineating the mechanism of protein biosynthesis have been described in previous reviews[16,17]. The very interesting relationship of peptide bond formation and polypeptide chain termination is to be discussed in the next article in this series by Caskey[2].

Step III involves elongation factor G

The translocation Step III (see Fig. 1) involving relative movement between the ribosome and mRNA clearly consumes much energy during protein biosynthesis.

Studies *in vitro* suggest that EF-G and GTP are together required for the movement of peptidyl-tRNA from the A-site to the P-site of 70 S ribosomes. Concomitantly, the discharged tRNA is ejected from the P-site, and EF-G may also be concerned with this process. Possibly, hydrolysis of GTP could provide the driving energy for the translocation process, since EF-G has been shown to have an associated GTPase activity in the presence of ribosomes (splitting GTP to GDP and inorganic phosphate). However, under certain conditions, slow translocation is possible in the absence of EF-G[18]. Thus, although the translocation process appears to be an inherent property of the monoribosome, EF-G does seem to be required for translocation at the normal rate. The precise role of EF-G in the translocation process is obscure, because GTPase activities (one coupled and one uncoupled) are associated with EF-G. The uncoupled activity has complicated the determination of the number of GTP molecules being split during the elongation step. Probably, the coupled GTPase activity is concerned with the affinity of EF-G for the 70 S ribosome. GTP cleavage after translocation could cause the release of EF-G from the ribosome. The presence of EF-G and GTP on the ribosome facilitates the translocation event but then EF-G must vacate its ribosomal site to allow the ternary complex to bind to the A-site, hence allowing a new elongation step to occur. EF-G and EF-Tu appear to bind at the same or overlapping sites on the 70 S ribosome. Furthermore, their ribosomal binding sites require the presence of ribosomal proteins L7 and L12[7]. *E. coli* EF-G's amino acid sequence (mol. wt 80,000) is being determined and preliminary results have shown that a stretch of about 100 amino acids near the N-terminus has a strong homology with EF-Tu (R. Laursen, personal communication) perhaps reflecting the common binding site for the ribosome or even GTP.

The translocation step is inhibited by a number of antibiotics and GTP analogues which have been useful for determining molecular events in the translocation process[16,17]. Fusidic acid, thiostrepton, viomycin and non-hydrolysable GTP analogues have been especially useful[17]. Fusidic acid appears to bind EF-G irreversibly to the ribosome and it has been possible to isolate mutants with fusidic acid-resistant EF-G. Since temperature sensitive mutants are also available, it should be possible to relate a genetic analysis to changes in structure and function.

Eukaryotic elongation factors

The elongation factors from eukaryotic sources are much less well characterized than from bacteria such as *E. coli*. In general, the eukaryotic factors equivalent to EF-T and EF-G are called EF-1 and EF-2. An EF-Ts-like activity is associated with a polypeptide EF-1β[19] (mol. wt $\simeq$ 30,000) which can be obtained from pig liver or rabbit reticulocytes. The EF-Tu activity is exhibited by another polypeptide EF-1α (mol. wt $\simeq$ 53,000). EF-1β is usually isolated in a complex containing another polypeptide EF-1γ (mol. wt $\simeq$ 53,000) of uncertain function. Interestingly, it has been established that organelles in eukaryotes contain elongation factors different from the cytoplasmic ones[20]. In this respect, mitochondrial elongation factors are synthesized in the cell cytoplasm whereas chloroplast elongation factors are synthesized in the organelle itself.

Earlier findings of a heavy, aggregated form of EF-1 in mammalian cells seem to have been dependent upon the type of source material. It has been reported that these aggregated forms designated EF-1$_H$ are held together by phospholipids and cholesterol esters (see Ref. 6), but so far the question of whether EF-1$_H$ (mol. wt of up to 5×10^5) plays a regulatory role in eukaryotic protein biosynthesis is unclear. For example, it is possible that EF-1H is a storage form of the active subunit EF-1α so that the amount of active subunit available for participating in protein biosynthesis can be controlled by the association into or dissociation of EF-1H.

Concluding remarks

Biological activities related to polypeptide chain elongation can be mentioned only very briefly. A possible coupling of translation and transcription involves the synthesis of unusual guanosine nucleotide derivatives (a tetraphosphate and pentaphosphate), also called magic spots MSI and MSII. Under normal conditions, the ternary complex binds to the ribosome during the decoding process. However, if the tRNA is not charged (as during starvation) it binds directly to the ribosome decoding the mRNA setting off a repetitive reaction synthesizing MSI or MSII which could inhibit RNA polymerase.

The elucidation of the detailed molecular mechanism of chain elongation is hampered by the lack of structural information on the ribosomal environments associated with the A- and P-sites or possible conformational changes in the ribosome or active components such as aminoacyl-tRNA. Conformational change might be used as a driving force in translocation and for the entry and exit of tRNA. Only recently has it been established[21,22] that the tRNA decodes the mRNA in both P- and A-sites.

The accuracy of translation is clearly maintained at the elongation step. Mistakes occur with less than 1% frequency. Whether the proposal that energy from GTP hydrolysis is used to ensure accuracy is correct is still unclear.

Much effort is being spent trying to decide whether control of protein biosynthesis occurs at the translation level, and if so, how[23]. The relation of initiation and control has been reviewed last month[1]. Clearly, a shortage of a particular essential aminoacyl-tRNA could slow the elongation rate, but it is not clear how and when this process is really involved. Recently, the very interesting finding of attenuation[24] of transcription has provided us with a feasible mechanism by which an aminoacyl-tRNA could act as a repressor-like molecule at the initiation of transcription.

References

1 Hunt, T. (1980) *Trends Biochem. Sci.* 5, 178–181
2 Caskey, T. (1980) *Trends Biochem. Sci.* 5, Sept. (in press)
3 Watson, J. D. (1964) *Bull. Soc. Chim. Biol.* 46, 1399–1425
4 Brimacombe, R. (1976) *Trends Biochem. Sci.* 1, 181–183
5 Bielka, H. (1978) *Trends Biochem. Sci.* 3, 156–158
6 Miller, D. L. and Weissbach, H. (1977) in *Molecular Mechanisms of Protein Biosynthesis* (Weissbach, H. and Pestka, S., eds), pp. 323–373, Academic Press, New York
7 Brot, N. (1977) in *Molecular Mechanisms of Protein Biosynthesis* (Weissbach, H. and Pestka, S., eds), pp. 375–411, Academic Press, New York
8 Kaziro, Y. (1978) *Biochim. Biophys. Acta* 505, 95–127
9 Reeh, S. and Pedersen, S. (1978) in *Gene Expression*, Vol. 43 of Federation of European Biochemical Societies (Clark, B. F. C., Klenow, H. and Zeuthen, J., eds), pp. 89–98, Pergamon Press
10 Wittinghofer, A., Frank, R., Gast, W. H. and Leberman, R. (1979) *J. Mol. Biol.* 132, 253–256
11 Miller, D. L. (1978) in *Gene Expression* (Clark, B. F. C., Klenow, H. and Zeuthen, J., eds), 11th FEBS Meeting 1977, Vol. 43, pp. 59–68, Pergamon Press
12 Rich, A. (1978) *Trends Biochem. Sci.* 3, 34–37
13 Clark, B. F. C. (1979) in *Nonsense Mutations and tRNA Suppressors* (Celis, J. E. and Smith, J. D., eds), pp. 1–46, Academic Press, London
14 Sprinzl, M. and Cramer, F. (1979) *Prog. Nucleic Acid Res. Mol. Biol.* 22, 1–69
15 Söll, D. and Schimmel, P. R. (1974) in *Enzymes*, 3rd edn, Vol. 10, 489–538
16 Jimenez, A. (1976) *Trends Biochem. Sci.* 1, 28–30
17 Modolell, J., Girbés, T. and Vazquez, D. (1978) in *Gene Expression* (Clark, B. F. C., Klenow, H. and

Zeuthen, J., eds), 11th FEBS Meeting 1977, Vol. 43, pp. 79–87, Pergamon Press
18 Spirin, A. S. (1978) in *Gene Expression* (Clark, B. F. C., Klenow, H. and Zeuthen, J., eds) 11th FEBS Meeting 1977, Vol. 43, pp. 69–78, Pergamon Press
19 Slobin, L. (1979) *Eur. J. Biochem.* 96, 287–293
20 Ciferri, O., Di Pasquale, G. and Tiboni, O. (1979) *Eur. J. Biochem.* 102, 331–335
21 Wurmbach, P. and Nierhaus, K. H. (1979) *Proc. Natl. Acad. Sci. U.S.A.*, 76, 2143–2147
22 Lührmann, R., Eckhardt, H. and Stöffler, G. (1979) *Nature (London)* 280, 423–425
23 Leader, D. P. (1979) *Trends Biochem. Sci.* 4, 205–208
24 Danchin, A. and Ullmann, A. (1980) *Trends Biochem. Sci.* 5, 51–52

Brian Clark is at the Department of Chemistry, Aarhus University, 8000 Aarhus C, Denmark.

Addendum – B. Clark

The products of *tufA* and *tufB* have now been shown by protein sequencing to differ only in the C-terminal amino acid, i.e. position 393 (Ref. 25). The *tufA* gene product contains Gly and the *tufB* Ser. This was confirmed by the determination of the DNA sequences of the *tufA*[26] and *tufB*[27] genes and their surroundings. The two genes differ in 13 bases, but 11 are in the third base positions of codons so that they are silent changes with respect to the gene products. Additionally, of the two differences at first base positions of codons one occurs in the initiation codon position so that GUG starts the *tufA* mRNA and AUG the *tufB* mRNA. Thus the only expressed difference comes from the C-terminal amino acid codon; in the *tufA* gene it is GGC and in the *tufB* it is AGC.

A knowledge of the EF-Tu primary structure[25] was needed to build our three-dimensional model. In addition to the difference at the C-terminal amino acid, interestingly, the N-terminal amino acid Ser-1 is acetylated in both gene products and Lys-56 is partly methylated so that the total bacterial EF-Tu preparation contains about 60% mono Me Lys. It is not known whether this is concerned with regulatory phenomena.

About 70% of the three-dimensional structure of EF-Tu has now been determined. The GDP-binding site is located on the surface of a highly structured part called the tight domain[28].

'Footprinting' of aminoacyl tRNAs in ternary complexes indicates that the aminoacyl and T stems also interact with EF-Tu[29] in addition to the aminoacyl-end region.

The preliminary indications that *E. coli* EF-G has about a 100 amino acid stretch of homology with EF-Tu has been confirmed[30].

EF-1α has been purified from a number of additional sources, especially the brine shrimp, *Artemia salina*. We have found a conservation of antigenic properties among various EF-1α preparations. This conservation does not appear to extend from eukaryote EF-1α to its prokaryote bacterial counterpart EF-Tu.

Finally, there is now reasonable evidence that the energy from GTP hydrolysis is used to ensure accuracy in translation.

References
25 Jones, M. D., Petersen, T. E., Nielsen, K. M., Magnusson, S., Sottrup-Jensen, L., Gausing, K. and Clark, B. F. C. (1980) *Eur. J. Biochem.* 108, 507–526
26 Yokota, T., Sugisaki, H., Takanami, M. and Kaziro, Y. (1980) *Gene* 12, 25–31
27 An, G. and Friesen, J. D. (1980) *Gene* 12, 33–39
28 Rubin, J. R., Morikawa, K., Nyborg, J., La Cour, T. F. M., Clark, B. F. C. and Miller, D. L. (1981) *FEBS Lett.* 129, 177–179
29 Boutorin, A. S., Clark, B. F. C., Ebel, J. P., Kruse, T., Petersen, H. U., Remy, P. and Vassilenko, S. (1981) *J. Mol. Biol.* 152, 593–608
30 Ovchinnikov, Yu, A., Alakhov, Yu, B., Bundulis, Yu, P., Bundule, M. A., Dovgas, N. V., Kozlov, V. P., Motuz, L. P. and Vinokurov, L. M. (1982) *FEBS Lett.* 139, 130–135

Peptide chain termination

C. Th. Caskey

In this, the third of our three-part series on protein synthesis, chain termination is reviewed. Termination differs markedly from initiation and elongation (reviewed in July and August TIBS) since the codon recognition molecules are proteins, not tRNA, and a perturbation of the peptidyltransferase leads to hydrolysis of the peptidyl-tRNA rather than the formation of peptide bonds.

Peptide chain termination is directed by one of three specific codons (UAA, UAG, or UGA) and results in the release of the completed peptide from ribosomal bound tRNA. In *E. coli* two proteins (release factors RF1 and RF2) have different codon specificities while a third factor, which stimulates the event, recognizes GDP and GTP but not codons. A single larger protein factor (RF) isolated from rabbit reticulocytes, functions with any of the three termination codons and recognizes GTP. The ribosomal peptidyltransferase is required for hydrolysis of peptidyl-tRNA at chain termination. Thus the process of peptide chain termination differs markedly from chain initiation and elongation since the codon recognition molecules are proteins not tRNA, and a perturbation of the peptidyltransferase leads to hydrolysis of the peptidyl-tRNA rather than the formation of peptide bonds[1].

Release factors

The participation of protein factors in peptide chain termination was initially proposed by Ganoza[2] and subsequently established by Capecchi[2]. Development of a simple assay *in vitro* for peptide chain termination led to our identification of two codon-specific protein factors in *E. coli*[3]. Using an initiation complex (f[³HMet-tRNA:AUG:ribosome) as substrate for the peptide chain termination event, release of f[³H]methionine (a peptide analogue) was found to require protein factors, termination codons, and, under certain conditions, guanine nucleotides.

Two codon-specific release factors (RF1 for UAA or UAG and RF2 for UAA or UGA) have been purified to homogeneity from *E. coli* (see Table I). Since the factors are free of oligo- or poly-nucleotides the codon specificity results from protein recognition mechanisms[1]. These acidic proteins have been found to have common immune determinants and therefore probably some structural homology although they differ significantly in their physical properties[6]. While some preliminary data are available on the structure of RF2, the amino acid sequence of RF1 and RF2 are not known and their genes have not yet been cloned.

During the purification of RF1 and RF2 we discovered a third release factor (RF3) which alone had no release activity in the presence of termination codons but stimulated the activities of RF1 and RF2[6]. This protein factor interacts with GTP or GDP and, as will be discussed subsequently, stimulates RF1 and RF2 binding to the release from ribosomes[1]. The RF3 molecule has not yet been further characterized (Table I). Capecchi and Klein found that a purified fraction of EF-Tu possessed this stimulatory activity and thus concluded that RF3 and EF-Tu are identical[3]. In our laboratory, however, preparations with high RF3 activity have been separated from EF-Tu activity[7] and we conclude that RF3 is a specific protein factor involved in peptide chain termination, not chain elongation.

Eukaryotic release factor was initially isolated from rabbit reticulocytes and subsequently from mammalian liver[8] and insect cells[9]. The molecular characteristics of rabbit reticulocyte release factor are given in Table I. The single protein factor will *in vitro* stimulate release of f[³H]methionine from the reticulocyte ribosomal substrate (f[³H]Met-tRNA: AUG:ribosome) in the presence of tetranucleotides UAAA, UAGA, or UGAA. The requirement for a tetranucleotide rather than trinucleotide is, we believe, a characteristic of the *in vitro* assay since substitution of different bases in the 3'-position of the tetranucleotide had no adverse effect on recognition. Thus it appears the 3' terminal base is not a part of the recognition process but confers an element of stability on the codon recognition complex. The reticulocyte RF interacts with GTP and GDP and promotes an active ribosomal dependent GTPase reaction[8] (see below).

Termination codon recognition

The recognition of peptide chain termination codons has been of considerable interest as it involves a deviation from the principle of the recognition of one nucleic acid by another by base pairing. The recognition of peptide chain termination codons on ribosomes is studied as a partial reaction utilizing ³H-labelled terminator codons for the detection of RF·[³H]terminator codon:ribosome complexes, which are retained on millipore filters[10]. In this assay, RF1 binds to ribosomes in response to UAA and UAG, while RF2 binds in response to UAA and UGA. Capecchi and Klein[3] attempted to study this recognition independent of ribosomes by equilibrium dialysis of RF and oligonucleotides containing terminator codons. Although the specificity of codon recognition was not absolute in the absence of ribosomes the studies supported the concept that RF molecules were responsible for codon recognition. More recently the recognition of specific nucleotide sequences of DNA by proteins, restriction endonucleases, has been extensively characterized[11].

The molecular structure of the RF 'anticodon' region is still unknown. The specificity of codon recognition has, however, been investigated indirectly using modified trinucleotides (3-Me-UAH, 3-Me-UAA, 5-Me-UAG, 5-Me-UAA, Br-UAG, h-UAH and UAI). These studies revealed a specificity for RF which is remarkably similar to that of the Watson–Crick and wobble base pairing[12].

TABLE I
Soluble chain termination factors

Designation	Molecular weight	Codon specificity	GTP recognition
E. coli			
RF1	47,000[a]	UAA and UAG	None
RF2	48,000[a], 35,000[b]	UAA and UGA	None
RF3	46,000[a]	None	Yes
Rabbit reticulocyte			
RF	56,500[a], 105,000[b]	UAAA, UAGA and UGAA	Yes

[a] By sodium dodecylsulphate–polyacrylimide gel electrophoresis.
[b] Equilibrium sedimentation under non-denaturing conditions.

Dalgarno and Shine[13] proposed, on the basis of the sequence of the 3' terminus of the 16 S rRNA, that recognition of termination codons may involve these sequences. They found a high frequency of UUA_OH terminal sequences as well as a UCA sequence within eight residues of the terminus. Such sequences could accommodate base pairing with UAA, UAG, and UGA. The importance of the 3' terminus of the 16 S rRNA in the recognition of termination codons is clear from our studies employing the specific nuclease cloacin DF13[14]. Cleavage with cloacin of 49 nucleotides from the 3' terminus of 16 S RNA in ribosomes inactivates the recognition of codons by RF but not the RF-mediated peptidyl-tRNA hydrolysis. It is unclear if this effect of the nuclease on codon recognition is the consequence of a perturbation of the ribosomal structure. Using another approach we found the recognition of codons on *B. subtilis* ribosomes by *E. coli* RF to be qualitatively unaltered, although the 3' terminus of the 16 S rRNA of *B. subtilis* does not contain sequences complementary to termination codons. Thus the preponderance of data favours the idea that RF1 and RF2 molecules recognize directly termination codons.

A single protein factor in rabbit reticulocyte extracts recognizes all three termination codons, a situation significantly different from that in *E. coli*. RF molecules from *E. coli* and rabbit reticulocytes are active only with the homologous ribosomes[15].

RF soluble factor and GTP requirements

In *E. coli* the protein factor RF3 promotes binding to and release from ribosomes of RF1 and RF2[7]. RF3 reduces the K_m for trinucleotide codons by 5–8-fold during peptide chain termination *in vitro* without affecting the V_{max} of peptidyl-tRNA hydrolysis, suggesting that the stimulatory activity is related to ribosomal binding[6], The RF3 stimulates formation of RF1 or RF2:[^{3}H]UAA:ribosomal intermediates supporting this concept. Furthermore, addition of GDP to such recognition complexes formed in the presence of RF3 leads to their rapid disruption. We have not succeeded in demonstrating a requirement for GTP for the binding of RF1 or RF2 to ribosomes in the presence of RF3, although RF3 is known to have a ribosomal dependent GTPase activity. It is clear that EF-G, another soluble factor that has a ribosomal dependent GTPase activity, cannot simultaneously occupy the ribosome with RF1 or RF2. These data indicate that both binding and release of RF1 and RF2 are stimulated by RF3 and suggest that GTP and its hydrolysis are involved. However, proof of this proposed cyclic series of events has not been possible *in vitro*.

In the case of peptide chain termination using reticulocyte ribosomes and RF, the situation is clear. The release factor possesses a ribosomal dependent GTPase[16] which is markedly stimulated by codons. The binding of RF to ribosomes is stimulated by GTP or the non-hydrolysable analog GDPCP, but not by GDP. Since GTP but not GDPCP stimulates the catalytic behavior of RF it appears likely that GTP facilitates RF binding to the ribosome whereupon GTP hydrolysis occurs, followed by displacement of the RF and GDP from the ribosome (Fig. 1).

Peptidyl-tRNA hydrolysis

The role of the peptidyltransferase activity of the 50 S ribosomal subunit in peptide chain termination was initially suggested from the study of antibiotics which inhibited both the peptidyltransferase and the peptidyl-tRNA hydrolysis step of chain termination[17]. This partial reaction of peptide chain termination could be studied independent of codons if RF molecules were made to bind to the ribosome by the addition of ethanol[1]. The antibiotics ami-

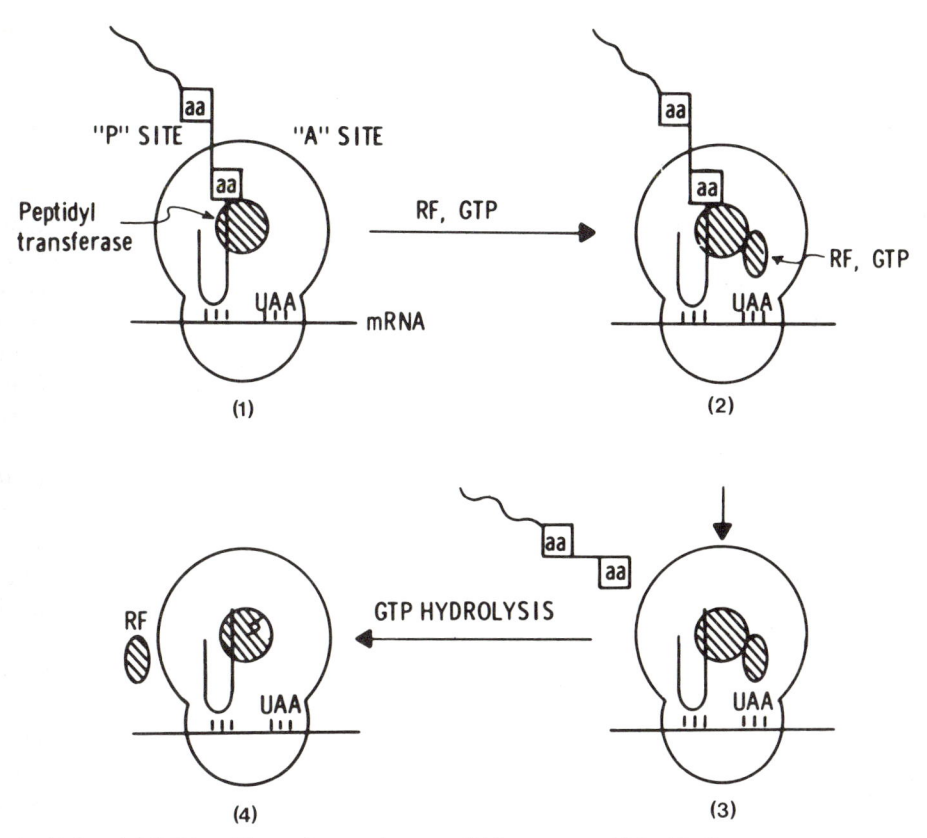

Fig. 1. (1) Peptidyl-tRNA in 'P' site after translocation. (2) Interaction of RF with ribosome in response to terminator codon. (3) Peptidyl-tRNA hydrolysis involving interaction of RF and peptidyltransferase. (4) Dissociation of RF from the ribosome.

cetin, lincocin, chloramphenicol, and sparsomycin inhibit peptidyltransferase and RF-mediated peptidyl-tRNA hydrolysis in parallel. These antibiotics do not inhibit the binding of RF to ribosomes in response to codons, indicating an inhibition of a specific partial reaction of peptide chain termination.

Peptidyltransferase has several catalytic activities which can be demonstrated *in vitro*, including hydrolysis of peptidyl-tRNA[18]. It catalyses peptide bond formation when an amino group is the nucleophilic agent attacking the aa-tRNA ester linkage; this occurs normally during peptide chain elongation by aa-tRNA. The peptidyltransferase center will also catalyse ester formation when the nucleophilic agent is an alcohol, and hydrolysis of the nascent peptidyl-tRNA when the nucleophilic agent is water. Since RF has no esterase activity and peptidyltransferase is required for peptidyl-tRNA hydrolysis, it appears most likely that RF interacts with the peptidyltransferase to stimulate hydrolysis of the peptidyl-tRNA. It is of interest that two antibiotics, lincocin (with *E. coli* ribosomes) and anisomycin (with rabbit reticulocyte ribosomes) have differential effects on the peptidyltransferase activities of peptide bond and ester formation, and peptidyl-tRNA hydrolysis. Each inhibits, on the respective ribosomes, peptide bond and ester formation while stimulating markedly peptidyl-tRNA hydrolysis[17]. Such modification of peptidyltransferase specificity might operate in chain termination, with RF, rather than lincocin or anisomycin, being the modifying component. Recently, using lincocin analogs the structural requirements for this effect have been determined[19].

Ribosomal site of RF binding

The ribosomal proteins required for peptide chain termination have been investigated with: (1) antibiotics, (2) antibodies directed against specific proteins, (3) ribosomes depleted of proteins, and (4) crosslinkage of RF to ribosomal proteins. The results, collectively, give an indication of the ribosomal domain(s) of contact.

The ribosomal proteins S4 and L4 are involved in peptide chain termination because the antibiotics streptomycin and erythromycin are inhibitors of peptide chain termination and mutations in these proteins lead to resistances to these antibiotics[17]. Using antibodies developed to specific ribosomal proteins, the proteins L7/L12, S2 and S3 were shown to be required for RF binding, while L16 and L11 affected peptidyl-tRNA hydrolysis[19]. These proteins have been implicated as part of the peptidyltransferase center and surrounding domain, respectively. Both L16 and L11 are required for chloramphenicol binding[20]. Ribosomes depleted of L7/L12 are defective for RF binding[21], thus confirming the antibody studies. Depleting ribosomes of L11 did not significantly affect the peptidyltransferase activity and had minimal effect on peptide chain termination[22]. Direct linkage of RF to ribosomal proteins through di-imido, bifunctional, crosslinking agents has provided direct evidence for close contact[23]. RF was found to crosslink to L2, L7/L12, L11, S17, S21 and S18. Pongs and Rossner[24], using an affinity probe, which is a derivative of the termination codon UGA, found crosslinkage to S18 and S4 upon codon recognition. It is not surprising that RF makes contact with many of the ribosomal proteins that also interact with EF-Tu[25]. The need for an intact 3' terminus of the 16S rRNA has been previously discussed.

Models of the 30 S and 50 S ribosomal subunits have been developed by Lake[26] and Stöffler[27] based upon immune electron microscopy which estimates shape and position of ribosomal proteins. The 30 S ribosomal subunit has an embryo shape with S4, S18, and S21 localized to the small head, the region involved in codon recognition for aa-tRNA and interaction with IF2 and IF3[28]. The 50 S ribosomal subunit is envisaged as a 'crowned seat' where L2, L7/L12, and L11 are located within the seat. This binding domain for RF is a region of the interface between the two ribosomal subunits[28].

Regulation at peptide chain termination codons

Mutations which generate chain termination codons lead to premature chain termination and release of protein fragments. Such mutations can be phenotypically suppressed in cells that carry suppressor tRNA mutations. The mutant suppressor aa-tRNA can recognize termination codons and translate them. RF and aa-tRNA$_{su}$ compete for the termination codon. Thus regulation of mutant gene expression is achieved. Nonsense mutations occur in prokaryotes and eukaryotes and a variety of suppressor tRNA mutations have been identified in both[1].

Concluding comments

A model for the intermediate events of peptide chain termination, which is based upon data obtained from studies *in vitro* of bacterial and mammalian peptide chain termination, is presented in Fig. 1. Not all intermediate events and requirements have been demonstrated for each cell type, and thus the model is the best fit of all the available information.

Both mammalian and bacterial cells utilize the same terminator codons (UAA, UAG, and UGA). The codons are recognized by two factors in *E. coli*, while a single factor from rabbit reticulocytes recognizes all three. A separate factor, RF3, identified in bacterial extracts, stimulates the binding and release of RF1 and RF2 to ribosomes and interacts with GDP and GTP. The RF of reticulocytes requires GTP and its hydrolysis for binding and release from ribosomes. The hydrolysis of peptidyl-tRNA at peptide chain termination is catalysed by the peptidyltransferase center as a consequence of RF binding to the ribosome.

These intermediate steps lead to the cyclic binding and release of RF in response to termination codons and result in the release of nascent peptides from ribosomal bound peptidyl-tRNA. Suppressor aa-tRNA molecules compete in this recognition leading to chain elongation rather than termination at such codons.

Acknowledgement

This work was supported by the Robert Welch Foundation.

References

1 Caskey, C. T. and Campbell, J. M. (1979) in *Nonsense mutations and tRNA suppressors* (Celis, J. E. and Smith, J. D., eds), p. 81, Academic Press, New York
2 Ganoza, M. C. (1966) *Cold Spring Habor Symp. Quant. Biol.* 31, 273
3 Capecchi, M. R. and Klein, H. A. (1969) *Cold Spring Habor Symp. Quant. Biol.* 34, 469
4 Scolnick, E., Tompkins, R., Caskey, T. and Nirenberg, M. (1968) *Proc. Natl. Acad. Sci. U.S.A.* 61, 768
5 Campbell, J. M. and Caskey, C. T. (1979) *Arch. Biochem. Biophys.* 90, 1032
6 Milman, G., Goldstein, J., Scolnick, E. and Caskey, T. (1969) *Proc. Natl. Acad. Sci. U.S.A.* 63, 183
7 Goldstein, J. L. and Caskey, C. T. (1970) *Proc. Natl. Acad. Sci. U.S.A.* 67, 537
8 Goldstein, J. L., Beaudet, A. L. and Caskey, C. T. (1970) *Proc. Natl. Acad. Sci. U.S.A.* 67, 99
9 Ilan, J. (1973) *J. Mol. Biol.* 77, 437
10 Scolnick, E. M. and Caskey, C. T. (1969) *Proc. Natl. Acad. Sci. U.S.A.* 64, 1235
11 Roberts, R. J. (1980) *Nucleic Acid Res.* 8, r63
12 Smut, J., Kemper, W., Caskey, T. and Nirenberg, M. (1970) *J. Biol. Chem.* 245, 2753
13 Dalgarno, L. and Shine, J. (1973) *Nature (London) New Biol.* 245, 261
14 Caskey, C. T., Bosch, L. and Konecki, D. S. (1977) *J. Biol. Chem.* 252, 4435
15 Konecki, D., Aune, K. C., Tate, W., Caskey, C. T. (1977) *J. Biol. Chem.* 13, 4514
16 Beaudet, A. L. and Caskey, C. T. (1971) *Proc. Natl. Acad. Sci. U.S.A.* 68, 619
17 Caskey, C. T. and Beaudet, A. L. (1971) in *Molecular Mechanisms of Antibiotic Action on Protein Biosynthesis and Membranes*, p. 326, Elsevier, Amsterdam
18 Scolnick, E., Milman, G., Rosman, M. and Caskey, C. T. (1970) *Nature (London)* 225, 152
19 Tate, W. P., Caskey, C. T. and Stöffler, G. (1975) *J. Mol. Biol.* 93, 375
20 Roth, H. E. and Nierhaus, K. H. (1975) *J. Mol. Biol.* 94, 111
21 Brot, N., Tate, W. P., Caskey, C. T. and Weissbach, H. (1974) *Proc. Natl. Acad. Sci. U.S.A.* 71, 89
22 Armstrong, I. L. and Tate, W. P. (1978) *J. Mol. Biol.* 120, 155
23 Stöeffler, G., Tate, W. P. and Caskey, C. T. (1982) *J. Biol. Chem.* 257, 4203
24 Pongs, O. and Rossner, E. (1975) *Hoppe-Seyler's Z. Physiol. Chem.* 356, 1297
25 José, C. S., Kurland, C. G. and Stöffler, G. (1976) *FEBS Lett.* 71, 133
26 Lake, J. A. (1976) *J. Mol. Biol.* 105, 131
27 Tischendorf, G. W., Zeichhardt, H. and Stöffler, G. (1979) *Mol. Gen. Genet.* 134, 209
28 Brimacombe, R., Stöffler, G. and Wittman, H. G. (1978) *Annu. Rev. Biochem.* 47, 217

C. Th. Caskey is at the Howard Hughes Medical Institute, Baylor College of Medicine, Houston, TX 77030, U.S.A.

Enzymic editing mechanisms in protein synthesis and DNA replication

Alan R. Fersht

The exquisite fidelity of the genetic coding process is maintained during the replication of DNA and the synthesis of proteins by a series of editing or proof-reading reactions which remove errors. Without these checks, mutation rates would be unacceptably high and proteins largely heterogeneous.

The specificity of enzymes is as remarkable as their catalytic power. Yet the high selectivity of the common metabolic enzymes for their substrates seems almost trivial when compared with the extraordinary fidelity of the enzymes involved in macromolecular synthesis. It is now known from experiment, for example, that the error rate in DNA replication in *Escherichia coli* is only one mistake in 10^8–10^{10} nucleotides polymerized and the overall error rate in transcribing the DNA and translating the message into protein is only about 1 in 10^4 amino acid residues incorporated. This specificity is, in fact, beyond the theoretical thermodynamic limits for simple enzymes and is only possible because of the evolution of *editing* or *proof-reading* mechanisms.

Editing or proof-reading

In addition to their synthetic active site, certain key enzymes involved in polymerization have evolved a second, hydrolytic, active site which is used to destroy incorrect intermediates as they are formed. The synthetic process is thus double-checked at each step so that errors may be removed before they are permanently incorporated. Just how this is done will be seen later.

Editing is distinct from *repair*. The double-stranded DNA molecule, because of the complementary nature of the base pairing, is capable of being post-replicatively repaired by the excision of lesions in one strand followed by patching and sealing by a DNA polymerase and ligase.

Selection of amino acids in protein synthesis

The problem: Amino acids are selected in protein synthesis during the aminoacylation of tRNA. This is a two-step reaction composed of activation of the amino acid (equation 1) followed by its transfer to tRNA (equation 2). Both steps are catalysed by the same aminoacyl-tRNA synthetase, each amino acid (AA) having its own specific enzyme (E^{AA}) and tRNAs ($tRNA^{AA}$).

$$E^{AA} + AA + ATP \rightarrow$$
$$E^{AA} \cdot AA\text{-}AMP + PPi \quad (1)$$
$$E^{AA} \cdot AA\text{-}AMP \xrightarrow{tRNA^{AA}}$$
$$AA\text{-}tRNA^{AA} + AMP + E^{AA} \quad (2)$$

In principle, two selections are made: recognition by the enzyme of first the correct amino acid and then the correct tRNA. In practice, tRNA recognition is not a problem; it is such a large molecule that there is adequate scope for distinctive structural variation. On the other hand, as pointed out by Pauling in 1958[1], the amino acids are generally so similar in structure that there are severe problems in distinguishing between them. The classic difficulty is the discrimination between an amino acid and its smaller homologue, for example, between alanine and glycine. Since the cavity at the active site of the alanyl-tRNA synthetase is large enough to accommodate alanine, it must also be able to bind the smaller glycine. Any discrimination between the two can arise only from the difference in their binding energies with the enzyme; in this case the binding energy of a methylene group, which can contribute only up to a factor of 100–200 to the specificity. This problem is repeated with the isoleucyl-tRNA synthetase and valine, and the threonyl-tRNA synthetase and serine. On the other hand, amino acids larger than the substrate molecule should be rejected more efficiently because of the steric hindrance resulting from too small an active site; the repulsive forces can often be far higher than the attractive ones.

It is found experimentally that the aminoacyl-tRNA synthetases do make the predicted errors by activating the smaller (or isosteric) amino acids (equation 1), but the overall reaction (equations 1–2), however, is very precise.

The solution: The answer to the paradox of how the accuracy of the overall reaction is higher than the partial reaction was found experimentally by Norris and Berg in 1965[2]. They isolated large quantities of the isoleucyl-tRNA synthetase and examined directly its complexes with isoleucyl adenylate in the 'correct' reaction, and with valyl adenylate formed in the 'incorrect' reaction. Whereas the addition of $tRNA^{Ile}$ to the correct complex leads to the formation of the Ile-$tRNA^{Ile}$ (equation 3), the addition to the 'incorrect' complex leads to its quantitative hydrolysis (equation 4).

$$E^{Ile} \cdot Ile\text{-}AMP \xrightarrow{tRNA^{Ile}}$$
$$Ile\text{-}tRNA^{Ile} + E^{Ile} + AMP \quad (3)$$

$$E^{Ile} \cdot Val\text{-}AMP \xrightarrow{tRNA^{Ile}}$$
$$Val + tRNA^{Ile} + E^{Ile} + AMP \quad (4)$$

The result of the activation (equation 1) and the hydrolytic (equation 4) reactions is that in the presence of valine and tRNA^{Ile}, the isoleucyl-tRNA synthetase acts as an ATP-pyrophosphatase, wastefully, and of necessity, hydrolysing ATP. In this experimental discovery of editing mechanisms, Norris and Berg established diagnostic tests of their occurrence: (a) the formation of an intermediate but not of the final product; (b) the consumption of a high energy compound (ATP, GTP, dNTPs etc.).

The editing pathway: It was subsequently found that the aminoacyl-tRNA synthetases are weak esterases towards aminoacyl-tRNA but the hydrolytic activity is higher with some mischarged tRNAs. This led to the suggestion that the esterase activity is responsible for editing[3,4].

The reaction pathway has now been rigorously established for the editing of amino acid activation by the valyl-tRNA synthetases[5,6]. The incorrect amino acid (e.g. threonine which is isosteric with valine, or α-aminobutyrate which is smaller by one methylene group) is activated and transferred to the tRNA^{Val} in the normal manner. But the misacylated tRNA is rapidly and specifically deacylated before it can dissociate from the enzyme (equation 5).

$$E^{Val} \cdot Thr-AMP \xrightarrow{tRNA^{Val}}$$
$$AMP + E^{Val} \cdot Thr-tRNA^{Val} \xrightarrow{fast}$$
$$E^{Val} + Thr + tRNA^{Val} \quad (5)$$

The hydrolytic activity towards misacylated tRNA is not inhibited by the normal substrates of the aminoacylation reaction which bind at the synthetic active site[4-6] suggesting the presence of a distinct and separate hydrolytic site.

Amino acids are sorted by size and chemical characteristics – the double sieve

It has been suggested that the two sites sort the whole range of amino acids according first to size and then, if need be, by chemical characteristics[7,8]. Consider first the case of an aminoacyl-tRNA synthetase that has no isosteric competitor with its correct substrate, e.g. the isoleucyl-tRNA synthetase. The activation site is constructed to be just large enough to accommodate isoleucine but the hydrolytic site is smaller and just large enough to accommodate valine. All amino acids larger than the correct substrate are rejected at a tolerable level from binding at the activation site by steric hindrance. On the other hand, all smaller amino acids have access to the activation site and are activated, albeit at lower rates. But these same amino acids (or rather their products) are accepted by the hydrolytic site. The correct amino acid is largely excluded from the hydrolytic site by steric hindrance. The prinicple of steric exclusion is thus used twice to obtain the desired sorting. The above case is relatively simple since sorting can be performed by size alone. Where isosteric amino acids occur, such as threonine competing with valine, sorting by chemical characteristics must be superimposed upon the sorting by size. For example, the hydrolytic site of the valyl-tRNA synthetase presumably has a hydrogen bond acceptor/donor site to bind specifically the hydroxy group of threonine and repel the corresponding hydrophobic methyl group of valine. Otherwise, the exclusion of all larger amino acids by the activation site and the acceptance of all smaller amino acids by the hydrolytic or editing site takes place as described in the first case. The feasibility and energetics of the double-sieving can be seen from the data in Table I. The double-sieve hypothesis not only rationalizes all the evidence but makes predictions: for example, only amino acids smaller than, or isosteric with, the correct substrate require editing; unnatural isosteres (e.g. norleucine for methionine) may cheat the editing since recognition is primarily by size.

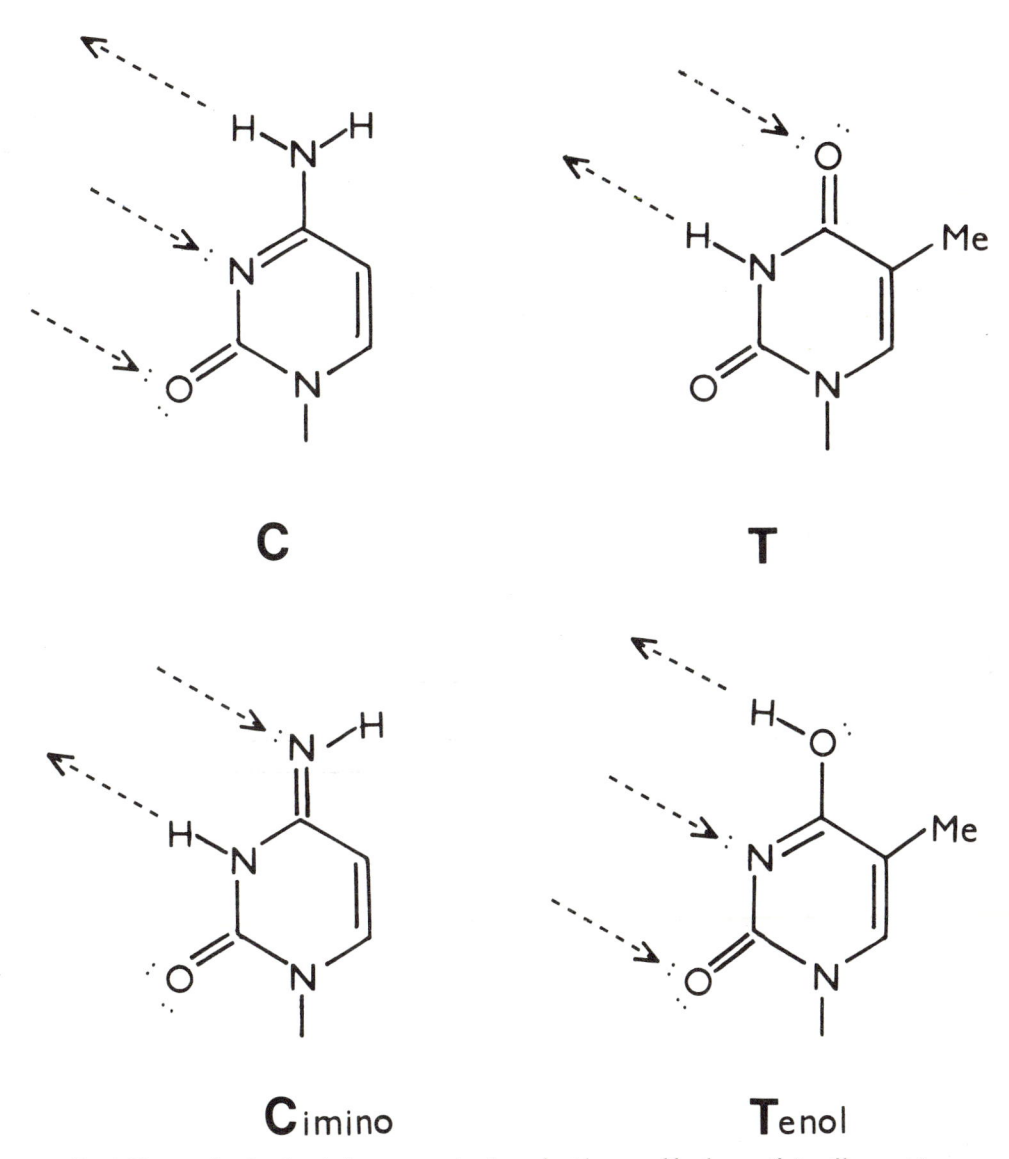

Fig. 1. Keto–enol and amino–imino tautomerism in nucleotides scrambles the specificity of base pairing.

Which aminoacyl-tRNA synthetases require editing activity?

Some amino acids are sufficiently different from their competitors that they are selected with a sufficient accuracy by the first site, without the need for subsequent editing. A rule of thumb for judging when editing mechanisms are required is that the error rate must be kept below about 5 × 10^{-4}, the maximum error rate observed so far for protein synthesis. For example, cysteine is bound to the cysteinyl-tRNA synthetase so much more tightly than its smaller or isosteric competitors (alanine, glycine, serine) that sufficient precision is attained without an editing site[9]. The binding energy of a methylene group is much lower and so the presence of one or two

TABLE I. Double-sieving by valyl-tRNA synthetase[a]

Amino acid	Relative rate of activation by E^{Val}	Rate constant for hydrolysis of AA-tRNA^{Val} catalysed by E^{Val} (s^{-1})
Ile	$\leq 2 \times 10^{-5}$	Not measured[b]
Val	1	4.5×10^{-3}
Thr	4×10^{-3}	40
αBut[c]	5×10^{-3}	50
Ala	1×10^{-4}	>1
Gly	9×10^{-8}	Not measured

[a] Data from Refs. 5 and 6.
[b] In similar examples AA-tRNA is relatively stable.
[c] α-Amino -L-butyric acid.

additional -CH$_2$- groups on the correct substrate is not generally sufficient to confer adequate selection without editing. Aminoacyl-tRNA synthetases that clearly need an editing site are: Ala (for Gly), Val (for Thr and Ala), Thr (for Ser), Ile (for Val), Met (for homocysteine) and probably also Phe and Leu (for other hydrophobic amino acids). The glycyl-tRNA synthetase presumably has no problems since all the competitive amino acids are larger and so can be excluded by steric hindrance.

DNA Replication

The problem: There is a fundamental difference between the selection of amino acids and the matching of complementary base pairs in the replication and transcription of DNA. Each amino acid has its own activating enzyme precisely tailored to its fit, but one DNA polymerase or RNA polymerase has to cope with all four combinations of base pairing. The overall size and shape of A:T, T:A, G:C and C:G are very similar and it appears that the polymerase has one synthetic active site which recognizes just this overall shape, i.e. whether a purine is matched with a pyrimidine. Specificity is provided by the complementarity of base pairing. Therin lies an inherent weakness: keto–enol and amino–imino tautomerism scramble the specificity. For example, as seen in Fig. 1, the enol form of T has base pairing characteristics identical to C, and the imino tautomer of C is equivalent to T. Since the overall shapes are the same, T$_{enol}$ *is* C and C$_{imino}$ *is* T as far as the polymerase is concerned. As the enol and imino tautomers are present at about 1 in 10^4–10^5, there is a threshold of about 10^{-4}–10^{-5} for the error rate. (Further errors can arise because there is a possibility of some purine–purine mismatching – see Ref. 10 for a detailed discussion.)

The (partial) solution: Again, specificity is enhanced by an editing mechanism, discovered by Brutlag and Kornberg with DNA polymerase I[11]. DNA is replicated in the 5' → 3' direction. But the enzyme also has an exonuclease activity working in the opposite direction (3' → 5') which removes mismatches as they are formed.

The polymerase believed to be responsible for the major replication of DNA in *E. coli* is DNA polymerase III, which also has a 3'–5' exonuclease activity. During 1979, whilst visiting Arthur Kornberg's laboratory, I was able to measure the fidelity of this enzyme during the copying of phage φX174 DNA *in vitro*[12] and found that (a) the kinetics are consistent with the active participation of an editing mechanism and (b) under conditions mimicking those in the cell, the error rate in copying the particular nucleotide studied is about 5 × 10^{-7}, close to that found from genetic studies of the mutation frequency of φX174 *in vivo*. Although the studies *in vitro* and *in vivo* match up nicely, the error rate is 2–3 orders of magnitude greater than that found for the replication of *E. coli* itself.

Post replicative mismatch correction – the additional factor: It has generally been thought that error correction during replication has to take place as soon as the nucleotide has been inserted and be-

TABLE II. Error rates and editing.

Process	Observed error rate	Predicted error rate in absence of editing	Experimental evidence for editing
Replication DNA → DNA	10^{-8}–10^{-10}	10^{-4}–10^{-5}	Yes
Protein Synthesis DNA → protein (overall)	$<10^{-3}$	10^{-1}–10^{-5}	Yes
DNA → mRNA (transcription)	$<10^{-3}$	10^{-4}–10^{-5}	No
E$_{AA}$:AA (amino acid recognition)	$<10^{-3}$	$\leq 10^{-1}$	Yes
E$_{AA}$:tRNA^{AA} (tRNA recognition)	$<10^{-3}$	10^{-6}–10^{-8}	No
mRNA:AA-tRNA (codon:anticodon)	$<10^{-3}$	?	Yes

fore further polymerization takes place. Otherwise, it would not be known which of the two nucleotides in a mismatched based pair is the 'incorrect' one. However, it now appears that methylation of DNA provides a label to identify which is the parent strand in a newly replicated duplex. E. coli DNA is methylated after its synthesis so that during replication the newly synthesized strand is temporarily different from its paired parent. Very recent data from Radman's laboratory show that there is an enzyme system which patrols the newly synthesized strand, excising mismatches and inserting the correct nucleotide[13]. Genetic studies show that this increases the fidelity of replication by a factor of 10^3–10^4. Since the overall error in replication is 10^{-8}–10^{-10} per base pair replicated, an accuracy of only 10^{-5}–10^{-7} is required for the DNA polymerase. This is just what was found for the Pol III-catalysed replication of ϕX174 DNA *in vitro* to give *single*-stranded viral DNA!

Editing elsewhere in protein synthesis

RNA polymerase does not have a 3' → 5' exonuclease activity and there appears to be no editing function. Studies *in vitro* give an error rate of about 10^{-4} per base transcribed[14] and this appears adequate for the overall accuracy. The specificity of codon:anticodon recognition during mRNA:AA-tRNA interactions on the ribosome in not known, but there is now evidence that an undefined editing mechanism operates[15,16]. A summary of where editing appears to be necessary and has been found is given in Table II.

Conclusion

A typical protein in E. coli has a mol. wt of about 110,000, containing some 1000 amino acid residues. Consider how the error rates listed in Table II for the absence of editing affect its composition. First the protein would be genetically unstable since the error rate for replicating its genome would be 0.03–0.3 per gene per generation. Second its construction would be too loosely defined, each molecule containing from a few to possibly 100 incorrect amino acids. When this is multiplied by the total number of proteins present in E. coli (~3000), it is realized that there would be a few, if any, viable progeny per generation and those produced would be inefficient. Thus, without the evolution of editing mechanisms, the present system of genetic code and amino acids would certainly not be able to maintain higher forms of life.

References

1. Pauling, L. (1958) in *Festschrift Arthur Stoll*, pp. 597–602, Birkhauser AG., Basel
2. Norris, A. and Berg, P. (1965) *Procl Natl. Acad. Sci. U.S.A.* 52, 330–337
3. Eldred, E. W. and Schimmel, P. R. (1972) *J. Biol. Chem.* 247, 2961–2964
4. Yarus, M. (1972) *Proc. Natl. Acad. Sci. U.S.A.* 69, 1915–1919
5. Fersht, A. R. and Kaethner, M. M. (1976) *Biochemistry* 15, 3342–3346
6. Fersht, A. R. and Dingwall, C. (1979) *Biochemistry* 18, 1238–1245
7. Fersht, A. R. (1977) *Enzyme Structure and Mechanism*, ch. 11, W. H. Freeman & Co., Reading and San Francisco
8. Fersht, A. R. and Dingwall, C. (1979) *Biochemistry* 18, 2627–2631
9. Fersht, A. R. and Dingwall, C. (1979) *Biochemistry* 18, 1245–1249
10. Topal, M. D. and Fresco, J. R. (1976) *Nature (London)* 263, 285–293
11. Brutlag, D. and Kornberg, A. (1972) *J. Biol. Chem.* 247, 241–248
12. Fersht, A. R. (1979) *Proc. Natl. Acad. Sci. U.S.A.* 76, 4946–4950
13. Glickman, B. W. and Radman, M. (1980) *Proc. Natl. Acad. Sci. U.S.A.* 77, 1063–1067
14. Springgate, C. F. and Loeb, L. A. (1975) *J. Mol. Biol.* 97, 577–591
15. Thompson, R. C. and Stone, P. J. (1977) *Proc. Natl. Acad. Sci. U.S.A.* 74, 198–202
16. Yates, J. L. (1979) *J. Biol. Chem.* 254, 11550–11554

Alan R. Fersht is at the Department of Chemistry, Imperial College of Science and Technology, London SW7 2AY, U.K.

RNA and DNA sequence recognition and structure–function of aminoacyl tRNA synthetases

Paul Schimmel, Scott Putney and Ruth Starzyk

Aminoacyl tRNA synthetases establish the rules of the genetic code through specific protein-RNA interactions which may have a covalent component. Structural features of these enzymes are now emerging and recently it was discovered that genes can be regulated through synthetase-DNA interactions.

Genes are decoded into proteins by rigid adherence to rules that assign specific trinucleotide sequences to specific amino acids. These relationships are made entirely by aminoacyl tRNA synthetases which are the biochemical agents responsible for correlating amino acids with trinucleotides.

Synthetases, which probably arose over one billion years ago, were discovered only 25 years ago during the research on coding[1-4]. These enzymes catalyze the attachment of an amino acid (AA) to a cognate tRNA in a two step reaction shown below:

$$E + AA + ATP \rightleftharpoons E \cdot AA \sim AMP + PPi \quad (1)$$
$$E \cdot AA \sim AMP + tRNA \rightleftharpoons AA\text{-}tRNA + AMP + E \quad (2)$$

In the first reaction, an amino acid is condensed with ATP to make firmly bound aminoacyl adenylate (AA~AMP); in the second reaction, adenylate reacts with tRNA to give aminoacyl tRNA (AA-tRNA). For each amino acid there are specific tRNA species and one specific aminoacyl tRNA synthetase. Within a single-stranded region of tRNAs known as the anticodon loop, there is a trinucleotide sequence (anticodon) which corresponds to the particular amino acid associated with that tRNA. It is in the reaction shown above that synthetases correlate amino acids with trinucleotides[2-4].

As proteins intimately associated with the genetic code and translation of genetic information in all living systems, synthetases have been and continue to be investigated extensively. Characterizations and structural interrelationships, the recognition of transfer RNA, and the regulation of biosynthesis are major current issues. Many early ideas about these enzymes are incomplete or wrong and unexpected features continue to be discovered. Some of the more recent concepts and ideas are summarized below.

Diverse quaternary structures and subunit sizes of synthetases

Seventeen amino acid specific synthetases from *Escherichia coli* have been characterized and many from other prokaryotes and eukaryotes have also been analyzed[3]. While each synthetase catalyzes the same general type of reaction, they are heterogeneous with respect to subunit size and quaternary structure. Subunit sizes vary from 37 to 120 kd and four quaternary structures are found: α, α_2, $\alpha_2\beta_2$, and α_4. Some data show that, while synthetases are heterogeneous amongst themselves with respect to quaternary structure, a particular enzyme's quaternary structure is the same in various organisms[3].

The smallest enzyme is *E. coli* Trp-tRNA synthetase, an α_2 protein with a subunit mol. wt of 37,000. Each subunit of this enzyme appears to have its own sites for amino acid, adenylate and tRNA[5]. Therefore, the minimum size polypeptide chain required to carry out reactions (1) and (2) above is about 330 amino acids. The longest polypeptide chains are associated with E. coli Ile- and Val-tRNA synthetases – each are α enzymes with mol. wts of about 110,000 (Ref. 3) and the largest synthetase quaternary structure characterized thus far is E. coli Ala-tRNA synthetase – an $α_4$ enzyme with a mol. wt of 380,000 (Ref. 6).

There are three explanations for heterogeneous subunit sizes and quaternary structures: First, these enzymes may have evolved by convergent evolution from unrelated ancestors and have different polypeptide sizes and quaternary structures although they all catalyse similar reactions. This is difficult to prove or disprove. Second synthetases may be more similar than is apparent by simply comparing their sizes. Several reports on large synthetases (subunits with mol. wts greater than 60,000) state that polypeptides contain repeated sequences[3]. If this is true, then the heterogeneity of subunit size and quaternary structure is of less significance. However, as explained below, doubt has been cast recently on the existence of repeated sequences. A third possibility is that synthetases have different sizes partly because they perform functions, in addition to aminoacylation, which vary from protein to protein. For example, one synthetase has been found to act as a regulator of transcription. This role is believed to depend upon the particular quaternary structure assumed by that synthetase[7].

Synthetase primary structures

Understanding the inter-relationships between synthetases and regions crucial for common functions requires a knowledge of primary structure. Bacillus stearothermophilus Trp-tRNA synthetase – a 327 amino acid polypeptide – was the first to be sequenced completely[8]. Its primary structure was determined by classical protein sequencing methods. The second synthetase to be sequenced was the 875 amino acid polypeptide of E. coli Ala-tRNA synthetase[9]. In this case the gene was sequenced by Maxam-Gilbert methods and was translated into a polypeptide. Because an error in the DNA sequence (such as a deleted or inserted base) can shift the reading frame to the wrong phase, and because such errors have led to deduction of incorrect polypeptide sequences derived from DNA sequences (see Ref. 9), the reading frame of this gene was checked over its entire coding region. Oligopeptide sequences from scattered random regions of Ala-tRNA synthetase were matched to the DNA sequence and gas chromatography–mass spectrometry provided multiple oligopeptide sequences derived from protein hydrolysates[9,10].

Sequences of other synthetases are in progress, or have recently been completed. These include E. coli Gln-, Phe-, Trp-, Tyr-, and Gly-tRNA synthetases, as well as yeast Met-tRNA synthetase.

Homologies between synthetases

There is considerable homology between B. stearothermophilus Trp-tRNA synthetase and its E. coli counterpart[8]. These homologies are more extensive in the N-terminal third than in the remainder of the polypeptide. Further refinement of this comparison is expected when the complete E. coli sequence is published.

E. coli Ala-tRNA synthetase (875 amino acids per subunit) has been compared with the partial sequence of E. coli Tyr-tRNA synthetase (about 420 amino acids per subunit). A significant homology (Fig. 1) occurs between residues 370 to 382 of Ala-tRNA synthetase and residues 9 to 25 of Trp-tRNA synthetase[9]. As shown, within a 13 residue stretch in the alanine enzyme there is a match of 11 with those in a 17 residue segment of the tyrosine enzyme. This homology is in a part of Ala-tRNA synthetase which is important for the formation of aminoacyl adenylate[9]. Data on the functional significance of this region in Tyr-tRNA synthetase are lacking.

Ala-tRNA Synthetase

Leu- - -GluArgGlyLeu- - -AlaLeuLeu- - -AspGluGlu- - -LeuAla
LeuGlnGluArgGlyLeu ValAlaGln Val Thr AspGluGlu Ala LeuAla

Tyr-tRNA Synthetase

Fig. 1.

As for *E. coli* Trp-tRNA synthetase, 15 of 18 consecutive nucleotides in the carboxyl-terminal half match with an 18 nucleotide stretch in the carboxyl-terminal half of Ala-tRNA synthetase. The resulting amino acid sequences are very similar, which means that the nucleotide sequence homologies are in the same reading frame[9]. However, the significance of this match, if any, remains unknown.

The homology between *E. coli* Ala- and Tyr-tRNA synthetases is clearcut and is the first example of homology between two different synthetases. However, the region of homology is small and little homology (see above) exists between alanyl and tryptophanyl enzymes. Sophisticated statistical comparisons of sequences may reveal greater homologies[11], but the problem is difficult because the great variations in chain lengths of the synthetases (described above) makes it hard to align them correctly.

Lack of sequence repeats in a large synthetase

Reports of repeated sequences in large molecular weight synthetase subunits ($M_r > 60,000$) led to speculations that synthetases evolved by gene duplication[3]. These assertions are not based upon complete sequence information, but rather upon peptide mapping data. Because of its size (875 amino acids), Ala-tRNA synthetase is an ideal candidate for sequence repeats. However, no repeated sequences longer than tetrapeptides are present and only five tetramers occur twice in the sequence. Thus, the alanine polypeptide cannot be viewed as two similar polypeptides covalently fused. Whether Ala-tRNA synthetase is an exception, or whether earlier assertions are incorrect, is not yet known. Note also that Ala-tRNA synthetase has roughly 550 more amino acids than Trp-tRNA synthetase and that these are in non-repetitious sequences. A possible role for a large mass of extra

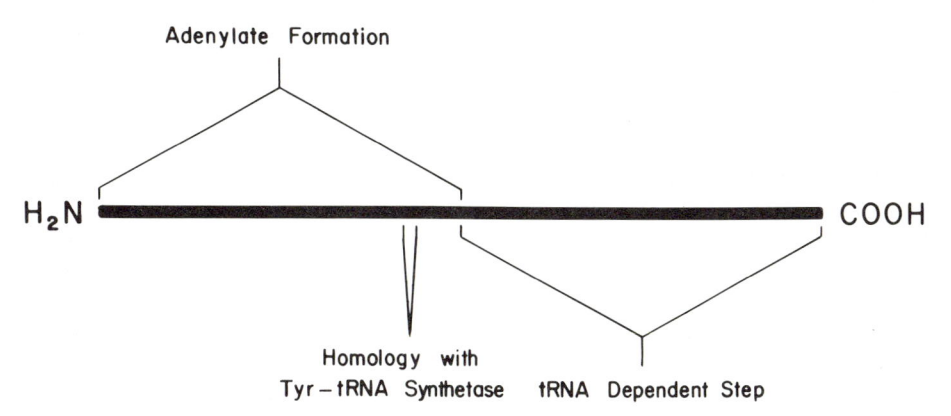

Fig. 2. Schematic structural map of E. coli *Ala-tRNA synthetase. The diagram is based on data described in the text.*

polypeptide is discussed below.

Structural domains with distinct functions

Proteases cut Ala-tRNA synthetase at the middle of its chain to give a 48 kd N-terminal fragment. This fragment has three properties: it is fully active in adenylate formation (a mole of fragment has the same specific activity as one-fourth the specific activity of a mole of tetrameric native enzyme); therefore, adenylate formation is confined entirely to an N-terminal domain. As this fragment is inactive in tRNA aminoacylation, the carboxyl-terminal half must be required for the tRNA-dependent step (equation 2). The N-terminal fragment exists as a monomer under conditions where native enzyme exists only as a tetramer; because the fragment has the same specific activity as one-fourth that of tetramer and it exists as monomer, the subunits in the native enzyme are evidently identical and independent with respect to adenylate formation[6].

These data establish the existence of a polypeptide domain with adenylate formation activity. They also suggest an additional domain for the tRNA-dependent step. Fig. 2 gives a schematic illustration. The segment which interacts with nucleic acid could be small. For example, cro protein, which binds a specific sequence in the lambda operon, is only 66 amino acids long[12] and lambda repressor, which binds a different sequence in the same operon, uses a segment of about 90 residues[13]. At some point during catalysis an interaction between the synthetase domains is required to transfer amino acid from adenylate to tRNA. Transfer of substrates between domains is suggested in other contexts[14].

Recognition sites on tRNA

Cytoplasmic tRNA molecules vary in length from 73 to 93 nucleotides[15]. A cloverleaf hydrogen bonding pattern characterizes their secondary structure which in turn is arranged in an L-shaped tertiary conformation accommodating virtually all tRNA sequences. Many studies have shown that synthetases bind in and around the inside of the L[16].

While tRNAs vary considerably in length and nucleotide sequence, all cytoplasmic tRNAs have uridine (or 4-thiouridine) at position 8 (Ref. 15). This uridine is located at the inside of the vertex of the L where it is well positioned to

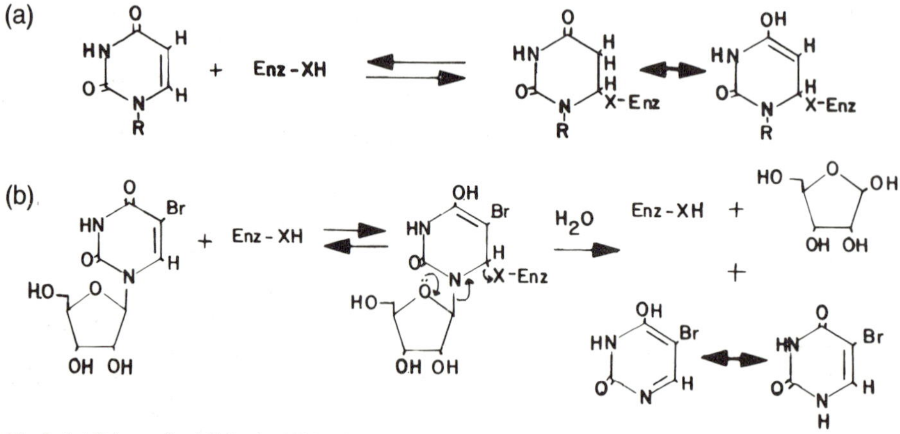

Fig. 3. (a) Scheme for Michael addition between a synthetase nucleophile and the uracil ring. The resulting adduct could exist in two tautomeric forms, as shown.

(b) Scheme for synthetase catalysed deribosylation of 5-bromouridine. A Michael adduct initially forms between an enzyme nucleophile and the uracil ring. The unstable adduct then decomposes to ribose, free synthetase and 5-bromouracil.

interact with bound synthetase; indeed synthetases have been shown to bind to this uridine[4]. This interaction results in a synthetase-catalyzed exchange of pyrimidine ring H-5 atoms on uridine-8. Observation of this exchange led to further work that proved the existence of a site on synthetases which covalently binds the uracil ring.

Synthetases covalently bind the uracil ring

Fig. 3a shows a Michael addition between an enzyme nucleophile and the uracil ring which leads to an H-5 exchange for solvent hydrogen atoms. Synthetases catalyze this exchange at uridine-8 in tRNA, and also in uridine mononucleoside. Further studies of interaction with mononucleoside have been conducted with 5-halouridines such as 5-bromouridine. The electron-withdrawing power of a halogen at C-5 renders C-6 more susceptible to nucleophilic attack, and increases the probability of trapping a covalent synthetase-pyrimidine adduct. Surprisingly, all the synthetases tested catalysed the slow deribosylation of 5-bromouridine to 5-bromouracil and ribose[17]. The deribosylation reaction undoubtedly proceeds through an unstable Michael adduct (Fig. 3b).

This unusual deribosylation reaction is common to many and perhaps all synthetases of prokaryotes and eukaryotes. While extensive primary structure homologies may not exist between enzymes (although this remains to be examined fully; see above), strong selective pressure has nevertheless preserved a site which interacts with the uracil ring.

Uridine interaction site is required for synthetase activity

If the site on a synthetase which interacts with uridine is required for activity, then covalent labeling of this site should inactivate the enzyme. Michael adducts between synthetases and uridine or 5-halouridines are labile and decompose according to the reversible reaction in Fig. 3a, or through deribosylation (Fig. 3b). These pathways prevent straightforward covalent labeling of the uridine interaction site. However, under certain circumstances the Michael adduct rearranges to yield a stable enzyme-nucleoside complex.

The stable adduct is catalytically inactive and this observation establishes a functional significance for the uridine interaction site[18]. For Ala-tRNA synthetase, attachment is to a region of polypeptide required for the tRNA-dependent step. One residue is bound per catalytic site.

The role in catalysis for the uridine interaction site is not known. However, covalent interaction with uridine-8 may be necessary to facilitate reaction of bound adenylate with the 3'-end of tRNA. Fig. 2 shows that adenylate formation is on a discrete domain in Ala-tRNA synthetase. Thus, co-ordination between this domain and the region which interacts with tRNA must occur at some point to achieve aminoacylation. Ebel and co-workers reported recently that yeast Phe-tRNA synthetase aminoacylates adenosine mononucleoside if the enzyme is conformationally triggered, or primed, by the binding of defective yeast tRNA^{Phe} which has been stripped of its 3'-terminal adenosine (the normal site of aminoacylation)[19]. An interaction between synthetase and uridine-8 in defective tRNA^{Phe} could trigger aminoacylation of adenosine mononucleoside.

In addition to uridine-8 there must be other nucleotide sequences which determine recognition of tRNA. These sequences would permit or prevent the enzyme's access to uridine-8, depending upon whether the enzyme contacts cognate or non-cognate tRNA, respectively. In addition to some preferential binding affinity for cognate tRNA, synthetases have long been known to discriminate at the level of V_{max}[3]. Binding of non-cognate tRNA does not result in aminoacylation, perhaps because covalent addition to uridine-8 does not

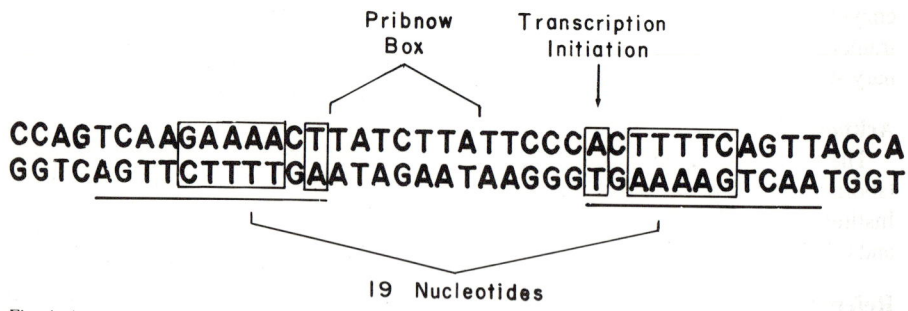

```
                    Pribnow      Transcription
                     Box          Initiation

CCAGTCAAGAAAACTTATCTTATTCCCACTTTTCAGTTACCA
GGTCAGTTCTTTTGAATAGAATAAGGGTGAAAAGTCAATGGT
```

                    19 Nucleotides

Fig. 4. A section of the 5'-non-coding region of alaS. The top sequence is the coding strand (same sequence as mRNA), and locations of Pribnow Box and point of transcription initiation are marked. The first codon of the polypeptide is 79 nucleotides downstream from the point of transcription initiation. Underlines denote sections on bottom strand protected from nuclease digestion by bound Ala-tRNA synthetase. Boxed base pairs form an interrupted palindrome whose centers are separated by 19 base pairs.

occur. The lack of H-5 exchange at uridine-8 in a non-cognate synthetase-tRNA mixture is consistent with this view[20].

DNA sequence recognition and transcription regulation

The concentrations of synthetases in cells are adjusted by metabolic regulation and by amino acid dependent repression of gene expression[21]. Numerous physiological studies have been reported on several synthetase systems, but only Ile- and Ala-tRNA synthetases have been examined in vitro[7,22]. A molecular explanation of amino acid dependent repression has only been derived for the alanine system.

The sequence of the 5'-non-coding region of alaS does not encode a leader peptide or have an attenuator sequence as do numerous operons involved in amino acid biosynthesis[23]. In vitro transcription of a restriction fragment encompassing the alaS promoter is repressed by Ala-tRNA synthetase, but not by other synthetases or other non-specific proteins. Moreover, Ala-tRNA synthetase does not repress transcription of another well studied promoter. These observations show that repression is specific. Furthermore, repression of transcription is enhanced greatly by concentrations of alanine comparable to the enzyme's K_m for that amino acid. This means that an enzyme-ligand complex is the most efficient repressor. Calculations show that only the complexed, and not the free enzyme, acts at in vivo enzyme concentrations[7].

DNA protection experiments show that repression occurs because Ala-tRNA synthetase binds to a specific DNA sequence which flanks the transcription start point. Fig. 4 shows the region of DNA protected by bound Ala-tRNA synthetase. Within the protected region there is an inverted repeat of six out of seven nucleotides. This sequence is: 5' GAAAANT. Nineteen base pairs separate the centers of the repeat and this is almost exactly two revolutions of a DNA B helix; therefore, the sequences bound by synthetase are on one face of the helix. This spacing and two-fold symmetry arrangement of nucleotide sequences requires that bound protein has a two-fold axis of symmetry. The pattern of products produced by inter-subunit chemical crosslinking of Ala-tRNA synthetase fits with this two-fold symmetry for the native tetramer[6] and the mass of the tetramer indicates that it is large enough to span 70–100 Å of DNA (Fig. 4).

These data suggest that one explanation for diverse subunit sizes and quaternary structures is that synthetases have roles in addition to the aminoacylation of tRNA. For example, Ala-tRNA synthetase not only has catalytic sites, but also has polypeptide sections required for tetramer assembly and DNA sequence recognition. These functions, which may vary from enzyme to enzyme, place further constraints on synthetase primary and quaternary structure.

Acknowledgements

The work summarized from the authors' laboratory has been supported by National Institutes of Health Grant Nos. GM15539 and GM23562.

References

1. Berg, P. (1961) *Ann. Rev. Biochem.* 30, 293–324
2. Ofengand, J. (1977) in *Molecular Mechanisms of Protein Biosynthesis* (Weissbach, H. and Pestka, S., eds), pp. 7–69, Academic Press, New York
3. Schimmel, P. R. and Söll, D. (1979) *Ann. Rev. Biochem.* 48, 601–648
4. Schimmel, P. R. (1979) *Advs. Enzymol. Rel. Areas Mol. Biol.* 49, 187–222
5. Muench, K. H. (1976) *J. Biol. Chem.* 251, 5195–5199
6. Putney, S. D., Sauer, R. and Schimmel, P. R. (1981) *J. Biol. Chem.* 256, 198–204
7. Putney, S. D. and Schimmel, P. R. (1981) *Nature (London)* 291, 632–635
8. Winter, G. P. and Hartley, B. S. (1977) *FEBS Lett.* 80, 340–342
9. Putney, S. D., Royal, N. J., Neuman de Vegvar, H., Herlihy, W. C., Biemann, K. and Schimmel, P. (1981) *Science* 213, 1497–1501
10. Herlihy, W. C., Royal N. J., Biemann, K., Putney, S. D. and Schimmel, P. R. (1980) *Proc. Natl Acad. Sci. U.S.A.* 77, 6531–6535
11. Doolittle, R. (1981) *Science* 214, 149–159
12. Hsiang, M. W., Cole, R. D., Takeda, Y. and Echols, M. (1977) *Nature (London)* 270, 275–277
13. Pabo, C. O., Sauer, R. T., Sturtevant, J. M. and Ptashne, M. (1979) *Proc. Natl Acad. Sci. U.S.A.* 76, 1608–1612
14. Eccleston, E. D., Thayer, M. L. and Kirkwood S. (1979) *J. Biol. Chem.* 254, 11399–11404
15. Gauss, D. H. and Sprinzl, M. (1981) *Nucl. Acid Res.* 9, r1–r23
16. Schimmel, P. R. (1980) *Crit. Rev. Biochem.* 9, 207–251
17. Koontz, S. W. and Schimmel, P. R. (1979) *J. Biol. Chem.* 254, 12277–12280
18. Starzyk, R. M., Koontz, S. W. and Schimmel, P. (1982) *Nature (London) (in press)*
19. Renaud, M., Bacha, H., Remy, P. and Ebel, J. P. (1981) *Proc. Natl Acad. Sci. U.S.A.* 78, 1606–1608
20. Schoemaker, H. J. P. and Schimmel, P. R. (1977) *Biochemistry* 16, 5454–5460
21. Umbarger, H. E. (1980) in *Transfer RNA: Biological Aspects* (Söll, D., Abelson, J. and Schimmel, P., eds), pp. 453–467, Cold Spring Harbor Laboratory, New York
22. Wirth, R., Kohles, V. and Böck, A. (1981) *Eur. J. Biochem.* 114, 429–437
23. Yanofsky, C. (1981) *Nature (London)* 289, 751–758

Paul Schimmel, Scott Putney and Ruth Starzyk are at the Department of Biology, Massachusetts Institute of Technology, Cambridge, MA 02139, U.S.A.

Feedback regulation of ribosomal protein synthesis in *Escherichia coli*

Masayasu Nomura, Dennis Dean and John L. Yates

The synthesis of ribosomal proteins and the assembly of ribosomes in Escherichia coli *are coupled. When the rate of ribosomal protein synthesis exceeds that of ribosome biosynthesis, certain key ribosomal proteins act as inhibitors that prevent the further translation of their own mRNA. This translational feedback mechanism ensures the balanced and co-ordinately regulated synthesis of many ribosomal proteins.*

Ribosomes are complex organelles responsible for all cellular protein synthesis (for reviews, see Refs 1, 2). An *Escherichia coli* ribosome contains three species of RNA (rRNA) and approximately 52 proteins (r-proteins). Each of the rRNAs and most of the r-proteins are found in the ribosome in stoichiometric amounts. Thus, the complexity of the ribosome can be viewed from two aspects: first, in terms of its structure and function in relation to protein synthesis; and, second, in terms of the genetic organization and physiological regulation of the genes coding for ribosome components.

Concerning the first aspect, one can ask how all these components are assembled together to form a functional ribosome. This problem has been studied by purifying each of the ribosomal components and reconstituting them *in vitro* to assemble active ribosomal subunits. Studies using this approach have demonstrated the sequential and co-operative interactions of these components during the *in vitro* assembly of ribosomes (for reviews, see Refs 3, 4). As regards the organization and regulation of genes coding for ribosomal components, we know that the synthesis rates of most if not all of the r-proteins reflect their accumulation rates, which in turn reflect the stoichiometric relationship of all these r-proteins in the ribosome[5]. In addition, we know that the rate of synthesis of ribosomes depends on the organism's growth rate; that is, when cells grow faster in nutritionally rich media, cells synthesize more ribosomes to satisfy the demand for increased rates of protein synthesis (for reviews, see Refs 6, 7). Thus, we can ask the following question: what mechanisms ensure the balanced and co-ordinately regulated synthesis rates of as many as 52 r-proteins? Although the answer to this question is still not complete, a model for translational feedback regulation of r-protein synthesis has been formulated and supported by a variety of experimental approaches. This model proposes that r-protein synthesis and ribosome assembly are coupled such that when r-protein synthesis exceeds the rate of ribosome biosynthesis certain key r-proteins act as inhibitors that prevent the further translation of their own mRNA[8,10]. In this article, we summarize these recent developments related to regulation. We note, to our satisfaction, earlier work on the assembly of ribosomes *in vitro* has now proved to be relevant to the regulation problem, giving insight into the mechanisms involved in the postulated feedback translational regulation.

Isolation of r-protein genes and their structural organization

Much of the groundwork, especially the isolation and characterization of r-protein genes, had to be done before the specific

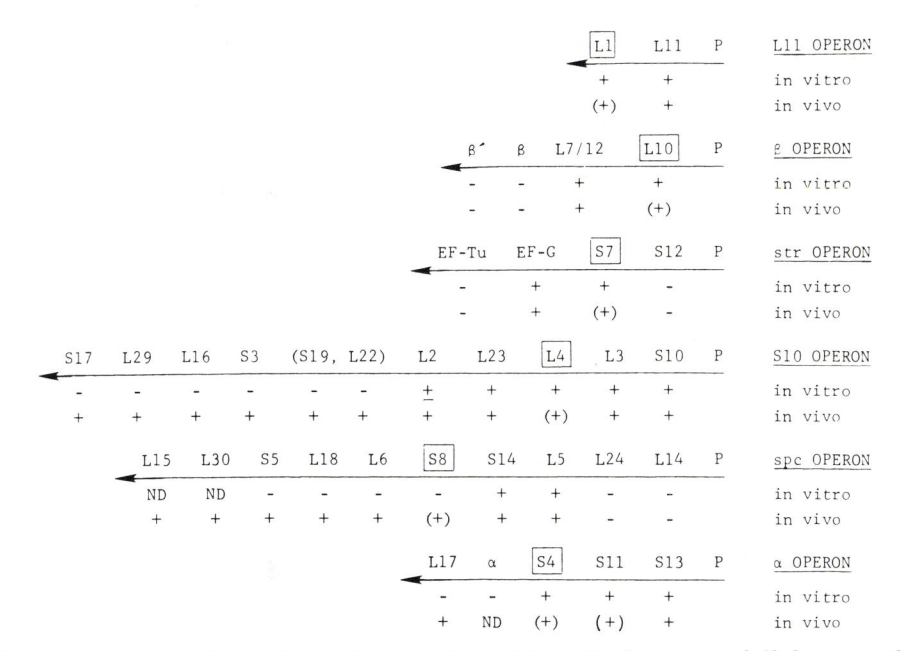

									L1	L11	P	L11 OPERON
									+	+		in vitro
									(+)	+		in vivo
						β´	β	L7/12	L10	P		β OPERON
						−	−	+	+			in vitro
						−	−	+	(+)			in vivo
						EF-Tu	EF-G	S7	S12	P		str OPERON
						−	+	+	−			in vitro
						−	+	(+)	−			in vivo
S17	L29	L16	S3	(S19, L22)	L2	L23	L4	L3	S10	P		S10 OPERON
−	−	−	−	−	±	+	+	+	+			in vitro
+	+	+	+	+	+	+	(+)	+	+			in vivo
L15	L30	S5	L18	L6	S8	S14	L5	L24	L14	P		spc OPERON
ND	ND	−	−	−	−	+	+	−	−			in vitro
+	+	+	+	+	(+)	+	+	−	−			in vivo
				L17	α	S4	S11	S13	P			α OPERON
				−	−	+	+	+				in vitro
				+	ND	(+)	(+)	+				in vivo

Fig. 1. Autogenous regulation of r-protein genes of E. coli *located in the* str-spc *and* rif *chromosomal regions. Genes are represented by the protein product. For each operon, the order of transcription from the promoter (P) is indicated by an arrow. Regulatory r-proteins are indicated by the boxes. Effects of the boxed protein on the* in vitro[15-20] *or* in vivo[19,20,24,25] *synthesis of proteins from the same operon are shown. Results obtained* in vivo *are those from experiments using hybrid plasmids to achieve overproduction of the regulatory r-protein. +, specific inhibition of synthesis; −, no significant effect on synthesis; ±, weak inhibition of synthesis observed* in vitro; *(+), inhibition presumed to occur* in vivo; *ND, not determined. It is not established how the regulation of synthesis of r-proteins S12, L14 and L24 takes place. Some indirect experiments suggest that L14 feedback regulates the translation of its own mRNA and L24 also regulates its own synthesis (our unpublished experiments).*

formulation of the translational regulation model and its experimental proof were achieved. Earlier studies on antibiotic-resistant mutants suggested that there was a cluster of r-protein genes in the 'str-spc region' of the *Escherichia coli* chromosome (72 min on the genetic map). The first success in the comprehensive analysis of r-protein genes was the isolation of a series of transducing phages (λspc1, λspc2 and λfus2) carrying r-protein genes[11,12]. More than half of the total r-protein genes were identified on these transducing phages and their organization was subsequently elucidated. Four r-protein genes were then found[13,14] on another transducing phage (λrif^d18) carrying a chromosome region derived from 89 min (see Ref. 7 for a review). The structures of the operons comprising these r-proteins are shown in Fig. 1. The availability of these transducing phages was important for the regulation studies not only because it made physical analysis of the r-protein genes possible, but also because: (a) measurements of r-protein mRNA became possible; (b) *in vitro* synthesis of r-proteins using these phage DNAs as template provided a system to study r-protein gene expression *in vitro*; (c) gene fusion techniques could be easily applied *in vitro* and; (d) nucleotide sequence analysis of the genes and their regulatory regions could be carried out (after the development of DNA sequencing techniques).

Origin of the model

An early indication that r-protein synthesis is regulated at the level of translation of r-protein mRNA came from measurements of r-protein synthesis rates and r-protein mRNA synthesis rates in *E. coli* strains merodiploid for a group of r-protein genes (in the *str-spc* region)[8,9]. It was found that the rates of synthesis of r-protein mRNA increase in proportion to the increase in gene dosage, but that the rates of r-protein synthesis do not increase relative to the synthesis rates of r-proteins whose genes exist in a single copy. Therefore, it appeared that *E. coli* cells could prevent the translation of an excess amount of r-protein mRNA. Since the apparent r-protein mRNA inactivation was specific only for the excess mRNA it seemed that the information required for inactivation might be contained within the excess mRNA. We suggested that r-proteins whose synthesis is directed by the excess mRNA might themselves inactivate the mRNA as a consequence of the accumulation of 'un-needed' r-proteins[8]. Indeed, such a mechanism could ensure the balanced and co-ordinate synthesis of all r-proteins.

Evidence *in vitro*

A DNA-directed *in vitro* protein synthesis system was used to test directly whether r-proteins could inhibit the translation of r-protein mRNA. Various DNA templates purified from transducing phages or as hybrid plasmids carrying r-protein operons were used to direct the synthesis of r-proteins *in vitro*. The synthesis of these r-proteins was measured in the presence or absence of exogenous purified r-proteins. When added to the *in vitro* system some r-proteins caused a clear and specific inhibition of DNA-directed r-protein synthesis. By uncoupling the transcription step from the translation step it was possible to demonstrate that the inhibition occurred only at the level of translation. The following proteins were found to have repressor activity: L1 (L11 operon)[15], L4 (S10 operon)[16], L10 (β operon)[17,18], S4 (α operon)[15], S7 (*str* operon)[19], and S8 (*spc* operon)[15,20]. S20 was also shown to have repressor activity using other *in vitro* procedures[21,22]. The observed inhibition was specific. For example, the addition of L1 to the *in vitro* system inhibited the synthesis of L1 and L11, both of which are contained within the L11 operon, and this inhibition was observed only with L1, and not with any other r-proteins tested. Similarly, specific inhibition of the synthesis of S13, S11 and S4 was observed only with S4, and not with any other r-proteins tested, including L1. As will be discussed later, not all r-proteins within an operon are inhibited by a particular 'repressor' r-protein from that operon.

Evidence *in vivo*

Evidence that the regulatory properties of r-proteins identified by the *in vitro* procedure actually reflect a regulatory role for these r-proteins *in vivo* came, independently, from a series of experiments by Lindahl and Zengel[23,24]. They placed the expression of individual r-proteins from the S10 operon under the control of the *lac* regulatory elements in a recombinant plasmid. Strains carrying plasmids constructed in this way over-synthesize r-proteins whose genes are contained on the plasmid when an inducer of the *lac* operon is added to the growth media. Thus, the physiological effects of the overproduction of specific r-proteins could be examined. The overproduction of r-protein L4 *in vivo* specifically inhibited the synthesis of other proteins whose genes were contained within the S10 operon. Using procedures similar to that of Lindahl and Zengel we have identified *in vivo* regulatory roles for r-proteins L1, L10, S4, S7 and S8[19,20,25]. The *in vivo* results, therefore, confirm the *in vitro* observation: certain r-proteins are capable of inhibiting the synthesis of other proteins encoded in the same operon (see Fig. 1). Also it should be noted that S7, S8 and L10 inhibit the *in vivo* synthesis of some, but not all of the proteins from their operon, that is, operons are sometimes subdivided into 'units of regulation'. This observation is consistent with the conclusion from *in*

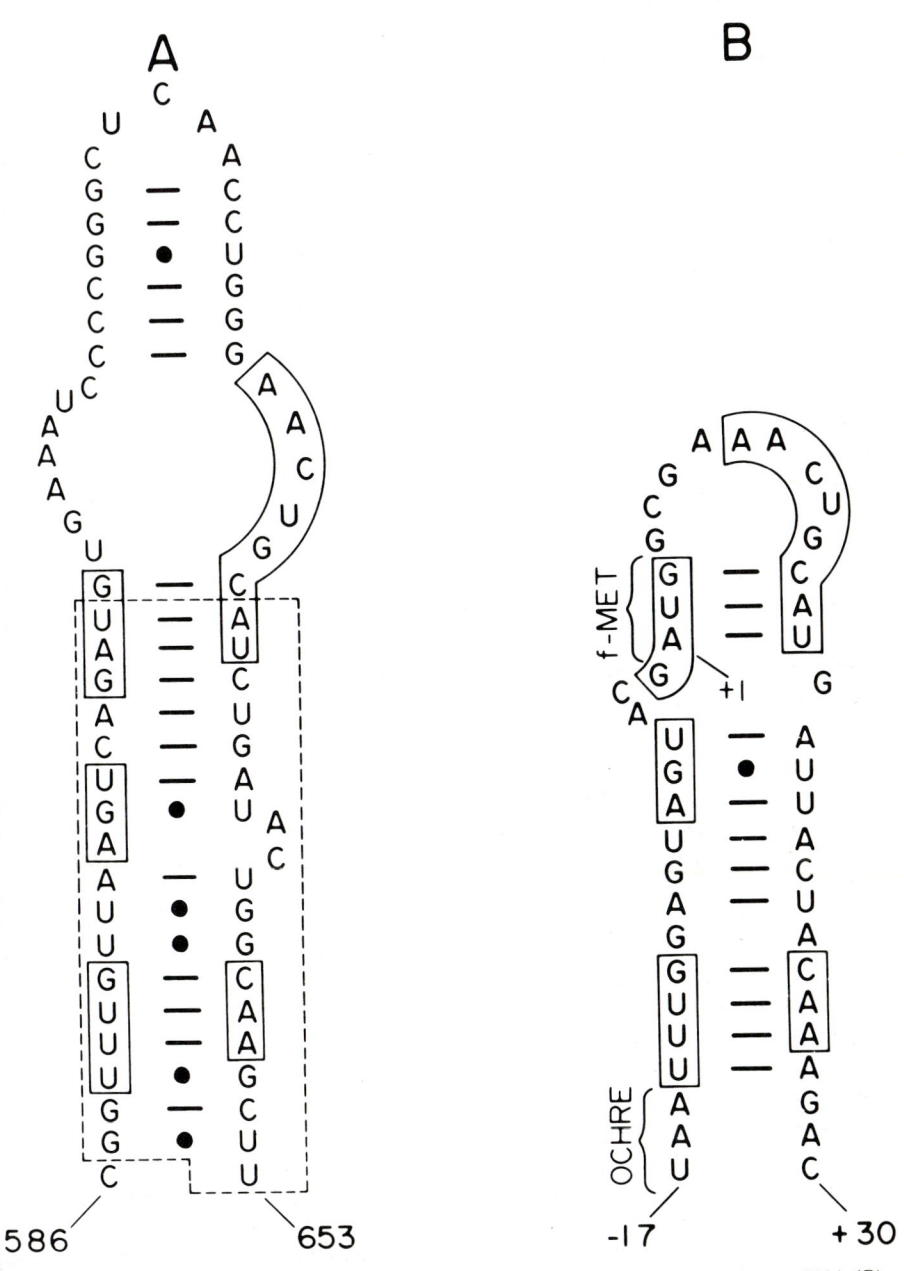

Fig. 2. Model of the secondary structure of S8 binding sites on 16S rRNA (A) and on mRNA (B). Homologies which are considered to be significant are boxed. (The structure shown in (A) is taken from Ref. 34.) The binding site for S8[35,36] is indicated by a broken line. In (B) the L24 coding region ends at −18, and the L5 coding region begins at +4. Since S8 inhibits translation of the L5 and distal cistrons, but not translation of L14 and L24 cistrons, the mRNA around the L24-L5 intercistronic region shown here is the likely target site for S8.

vitro experiments that these repressor r-proteins act at translation, and not at transcription.

However, a comparison of the *in vitro* and *in vivo* results shows that, in some instances, synthesis of r-proteins from promoter–distal genes was inhibited *in vivo* but not *in vitro*. For example, the overproduction of L4 *in vivo* caused inhibition of all the proteins encoded by the S10 operon, whereas L4 inhibited the synthesis of only the first four r-proteins in the operon *in vitro*. We believe that this regulation of promoter distal genes *in vivo* is due to a 'polar effect' caused by the direct inhibition of the translation of upstream mRNA, and that this polar effect is observed incompletely *in vitro*. Three mechanisms could account for the polar effect: (a) inhibition of translation of promoter–proximal mRNA leads to very rapid degradation of distal mRNA; or (b) translation inhibition leads to termination of transcription; or (c) translation of distal cistrons in a unit of regulation requires the translation of the first cistron in that unit. According to the third mechanism, the target sites of repressor r-proteins are at the beginning regions of mRNA corresponding to these first cistrons. The apparent escape of the distal cistrons from the inhibition *in vitro* is perhaps due to physical separation of distal mRNA from the target site, for example, by non-specific nucleolytic cleavages of mRNA in the *in vitro* system (see the discussion in Refs 16, 26). Experimental evidence to support this third mechanism was obtained with respect to the regulation of L11 and L1 synthesis by L1[26].

Homology between mRNAs for r-proteins and rRNA

How do r-proteins recognize their own mRNAs and thus prevent their translation? A striking feature of the identified regulatory r-proteins is that they are all r-proteins that are known to bind to rRNA early during ribosome assembly. It is possible that the regulation of r-protein translation can be regarded as a consequence of competition between rRNA and r-protein mRNA for repressor r-proteins[10]. This notion is supported by structural homologies between the known r-protein binding sites on rRNA and putative target sites for binding of regulatory r-proteins on their respective mRNAs. One such homology observed for r-protein S8[27] is shown in Fig. 2. Similar homologous structures have been drawn between rRNA binding sites and mRNA target sites for r-proteins L1[28,29], S4, and S7[10]. It is possible that the repressor r-proteins recognize the same general structural features on their mRNA that they recognize on rRNA during ribosome assembly. According to our model the affinity of repressor r-proteins is much higher for rRNA than r-protein mRNA, and repressor-mRNA interaction only occurs when the rate of r-protein synthesis exceeds the rate of ribosome accumulation.

We have summarized evidence that supports a model for the balanced and coordinate synthesis of r-proteins. The essential feature of this model is that r-protein synthesis is coupled, by a regulatory feedback mechanism, to the assembly of mature ribosomes. Experiments using mutant strains[30-32] and gene fusion techniques[33] indicate that the translational feedback mechanism actually plays an important physiological role in exponentially growing *E. coli* cells.

Acknowledgements

The work from the authors' laboratory was supported in part by the College of Agricultural and Life Sciences, University of Wisconsin–Madison, by Grant GM-20427 from the National Institutes of Health, and by Grant PCM79-10616 from the National Science Foundation. D.D. is supported by National Institutes of Health Postdoctoral Fellowship GMO-7553. This is paper No. 2540 from the Laboratory of Genetics, University of Wisconsin–Madison.

References

1. Nomura, M., Tissières, A. and Lengyel, P. (1974) *Ribosomes*, Cold Spring Harbor Laboratory, Cold Spring Harbor, New York
2. Chambliss, G. H., Craven, G. R., Davies, J. E., Kahan, L. and Nomura, M., eds (1980) *Ribosomes: Structure, Function and Genetics*, University Park Press, Baltimore
3. Nomura, M. and Held, W. A. (1974) in *Ribosomes* (Nomura, M., Tissières, A. and Lengyel, P., eds), pp. 193–223, Cold Spring Harbor Laboratory, Cold Spring Harbor, New York
4. Nierhaus, K. H. (1980) in *Ribosomes: Structure, Function and Genetics* (Chambliss, G., Craven, G. R., Davies, J., Davis, K., Kahan, L. and Nomura, M., eds), pp. 267–294, University Park Press, Baltimore
5. Dennis, P. P. (1974) *J. Mol. Biol.* 88, 25–41
6. Kjeldgaard, N. O. and Gausing, K. (1974) in *Ribosomes* (Nomura, M., Tissières, A. and Lengyel, P., eds), pp. 369–392, Cold Spring Harbor Laboratory, Cold Spring Harbor, New York
7. Nomura, M., Morgan, E. A. and Jaskunas, S. R. (1977) *Ann. Rev. Genet.* 11, 297–347
8. Fallon, A. M., Jinks, C. S., Strycharz, G. D. and Nomura, M. (1979) *Proc. Natl Acad. Sci. U.S.A.* 76, 3411–3415
9. Olsson, M. O. and Gausing, K. (1980) *Nature (London)* 283, 599–600
10. Nomura, M., Yates, J. L., Dean, D. and Post, L. E. (1980) *Proc. Natl Acad. Sci. U.S.A.* 77, 7084–7088
11. Jaskunas, S. R., Lindahl, L. and Nomura, M. (1975) *Proc. Natl Acad. Sci. U.S.A.* 72, 6–10
12. Jaskunas, S. R., Fallon, A. M. and Nomura, M. (1977) *J. Biol. Chem.* 252, 7323–7336
13. Lindahl, L., Jaskunas, S. R., Dennis, P. P. and Nomura, M. (1975) *Proc. Natl Acad. Sci. U.S.A.* 72, 2743–2747
14. Watson, R. J., Parker, J., Fiil, N. P., Flaks, J. G. and Friesen, J. D. (1975) *Proc. Natl Acad. Sci. U.S.A.* 72, 2765–2769
15. Yates, J. L., Arfsten, A. E. and Nomura, M. (1980) *Proc. Natl Acad. Sci. U.S.A.* 77, 1837–1841
16. Yates, J. L. and Nomura, M. (1980) *Cell* 21, 517–522
17. Brot, N., Caldwell, P. and Weissbach, H. (1980) *Proc. Natl Acad. Sci. U.S.A.* 77, 2592–2595
18. Fukuda, R. (1980) *Mol. Gen. Genet.* 178, 483–486
19. Dean, D., Yates, J. L. and Nomura, M. (1981) *Cell* 24, 413–420
20. Dean, D., Yates, J. L. and Nomura, M. (1981) *Nature (London)* 289, 89–91
21. Wirth, R. and Böck, A. (1980) *Mol. Gen. Genet.* 178, 479–481
22. Wirth, R., Kohler, V. and Böck, A. (1981) *Eur. J. Biochem.* 114, 429–437
23. Lindahl, L. and Zengel, J. (1979) *Proc. Natl Acad. Sci. U.S.A.* 76, 6542–6546
24. Zengel, J. M., Mueckl, D. and Lindahl, L. (1980) *Cell* 21, 523–535
25. Dean, D. and Nomura, M. (1980) *Proc. Natl Acad. Sci. U.S.A.* 77, 3590–3594
26. Yates, J. L. and Nomura, M. (1981) *Cell* 24, 243–249
27. Olins, P. O. and Nomura, M. (1981) *Nucleic Acids Res.* 9, 1757–1764
28. Gourse, R. L., Thurlow, D. L., Gerbi, S. A. and Zimmermann, R. A. (1981) *Proc. Natl Acad. Sci. U.S.A.* 78, 2722–2726
29. Branlant, C., Krol, A., Machatt, A. and Ebel, J.-P. (1981) *Nucleic Acids Res.* 9, 293–307
30. Olsson, M. O. and Isaksson, L. A. (1979) *Mol. Gen. Genet.* 169, 271–278
31. Jinks-Robertson, S. and Nomura, M. (1981) *J. Bacteriol.* 145, 1445–1447
32. Stöffler, G., Hasenbank, K. and Dabbs, E. R. (1981) *Mol. Gen. Genet.* 181, 164–168
33. Miura, A., Krueger, J. H., Itoh, S., de Boer, H. A. and Nomura, M. *Cell* 25, 773–782
34. Woese, C. R., Magrum, L. J., Gupta, R., Siegel, R. B., Stahl, D. A., Kop, J., Crawford, N., Brosius, J., Gutell, R., Hogen, J. J. and Noller, H. F. (1980) *Nucleic Acids Res.* 8, 2275–2293
35. Ungewickell, E., Garrett, R., Ehresmann, C., Stiegler, P. and Fellner, P. (1975) *Eur. J. Biochem.* 51, 165–180
36. Zimmermann, R. A., Mackie, G. A., Muto, A., Garrett, R. A., Ungewickell, E., Ehresmann, C., Stiegler, P., Ebel, J.-P. and Fellner, P. (1975) *Nucleic Acids Res.* 2, 279–302

Masayasu Nomura, Dennis Dean and and John L. Yates are at the Institute for Enzyme Research and the Departments of Genetics and Biochemistry, University of Wisconsin, Madison, Wisconsin 53706, U.S.A.

Structure and role of 5S RNA–protein complexes in protein biosynthesis

R. A. Garrett, S. Douthwaite and H. F. Noller

Rapid progress is being made in elucidating the structure of pro- and eukaryotic 5S RNAs and their protein complexes. Increasing evidence suggests that dynamic effects may occur in the 5S RNA, possibly mediated by proteins. Although the RNA appears to be situated in a functionally important part of the 50S subunit, and may play a dynamic role, it cannot yet be assigned any particular function.

When first contemplating the ribosome one is overwhelmed by the complexity of the structure, the large number of interrelated components and by the fact that it is a dynamic particle which will endure cooperative and allosteric interactions during a cycle of protein biosynthesis. One way to simplify such a problem, conceptually and experimentally, is to consider one part of the ribosome and to find out how it works – and then try to extrapolate any principles that emerge governing the mechanisms of protein–RNA recognition, conformational changes and co-operative interactions, and functional roles of the components, to the ribosome as a whole. The one part of the ribosome which may satisfy this requirement is the complex of 5S RNA and a small group of proteins.

Structure of 5S RNA

A base-pairing scheme, common to all prokaryotic 5S RNAs was first elucidated from comparative sequence studies by Fox and Woese[1]. It contains four base-paired regions constituting the sequences 1–10/110–120, 18–23/60–65, 31–34/48–51 and 82–86/90–94, which form part of the secondary structure drawn in Fig. 1. This was considered a minimal secondary structure by the authors and in deriving it they eliminated the bulges from the helices, which were prevalent in earlier models. The power of this comparative approach to secondary structure determination, pioneered by Fox and Woese, is only just beginning to be realized.

Experimental evidence supporting this basic secondary structure is now manifold and derives primarily from limited ribonuclease digestion and chemical modification studies using nucleotide specific reagents[2–5]. These approaches have provided direct support for three of the four helical regions, moreover, helices II and IV have already been extended as shown in Fig. 1[2,3]; but, the evidence for helix III is only negative, namely that no nuclease or reagent with a preference for single-stranded regions reacts there. There are three very accessible regions in the RNA structure around G_{13}, U_{40}–C_{43} and U_{87}–U_{89}; most of the remaining sequence, which is not drawn as helices, is inaccessible to nucleases and chemical reagents and is likely to be involved in tertiary interactions.

We have recently made a similar comparative sequence analysis of eukaryotic 5S RNAs and find that a universal structure can be drawn (as shown in Fig. 2) which is close to one of the early models[6]; a similar conclusion has been reached, independently, by Luehrsen and Fox[7]. This model differs from the prokaryotic one (Fig. 1) mainly in that it contains an extra helical region (helix V).

Location of protein binding sites

Of the three ribosomal proteins L5, L18 and L25 which assemble with *E. coli* 5S RNA, only the latter two bind strongly. L25 protects a fragment constituting nucleotides 69–110 against ribonuclease A attack (Ref. 8; see Fig. 1). This RNA-binding region is highly structured in free RNA. Analogous structured regions containing protein binding sites have been found in other pro- and eukaryotic 5S RNAs[9,10].

Although protein L18 tends to dissociate during the ribonuclease digestion[8], very mild treatment and selection for the intact L18–5S RNA complex in Mg^{2+}-containing gels, revealed sites around helix II which were protected by L18 against ribonucleases A and T_2 and a cobra venom nuclease specific for double helices (see Fig. 1). Chemical modification studies, using the reagent kethoxal, implicated the highly conserved guanines-24 and 69 in the RNA binding site[11]. In addition, L18 protected guanines-16, 23, 24, 54, 56, 64 and 67 against dimethylsulphate methylation[3]. These data are mutually consistent with the concept that nucleotides 15–24 and 52–69,

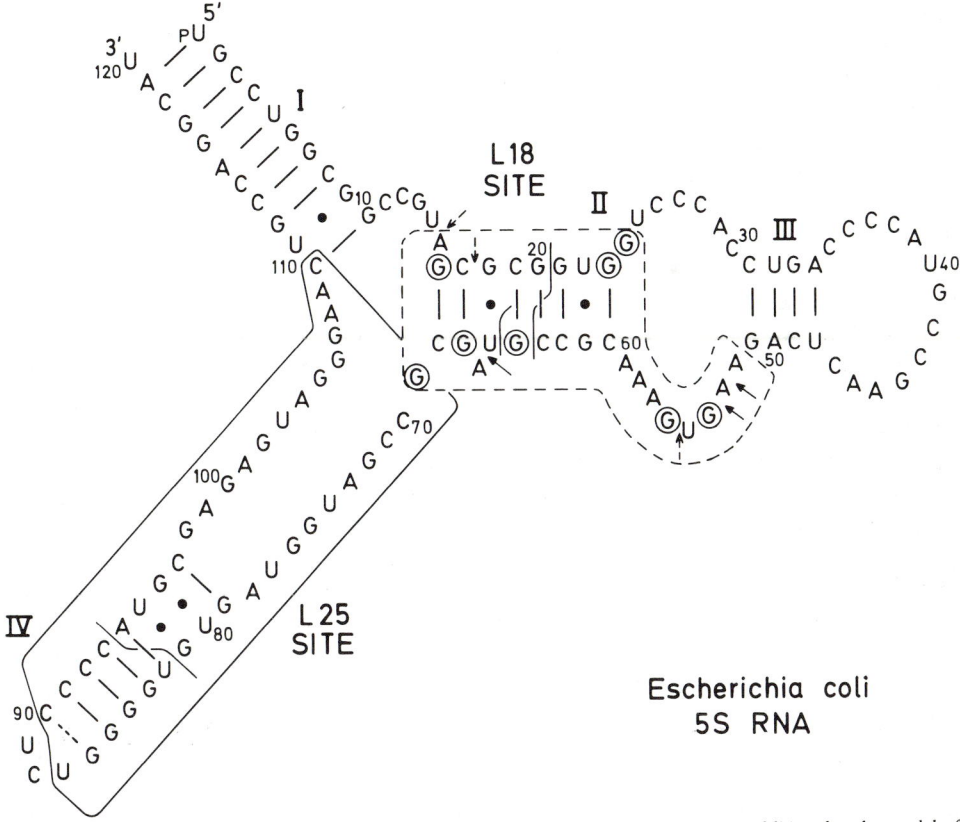

Fig. 1. Secondary structural model of Escherichia coli *5S RNA containing five base pairs additional to the model of Fox and Woese*[1]. *The binding regions of the proteins L18 and L25 are shown. The arrows indicate the ribonuclease A or T_2 cuts which occur in the free RNA but not in the protein–RNA complexes*[4]; *full arrows indicate strong cuts and broken arrows show weak cuts. Lines drawn through helices II and IV show the cobra venom ribonuclease cuts that are inhibited by the bound proteins*[4]. *The circled guanines are protected against modification by kethoxal*[11] *and/or dimethylsulphate*[3] *by L18.*

contain the binding site of L18; (Refs 3, 4, and 11) there are also some indications that the binding site may extend to the sequence 38–44 (Ref. 4). Little progress has been made in identifying the binding site of L5, although we know that strong co-operative RNA binding effects occur between L5 and L18, and that the two proteins are essential for 5S RNA–23S RNA assembly[12,13].

To date, no clear equivalence can be drawn between the pro- and eukaryotic proteins. There is weak evidence from protein sequence studies that gene fusion may have occurred such that one protein, YL3 from yeast, corresponds to L18 and L5 of E. coli[14]. Further, it is this protein which appears to effect a 5S RNA–5.8S RNA interaction[15]. However, this leaves us with a paradox, since the RNA binding site of YL3[10] corresponds to that of the other E. coli protein L25.

Protein–RNA recognition

The binding site of L18 has been sufficiently well characterized to allow speculation on the mode of protein–RNA recognition. L18 appears to make contact in the major groove of the central helix (positions 16–23/60–68); a result compatible with its ability to chase about half of the ethidium bromide from 5S RNA[12]. Although this helical region has a variable primary sequence in prokaryotes the protected guanines – 16, 23, 24, 54 and 69 – are universal[16] and could provide attachment points for the protein. The bulged adenosine-66 is also universal[16]; and it is the most accessible adenine to diethylpyrocarbonate modification in 5S RNA[3]. The internucleotide bond preceding it is also sensitive to ribonuclease digestion in free 5S RNA and is protected by L18[3,4]. It has recently been suggested, that this bulged nucleotide may provide a recognition signal for protein L18[3] and that such a mechanism might be common to other protein–RNA recognition systems[17]. In eukaryotic 5S RNAs a closely analogous central helix is found with a bulged nuc-

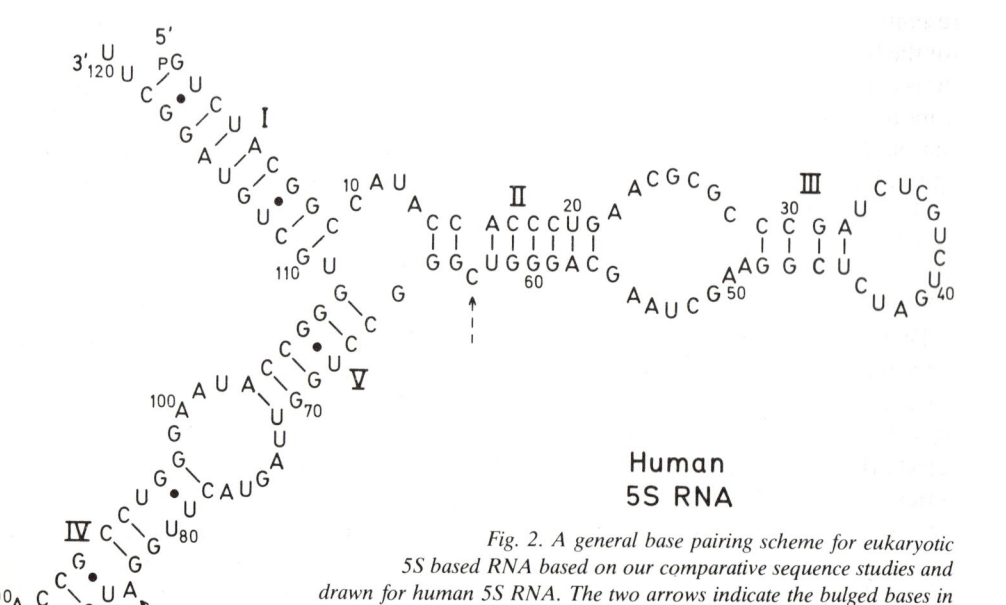

Human 5S RNA

Fig. 2. A general base pairing scheme for eukaryotic 5S based RNA based on our comparative sequence studies and drawn for human 5S RNA. The two arrows indicate the bulged bases in helices II and IV which are putative protein recognition sites.

leotide (helix II in Fig. 2), although no protein has yet been identified which binds to it. The bulged nucleotide varies from adenosine in yeast, to uridine in plants and cytosine in animals (see Fig. 2).

The RNA binding site of L25 is also shown in Fig. 1. The protein appears to bind to helix IV and prevent a double helical cut there. The remainder of the structure is less certain. It can be drawn as an unstable helix but it is doubtful whether that would account for the ribonuclease resistance and low chemical reactivity of the RNA region; although there are examples of proteins recognizing unstable or distorted helices[17]. It is more likely that the RNA region forms a stable tertiary structure which is recognized by the protein. This might explain the exceptional property of L25 – that it remains bound to the RNA in the absence of magnesium ions[8].

The corresponding region in eukaryotic 5S RNA is protected by a protein and this complex is also stable in the absence of magnesium[10]. As shown in Fig. 2, it contains two helical regions. Helix IV closely resembles that found in prokaryotes except for the bulged adenosine-83; in some structures a bulged guanosine-84 seems to offer a more stable alternative but the former can be drawn for all eukaryotes. By analogy with the L18 site the bulged nucleotide could constitute a protein recognition signal.

Conformational changes

There is an implied relation between conformational changes and function: The ribosome is a dynamic particle which moves relative to mRNA, receives and rejects tRNAs and factors, and opens and closes its subunits. It has been pointed out for 5S RNA, and this is also applicable to the large ribosomal RNAs, that a coiling–uncoiling of an unstable helical region could readily be converted into linear motion of, for example, the mRNA[1]. Such an effect could be catalyzed by a ribosomal protein like L18 (see Fig. 1).

At present, we have no proof that such mechanisms operate; we simply have evidence for the occurrence of dynamic effects. For example, protein L18 produces a large increase in the circular dichroism band of 5S RNA at 267 nm[18,19] and a similar effect has been observed for the yeast protein YL3–5S RNA complex[20]. Whilst this could derive from an electronic perturbation resulting from the direct chemical interaction of the protein and RNA, a more likely explanation is that a conformational change occurs possibly involving a switch between different helical conformers. So far chemical modification studies have revealed only minor structural changes[11]. L25 produces a small change in the circular dichroism spectrum of 5S RNA[18], which may correlate with changes in susceptibility of nucleotide bonds 68–69 and 89–90 to ribonuclease A attack and of guanine-69 to kethoxal modification[8,11].

Crothers and co-workers[21] have recently identified a proton-coupled conformational change that probably occurs in the 5S RNA tertiary structure at neutral pH. They suggest that since this is base-catalyzed it could be induced by the basic ribosomal proteins. Whether this is related to the aforementioned effects or not, is unknown.

Function of 5S RNA

When attempting to identify a function for 5S RNA one should perhaps ask: Why does the ribosome need a small RNA? One possible answer is that a small RNA can do something that a large RNA cannot. If one thinks in terms of the static structure, in so far as we understand it at present, there is probably nothing special about 5S RNA. However, the answer may lie in the dynamic properties of the molecule. It may be possible, for example, for a small RNA to produce some kind of alternating conformational change within a local region without influencing the rest of the ribosome. This may not be possible in the large RNAs which contain very large complex structural domains where a local change is more likely to be transmitted through the structure. It is likely, therefore, that the 5S RNA has a very special function.

The close structural homology between 5S RNA in pro- and eukaryotes suggests that they have the same functional role, although in eukaryotes the situation seems to be more complex since 5S RNA may form a complex with 5.8S RNA[15]. Although no equivalent to 5.8S RNA has been found in prokaryotes a partly homologous sequence occurs at the 5'-end of 23S RNA[22].

One proposed function is that 5S RNA acts as a reversible link between the small and large subunits[23]. Although base complementarity has been identified between conserved sequences in 5S RNA and sequences in both of the large ribosomal RNAs[23,24] there is no experimental evidence to support such RNA–RNA interactions.

Another hypothesis has been that the conserved complementary sequences T-ψ-C-G and C-G₄₄-A-A of tRNA and 5S RNA, respectively, interact in the ribosomal A-site. At one time this was supported by a wealth of circumstantial evidence, summarized by Erdmann[25], based mainly on the contention that the oligonucleotide T-ψ-C-G, or its analogues, could replace tRNA in certain ribosomal functions. It was also claimed that the 5S RNA–protein complex exhibited both GTPase and ATPase activities and inferred that these might be required for tRNA binding[25]. Recently, however, most of this evidence has been brought into serious question and much of it refuted[11,26]. In particular, it has been shown that when the sequence C-C-G₄₄-A-A is excised from the 5S RNA, and the ends rejoined by ligase, the smaller RNA can still be reconstituted into a 50S subunit which is active in protein biosynthesis[27].

What we know about the 5S RNA, from immunoelectron microscopy studies[28] and reconstitution experiments[29], is that it occurs at the subunit interface within a cluster of functionally important proteins which include L16, a protein essential for peptidyl transferase activity; protein L11 which stimulates GTPase activity, and the L7/12–L10 protein complex which forms part of the binding site for some of the ribosomal factors[30]. Its precise function remains to be determined.

References

1 Fox, G. E. and Woese, C. R. (1975) *Nature (London)* 256, 505–507
2 Noller, H. F. and Garrett, R. A. (1979) *J. Mol. Biol.* 132, 637–648
3 Peattie, D. A., Douthwaite, S., Garrett, R. A. and Noller, H. F. (1981) *Proc. Natl Acad. Sci. U.S.A.* 78, 7331–7335
4 Douthwaite, S., Christensen, A. and Garrett, R. A. (1982) *Biochem.* 21, 2313–2320
5 Douthwaite, S. and Garrett, R. A. (1981) *Biochem.* 20, 7301–7307
6 Nishikawa, K. and Takemura, S. (1974) *FEBS Lett.* 40, 106–109
7 Luehrsen, K. R. and Fox, G. R. (1981) *Proc. Natl Acad. Sci. U.S.A.* 78, 2150–2154
8 Douthwaite, S., Garrett, R. A., Wagner, R. and Feuteun, J. (1979) *Nucleic Acids Res.* 6, 2453–2470
9 Nazar, R. N., Willick, G. E. and Matheson, A. T. (1979) *J. Biol. Chem.* 254, 1506–1512
10 Nazar, R. N. (1979) *J. Biol. Chem.* 254, 7724–7729
11 Garrett, R. A. and Noller, H. F. (1979) *J. Mol. Biol.* 132, 637–648
12 Feuteun, J., Monier, R., Garrett, R., Le Bret, M. and Le Pecq, J. B. (1975) *J. Mol. Biol.* 93, 535–541
13 Spierer, P., Wang, C.-C., Marsh, T. L. and Zimmermann, R. A. (1979) *Nucleic Acids Res.* 6, 1669–1682
14 Nazar, R. N., Yaguchi, M., Willick, G. E., Rollin, C. F. and Roy, C. (1979) *Eur. J. Biochem.* 102, 573–582
15 Metspalu, A., Toots, I., Saarma, M. and Willems,

R. (1980) *FEBS Lett.* 119, 81–84
16 Hori, H. and Osawa, S. (1979) *Proc. Natl Acad. Sci. U.S.A.* 76, 381–385
17 Noller, H. F. and Woese, C. R. (1981) *Science* 212, 403–411
18 Bear, D. G., Schleich, T., Noller, H. F. and Garrett, R. A. (1977) *Nucleic Aids Res.* 4, 2511–2526
19 Spierer, P., Bogdanov, A. A. and Zimmermann, R. A. (1978) *Biochemistry*, 17, 5394–5342
20 Willick, G. E., Nazar, R. N. and Nguyen, T. V. (1980) *Biochemistry*, 19, 2738–2742
21 Kao, T. H. and Crothers, D. M. (1980) *Proc. Natl Acad. Sci. U.S.A.* 77, 3360–3364
22 Nazar, R. N. (1980) *FEBS Lett.* 119, 212–214
23 Azad, A. A. (1979) *Nucleic Acids Res.* 7, 1913–1929
24 Herr, W. and Noller, H. F. (1975) *FEBS Lett.* 53, 248–252
25 Erdmann, V. A. (1977) *Prog. Nucleic Acid Res. Mol. Biol.* 18, 45–90
26 Chinali G., Horowitz, J. and Ofengand, J. (1978) *Biochem.* 17, 2755–2760
27 Pace, B., Matthews, E. A., Johnson, K. D., Cantor, C. R. and Pace, N. R. (1982) *Proc. Natl Acad. Sci. U.S.A.* 79, 36–40
28 Stöffler, G., Bald, R., Kastner, B., Lührmann, R., Stöffler-Meilicke, M. and Tischendorf, G. (1980) in *Ribosomes* (Chambliss, G. *et al.*, ed.), pp. 171–205, University Park Press, Baltimore
29 Dohme, F. and Nierhaus, K. H. (1976) *Proc. Natl Acad. Sci. U.S.A.* 73, 2221–2225
30 Garrett, R. A. (1979) *Int. Rev. Biochem.* (Offord, R. E., ed.), Vol. 25, pp. 121–177, University Park Press, Baltimore

R. A. Garrett and S. Douthwaite are in the Division of Biostructural Chemistry, Chemistry Institute, Aarhus University, Denmark. H. F. Noller is in the Thimann Laboratories, University of California at Santa Cruz, CA 95064, U.S.A.

The organization of 16S ribosomal RNA
by R. A. Garrett

Models of the secondary structure of 16S ribosomal RNA from *Escherichia coli* have been rapidly developed since the correct RNA sequence became known. The latest versions of the models from Noller and Woese[1], Brimacombe[2] and Ehresmann[3] and their co-workers show a strong similarity; a cursory count reveals that about 30–35 of the 60 or so short helical segments are common to the models. Whilst this gives no room for complacency, the agreement is more encouraging than the painful experience with 5S RNA. (Who *didn't* propose a secondary structural model?). For this a tremendous debt must go to Woese, Fox and their co-workers who have emphasized that phylogenetic comparisons of sequences are absolutely essential for predicting reliable secondary structures in ribosomal RNAs (phylogenetic evidence for the existence of a helical region is provided when a putative base pair in one organism is found to be replaced by a different base pair in another). This has proved to be a critical test of secondary structure; all the other experimental data including chemical cross links, enzyme cuts and chemical modification sites are important but rarely conclusive, especially when, as for the large ribosomal RNAs, they are often obtained with either fragments of RNA or partially denatured RNA.

Several interesting features arise. The models confirm earlier protein binding studies which identified 3 major domains, interrupted by base pairing interactions between widely separated sequence regions (Fig. 1). The basic domain structures of 12S mitochondrial and 17–18S eukaryotic RNAs are conserved, but the size of certain regions (generally those of variable sequence) is diminished in the former and expanded in the latter, suggesting an interesting evolutionary flexibility. Especially well conserved are the invariant sequences which occur on the ribosome surface. These functionally important sequences, which appear in single stranded loops in the models, have been characterized mainly by Noller's group. They also point out that the introns in the eukaryotic ribosomal RNA genes occur in these regions only and do not separate the structural domains, as has been found for proteins.

Dynamic aspects of the RNA structure which are presumably essential for its function are still not understood; it is suggested that major rearrangements in base pairing

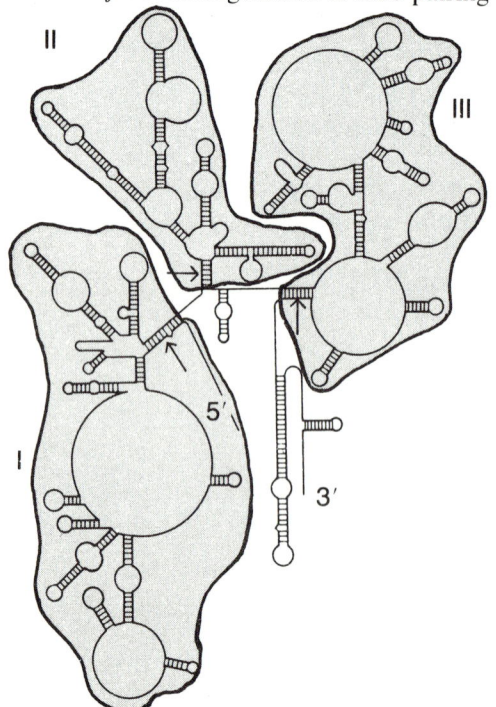

Fig. 1. Tentative secondary structure of 16S RNA according to Noller and Woese[1]. The three major domains I, II and III are encircled in grey. The double helices, composed of widely separated sequence regions, which define the limits of the domains, are indicated by arrows. The 3'-200 nucleotides (bottom right) constitute an additional minor domain which probably interacts primarily with domain III. (Modified figure reproduced with permission from Science.)

occur in both the small and large RNAs in different functional states of the ribosome. Whilst limited experimental evidence is presented, there is no convincing phylogenetic proof and the changes may be more subtle. As was pointed out some years ago, the large compact RNA domains probably have a higher potential than proteins for transmitting allosteric effects, especially along their double helical regions, without necessarily disrupting base pairing.

Many experimental problems lie ahead, but recent methodological developments should help us to revise and refine the models, infer higher order structure from a minimal secondary structure, deduce the significance of the many deviant helices containing mismatched and looped out bases and define the structural basis of RNA function.

References

1 Noller, H. F. and Woese, C. R. (1981) *Science*, 212, 403–411
2 Zweib, C., Glotz, C. and Brimacombe, R. (1981) *Nucleic Acids Res.* 9, 3621–3640
3 Stiegler, P., Carbon, P., Zuker, M., Ebel, J. P. and Ehresmann, C. (1981) *Nucleic Acids Res.* 9, 2153–2172

R. A. GARRETT

Division of Biostructural Chemistry, Aarhus University, Denmark.

Putting proteins in their place

Protein biosynthesis on membrane-bound ribosomes

David P. Leader

The distribution of messenger RNA between free and membrane-bound ribosomes is necessary for the correct cellular segregation of newly synthesized proteins. Molecular mechanisms for this partition of messengers, and for the compartmentation of their protein products, are now beginning to emerge.

Studies on the molecular details of protein biosynthesis in the past 20 years or so have tended to overshadow those concerned with the cellular aspects of this process. However, there have recently been exciting developments in our understanding of the way in which proteins synthesized in the intact mammalian cell are directed towards specific subcellular and, especially, extracellular locations. These developments will be briefly reviewed here in the context of earlier work, a more detailed account of which can be found elsewhere [1].

Ribosomes on membranes

The key to the study of the subcellular segregation of newly synthesized proteins was the electron microscope. This revealed that in secretory cells, such as those of liver and pancreas, there is an extensive system of membranes, termed endoplasmic reticulum, much of which is studded with knob-like protrusions giving it a 'rough' appearance. The protrusions on the 'rough endoplasmic reticulum' were shown to be ribosomes, some of the latter being also observed in areas of the cytoplasm from which membrane was absent. These two classes of ribosomes, 'free' and 'membrane-bound', could also be demonstrated by cell-fractionation techniques, the membrane-bound ribosomes in disrupted cells lying on the external face of small membranous vesicles.

The role of the ribosomes of the rough endoplasmic reticulum in the synthesis of secretory proteins was established through the efforts of many different workers; but the most outstanding contribution was from G. Palade, in recognition of which he received the Nobel Prize in 1974. By means of pulse-chase labelling of proteins with radio-active amino acids, combined with analysis by cell fractionation and autoradiography, he and his co-workers established that proteins synthesized on the rough endoplasmic reticulum of the exo-

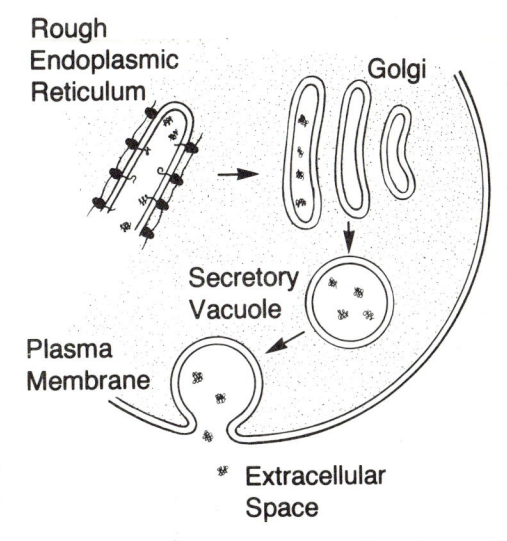

Fig. 1. *Passage of secretory proteins from the site of synthesis on the rough endoplasmic reticulum through various cell compartments to the extracellular space. The single diagram shows the successive locations of the protein.*

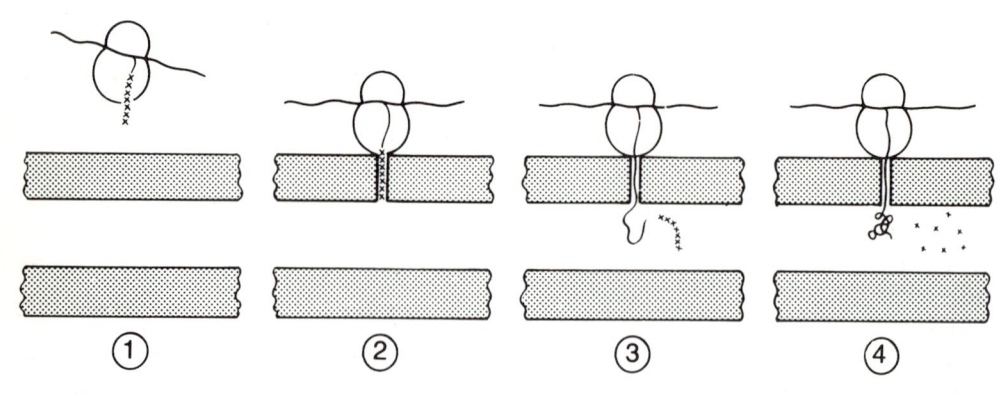

Fig. 2. *The Signal Hypothesis. The figure indicates four stages in the synthesis of secretory proteins:* (1) *free ribosomes initiate the synthesis of secretory protein, commencing with an N-terminal signal peptide (x-x-x-x-x-x);* (2) *the hydrophobic signal peptide attaches to the membrane of the rough endoplasmic reticulum, passing through it towards its lumen, and the ribosome attaches to the membrane via its 60 S subunit;* (3) *at some time before completion of the chain the signal peptide is cleaved;* (4) *the cleaved peptide is presumed to be subsequently degraded (although there is no direct evidence for this), the authentic protein continuing to be extruded until it is terminated, when it will start its passage out of the cell as indicated in Fig. 1.*

crine pancreas of the guinea-pig were extruded into its cisternal space and then transferred *via* the Golgi complex to secretory vacuoles before being eliminated from the cell [2]. This sequence of events is shown in Fig. 1, which indicates how fusion of the membrane of the secretory vacuole with the plasma membrane achieves the final discharge of the secretory proteins from the cell.

These findings led to the idea that there might be a general difference between the proteins synthesized on membrane-bound and on free ribosomes. Not only would the synthesis of secretory proteins be restricted to membrane-bound ribosomes, but that of intracellular proteins would be restricted to free ribosomes. Accordingly, a large number of studies were performed to identify the site of synthesis of specific proteins, usually employing immunochemical techniques to identify the latter (see Rolleston [3] for a detailed account of this work). These provided formidable evidence that ribosomes synthesizing a wide spectrum of secretory proteins are located almost exclusively on membranes; other examples being albumin (in liver), immunoglobulins (in plasma cells) and β-lactoglobulin (in the mammary gland). Moreover, certain intracellular proteins were, as anticipated, shown to be synthesized predominantly on free ribosomes (e.g. ferritin in liver, and globin in reticulocytes).

However, it was also clear that the original generalization was an oversimplification, and a certain proportion of intracellular proteins are synthesized on membrane-bound ribosomes. Some of these proteins are extruded into the cisternal space of the rough endoplasmic reticulum and may be directed to specific subcellular compartments (see below), but others are released into the cytosol [4] and the membrane-bound ribosomes on which these are synthesized may represent a distinct subclass, of unknown function.

Mode of attachment

The information specifying the endoplasmic reticulum as the site of synthesis of secretory proteins must clearly reside in the mRNA for such proteins. However, there are three main ways by which this information might be expressed. These are: (1) through direct interaction of mRNA with the membrane; (2) as an indirect consequence of the mRNA selecting a specific class of ribosomes that bind to membranes; (3) by the mRNA directing the synthesis of a particular type of polypeptide which could attach to the membrane. Knowledge of the nature of the attachment of ribosomes to membranes was clearly essential to distinguish between these possibilities.

Extensive studies of isolated membrane vesicles, especially those by D. Sabatini, G. Blobel and co-workers, lead to the conclusion that in the liver of the rat the ribosomes are bound to specific sites on the membranes through a salt-sensitive link to the large (60 S) subunit, the nascent peptide forming a firmer anchor when it has grown sufficiently to penetrate the membrane [5]. Most important, it was found that ribonuclease was neither sufficient nor necessary for the release of ribosomes from membranes, making it unlikely that attachment was through mRNA. It should be mentioned that there have been reports of release of membrane-bound ribosomes by ribonuclease in other cells, such as fibroblasts and HeLa cells [1], and that in the latter case the 3' polyadenylic acid region of the mRNA appears to interact directly with the membrane to cause attachment of ribosomes [6]. However, the cells used in these studies are clearly not typical of secretory tissues, and the different characteristics of their membrane-bound ribosomes suggests that these latter may be members of the separate functional subclass, alluded to above.

If the mRNA for secretory proteins does not initiate the attachment of ribosomes to membranes directly, it might be possible that it does so indirectly by selecting a specific subclass of ribosomes which can bind to membranes. Moreover, an extra protein has frequently been found associated with membrane-bound ribosomes [5], and this might conceivably be the basis for specific attachment of these ribosomes to the membrane. There is, however, little evidence to support this idea, and 60 S ribosomal subunits derived from free or membrane-bound ribosomes have an equal capacity to bind to the membranes of the rough endoplasmic reticulum [7]. While there seem to be specific attachment sites for ribosomes on the membranes of rough, but not smooth, endoplasmic reticulum, the binding of the ribosomes to these sites would seem to require initial direction to the membrane by some other, more specific, means.

It is evident that the interaction between membrane and nascent peptide during protein synthesis on membrane-bound ribosomes does not necessarily imply that such an interaction also initiates the attachment of ribosomes to membranes. Nevertheless, Blobel and Sabatini were strongly attracted to this possibility, and predicted that there must be some common feature shared by the nascent polypeptide chains of secretory proteins that would cause them to attach to membranes [8].

The signal peptide

Blobel and Sabatini did not specify the nature of the common feature which they predicted would modulate the attachment of the nascent polypeptide to membranes. However, it soon became apparent from studies by Milstein *et al.* [9] that this might be an extra sequence at the amino-terminus of the polypeptide. These workers observed that when the mRNA for an immunoglobulin light chain (a secreted protein) was translated in a cell-free system, lacking membranes, the product was about 15 residues larger than the authentic protein. Blobel and co-workers [10] showed that similar extensions can be detected during the synthesis of several other secretory proteins *in vitro* and established the amino-terminal and hydrophobic nature of these extensions by partial sequencing of the labelled precursor pro-

	−30	−20	−10	+1
Pre-proAlbumin (12)			Met.Lys.Trp.Val.Thr.Phe.Leu.Leu.Leu.Phe.Ile.Ser.Gly.Ser.Ala.Phe.Ser.	Arg...
Pre-IgG light chain (13)		Met.Asp.Met.Arg.Ala.Pro.Ala.Gln.Ile.Phe.Gly.Phe.Leu.Leu.Leu.Phe.Pro.Gly.Thr.Arg.Cys.	Asp...
Pre-Lysozyme (14)			Met.Arg.Ser.Leu.Leu.Ile.Leu.Val.Leu.Cys.Phe.Leu.Pro.Leu.Ala.Ala.Leu.Gly.	Lys...
Pre-Prolactin (15)	Met.Asn.Ser.Gln.Val.Ser.Ala.Arg.Lys.Ala.Gly.Thr.Leu.Leu.Leu.Leu.Met.Met.Ser.Asn.Leu.Leu.Phe.Cys.Gln.Asn.Val.Gln.Thr.	Leu...
Pre-penicillinase (E. coli) (16)		Met.Ser.Ile.Gln.His.Phe.Arg.Val.Ala.Leu.Ile.Pro.Phe.Phe.Ala.Ala.Phe.Cys.Leu.Pro.Val.Phe.Ala.	His...
Pre-Vesicular Stomatitis Virus glycoprotein (17)			Met.Lys.Cys.Leu.Leu.Tyr.Leu.Ala.Phe.Leu.Phe.Ile.His.Val.Asn.Cys.	Lys...
Pre-Lipoprotein (E. coli) (18)		Met.Lys.Ala.Thr.Lys.Leu.Val.Leu.Gly.Ala.Val.Ile.Leu.Gly.Ser.Thr.Leu.Leu.Ala.Gly.	Cys...
PRE-PROTEIN	SEQUENCE OF SIGNAL PEPTIDE (PRE-PIECE)			CLEAVAGE POINT

Fig. 3. Sequences of some signal peptides. The sequences above the horizontal line are for secretory proteins, and those below the line are for membrane proteins. The N-terminal amino acid of the authentic protein or pro-protein is numbered +1, and the amino acids of the signal peptide are given negative numbers starting from the cleavage point. The name 'pre-lipoprotein' is used, rather than 'pro-lipoprotein' [18], for consistency of usage of the terms 'pre-' and 'pro-'; and capitalization of the mature proteins is used to distinguish Prolactin from the various pro-Proteins. Different IgG light chains have different precursor sequences, the one presented being for MOPC-41 [13]. This is not intended to be a full list of published sequences of signal peptides, which at the time of writing number about a dozen.

teins. From these and other experiments Blobel and Dobberstein proposed the 'Signal Hypothesis' [11], the main tenet of which is that segregation of secreted proteins occurs by the binding of a hydrophobic amino-terminal sequence (signal peptide) to the membrane, this peptide being subsequently cleaved and, presumably, degraded within the cisternal space of the endoplasmic reticulum (Fig. 2).

Signal peptides have now been detected in many more eukaryotic secretory protein precursors; and also in a number of bacterial periplasmic proteins, which are synthesized on ribosomes bound to the inner face of the bacterial cell membrane and secreted into the periplasmic space. The complete amino acid sequences of several of these signal peptides have been determined, and some of them are shown in Fig. 3. From this it can be seen that their similarity is limited to a general hydrophobicity; the lengths (approx. 15-30 amino acids) and sequences showing considerable variation, despite earlier indications to the contrary [10]. The signal sequences on the initial protein precursors are designated 'pre' to distinguish them from the earlier discovered 'pro' sequences of intermediate protein precursors such as proinsulin and proalbumin. The fact that pre-proinsulin and pre-proalbumin have now been identified raises the question of whether these pre-sequences really are the initial amino-termini. This has in fact been demonstrated in several cases, either by cell-free protein synthesis with initiator met-tRNA [13,14] or from examination of the sequence of the mRNA [19]. Similar types of experiments also demonstrate unequivocally that the α- and β-chains of the non-secreted protein, haemoglobin, have no signal peptide.

Membrane proteins

Although discussion of the synthesis of proteins destined for specific cell compartments has so far centred on secretory proteins, it had, in fact, long been thought possible that membrane proteins might also be segregated by synthesis on membrane-bound ribosomes, a proposal which would help to account for the existence of the latter in non-secretory cells. However, when it became clear that some membrane proteins are synthesized on free ribosomes [3] this idea, in its original form, became untenable. Subsequently, though, it was pointed out that only proteins that are located – at least in part – on the external surface of the plasma membrane (which arises from the internal surface of the membrane of the endoplasmic reticulum – Fig. 1) would be expected to be positioned through synthesis on, and extrusion into, the membrane in a manner comparable to secretory proteins [20]. This now appears to be the case, and some of the membrane proteins of both eukaryotes and prokaryotes that are synthesized on membrane-bound ribosomes have been shown to have transient signal peptides, similar to those in secretory proteins (Fig. 3).

The segregation process

The signal hypothesis has at its centre the solid chemical evidence of the signal sequence; but not all the other details of the segregation process are quite so clear as this. There has been some experimental support for the idea that, in contrast to the scheme in Fig. 2, ribosomes bind to the membrane before the start of protein synthesis, and are displaced if there is no hydrophobic signal peptide [21]. Blobel, however, has presented persuasive evidence against this idea and in favour of his original concept [11] that the synthesis of secretory proteins is initiated on free ribosomes, which are then directed to the membrane by the signal peptide [22]. In either case there remains the thermodynamic problem of getting the rest of the protein (including the hydrophilic sections) through the membrane.

Work has also been done to characterize the specific sites on the membrane that bind the ribosomes following the initial event of attachment of the signal peptide. Sabatini and co-workers [23] have described two membrane proteins that interact with the ribosomes on the rough endoplasmic reticulum, and these appear to be absent from smooth endoplasmic reticulum. The detailed model of Blobel and Dobberstein [11] implies that the signal peptide interacts with the same component of the membrane, but it is not yet known whether this is in fact so.

Cleavage of the signal sequence seems to occur before completion of the polypeptide chain and to be catalysed by a peptidase associated with the cisternal faces of the membranes of the rough endoplasmic reticulum [22]. Correct processing has been observed with mixtures of mRNAs and membranes from different tissues and phylogenically distant eukaryotes, although it is not yet known whether the eukaryotic and prokaryotic signal peptidases can be interchaged. In view of the wide diversity of the sequences of the signal peptides (Fig. 3) it is very difficult to see what is the unique feature of the signal sequences recognized by the peptidase. Although there is usually a small amino acid at the immediate amino side of the cleavage site, this is clearly not sufficient, and something more subtle must also be important. Whether this involves more distal amino acids or some feature of secondary structure remains to be seen. The isolation of bacterial mutants in which the signal peptide is not cleaved offers a possible approach to this question [24].

Ovalbumin

There has been found, so far, one exception to the rule that secretory proteins have a transient signal peptide. Ovalbumin, unlike three other egg-white proteins synthesized in the same oviduct cells, is synthesized directly, without any precursor [25]. It has, however, been shown that ovalbumin is secreted by the same route as the other proteins, and that nascent ovalbumin will compete with nascent pre-prolactin *in vitro* for sites on microsomal membranes [26]. Thus it appears that the moderately hydrophobic amino-terminal sequence of ovalbumin acts as a non-cleavable signal peptide. This suggestion that secretion is possible without cleavage of the signal peptide is consistent with the results of studies on the mutant of *E. coli* pre-lipoprotein, where the signal peptide is not cleaved [24].

Other signals

Finally, it should be mentioned that other signals must be involved in directing proteins to other cell compartments. Certain proteins destined for organelles such as chloroplasts [27] and mitochondria [28] have been shown to be synthesized on free ribosomes, and their passage across the double-membranes of the organelles is associated with a proteolytic cleavage, not unlike that accompanying entry of diphtheria and other toxins into mammalian cells [29]. At least two proteins destined for peroxisomes are also synthesized on free ribosomes [30], a rather surprising result in view of the fact that peroxisomes appear to be formed by 'budding' from the endoplasmic reticulum. The mechanism by which such proteins reach the peroxisomes is not yet known; and similar ignorance exists over the transport of lysozomal and nuclear proteins. However, the rate of progress in this area makes one hopeful that answers to these problems will presently emerge.

Acknowledgements

I should like to thank Dr J. D. Pitts for helpful discussions and critical comments on the manuscript. I also gratefully acknowledge the generosity of the many authors who sent me details of their work prior to publication.

References

1. Sabatini, S. S. and Kreibich, G. (1976) in *The Enzymes of Biological Membranes* (Martonosi, A., ed.) Vol. 2, pp. 531–579, Plenum Press, New York
2. Palade, G. (1975) *Science* 189, 347–358
3. Rolleston, F. S. (1974) *Sub-Cell. Biochem.* 3, 91–117
4. Andrews, T. M. and Tata, J. R. (1971) *Biochem. J.* 121, 683–694
5. Sabatini, D. D., Borgese, N., Adelman, M., Kreibich, G. and Blobel, G. (1972) in *FEBS Symposium*, Vol. 27, pp. 147–171, North-Holland, Amsterdam
6. Milcarek, C. and Penman, S. (1974) *J. Mol. Biol.* 89, 327–338
7. Rolleston, F. S. (1972) *Biochem. J.* 129, 721–731
8. Blobel, G. and Sabatini, D. D. (1971) in *Biomembranes* (Manson, L. A., ed.) Vol. 2, pp. 193–5, Plenum Press, New York
9. Milstein, C., Brownlee, G. G., Harrison, T. M. and Mathews, M. B. (1972) *Nature (London), New Biol.* 239, 117–120
10. Devillers-Thiery, A., Kindt, T., Scheele, G. and Blobel, G. (1975) *Proc. Nat. Acad. Sci. U.S.A.* 72, 5016–5020
11. Blobel, G. and Dobberstein, B. (1975) *J. Cell Biol.* 67, 835–851
12. Strauss, A. W., Bennett, C. D., Donohue, A. M., Rodkey, J. A. and Alberts, A. W. (1977) *J. Biol. Chem.* 252, 6846–6855
13. Zemell, R., Burstein, Y. and Schechter, I. (1978) *Eur. J. Biochem.* 89, 187–193
14. Palmiter, R. D., Gagnon, J., Ericsson, L. H. and Walsh, K. A. (1977) *J. Biol. Chem.* 252, 6386–6393
15. McKean, D. J. and Maurer, R. A. (1978) *Biochemistry* 17, 5215–5219
16. Ambler, R. P. and Scott, G. K. (1978) *Proc. Nat. Acad. Sci. U.S.A.* 75, 3732–3736
17. Lingappa, V. R., Katz, F. N., Lodish, H. F. and Blobel, G. (1978) *J. Biol. Chem.* 253, 8667–8670
18. Inouye, S., Wang, S., Sekizawa, J., Halegoua, S. and Inouye, M. (1977) *Proc. Nat. Acad. Sci. U.S.A.* 74, 1004–1008
19. Seeburg, P. H., Shine, J., Martial, J. A., Baxter, J. D. and Goodman, H. M. (1977) *Nature (London)* 270, 486–494
20. Lodish, H. F. and Small, B. (1975) *J. Cell Biol.* 65, 51–64
21. Harrison, T. M., Brownlee, G. G. and Milstein, C. (1974) *Eur. J. Biochem.* 47, 613–620
22. Blobel, G. (1978) in *FEBS Symposium*, Vol. 43, pp. 99–108, Pergamon (Oxford)
23. Kreibich, G., Czakó-Graham, M., Grebenau, R., Mok, W., Rodriguez-Boulan, E. and Sabatini, D. D. (1978) *J. Supramol. Struct.* 8, 279–302
24. Lin, J. J. C., Kamazawa, H., Ozols, J. and Wu, H. C. (1978) *Proc. Nat. Acad. Sci. U.S.A.* 75, 4891–4995
25. Palmiter, R. D., Gagnon, J. and Walsh, K. A. (1978) *Proc. Nat. Acad. Sci. U.S.A.* 75, 94–98
26. Lingappa, V. R., Shields, D., Woo, S. L. C. and Blobel, G. (1978) *J. Cell Biol.* 79, 567–572
27. Dobberstein, B., Blobel, G. and Chua, N. H. (1977) *Proc. Nat. Acad. Sci. U.S.A.* 74, 1082–1085
28. Maccecchini, M. L., Rudin, Y., Blobel, G. and Schatz, G. (1979) *Proc. Nat. Acad. Sci. U.S.A.* 76, 343–347
29. Pappenheimer, A. M. (1978) *TIBS* 3, N220–223
30. Goldman, B. M. and Blobel, G. (1978) *Proc. Nat. Acad. Sci. U.S.A.* 75, 5066–5077

David P. Leader is at the Department of Biochemistry, University of Glasgow, Glasgow G12 8QQ, Scotland, U.K.

Addendum – *D. P. Leader*

The three years that have elapsed since this review first appeared have seen the accumulation of much information confirming the importance of the signal peptide for secretion, but most of the problems remaining at that time are still unsolved (reviewed in [31]). There have, however, been developments regarding the precise mechanism by which the signal peptide interacts with and passes into the membrane of the rough endoplasmic reticulum. In contrast to the original concept of linear secretion of the protein across the membrane (see Fig. 2), it may well be that the signal peptide folds back on itself, and the loop formed by this fold, rather than the amino-terminus, is first to enter the membrane.

This 'loop model' was first proposed for prokaryotes, which have as a common feature a charged residue near the amino terminus of the signal peptide. It was suggested that the function of this charge might be to anchor the peptide to the negatively charged inner surface of the bacterial cell membrane, the hydrophobic loop then being inserted into the membrane [18]. The idea was later thought also to apply to eukaryotes when it was reported that the non-cleaved signal peptide of ovalbumin was internal, rather than at the amino-terminal end of the protein. Although it is now known that this report was incorrect [32], there does appear to be an internal signal peptide for the anion transport protein of erythrocyte membranes [33]. However, it would seem premature to draw any general conclusions about the mode of secretion of eukaryotic proteins on the basis of the results obtained with a single membrane protein. Furthermore, there are complexities to secretion in eukaryotes (not encountered in prokaryotes), perhaps occasioned by the need to ensure recognition of the membrane of the rough endoplasmic reticulum, rather than that of, say, the mitochondrion. Evidence for the involvement of at least two proteins, or protein complexes (a 'signal recognition protein' [34] and a 'transfer protein' [35]) is emerging, and a role here also for 7S RNA [36], hitherto of unknown function, seems likely.

References

31 Strauss, A. W. and Boime, I. (1982) *CRC Critic. Rev. Biochem.* 12, 205–235
32 Braell, W. A. and Lodish, H. F. (1982) *J. Biol. Chem.* 257, 4578–4582
33 Braell, W. A. and Lodish, H. F. (1982) *Cell* 28, 23–31
34 Walter, P., Ibrahimi, I. and Blobel, G. (1981) *J. Cell Biol.* 91, 545–550
35 Meyer, D. I., Krause, E. and Dobberstein, B. (1982) *Nature* 297, 647–650
36 Ullu, E. (1982) *Trends Biochem. Sci.* 7, 216–219

The one and only missing link in protein secretion?
by A. M. Tartakoff

The N-terminal sequence of nascent secretory proteins consists of 20–40 hydrophobic amino acid residues. It is thought that one of the functions of this 'signal' sequence is to ensure that polysomes synthesizing these proteins bind to the rough endoplasmic reticulum (RER). However, as a hydrophobic sequence should interact with a range of organellar membranes, what governs its apparent selectivity for the RER?

The key factors may be ribosome-binding sites, e.g. the 'ribophorins', which are known to be confined to the RER[1]. Recent research suggests that this specificity may also be due to other proteins of the membranes of rough microsomes (fragmented RER). These proteins which are extracted with high concentrations of salt and/or elastase are essential for vectorial transport of nascent polypeptides across the membrane[2,3]. An 11 S protein complex extracted from dog pancreas microsomes has a number of remarkable properties which suggest that it binds both the hydrophobic N-terminus of the nascent chain and the microsomal membrane, thereby mediating vectorial transport[4,5,6].

(1) It binds with high affinity to free polysomes programmed with a message for the secretory protein prolactin – but not when programmed for a non-secretory protein, globin.

(2) In a cell-free protein synthetic system containing microsomes, this protein complex causes the prolactin message (but not the globin message) to bind to the microsomal membrane.

(3) In a cell-free protein synthetic system lacking membranes the complex inhibits the elongation of pre-prolactin (but not globin). The addition of microsomal membranes stimulates the synthesis of prolactin and pre-prolactin.

(4) If the microsomes in the system have been treated with high salt only an (apparently N-terminal) 8 kilodalton pre-prolactin fragment is synthesized. In the presence of the complex, protein synthesis is completed and vectorial transport of the chain occurs.

Hence, in the absence of the 11 S complex the elongation of pre-prolactin stops with a nascent chain segment equivalent to the hydrophobic extremity plus the stretch of polypeptide 'concealed' with the ribosome. Such chains 'destined' for export could not possibly be released to the cell sap and should not even be terminated in the absence of the normally microsome-associated 11 S complex.

This work raises two questions. Since the signal sequence seems to interact directly with protein, why is it hydrophobic in the first place? Second, what links the 11 S complex to the microsome?

References

1 Kreibich, G., Czako-Graham, M., Grebenau, R., Mok, W., Rodriguez-Boulan, E. and Sabatini, D. (1978) *J. Supramolec. Struct.* 8, 279–302
2 Meyer, D. and Dobberstein, B. (1980) *J. Cell Biol.* 87, 498–508
3 Walter, P. and Blobel, G. (1980) *Proc. Natl Acad. Sci. U.S.A.* 77, 7112–7116
4 Walter, P., Ibrahimi, I. and Blobel, G. (1981) *J. Cell Biol.* 91, 545–550
5 Walter, P. and Blobel, G. (1981) *J. Cell Biol.* 91, 551–556
6 Walter, P. and Blobel, G. (1981) *J. Cell Biol.* 91, 557–561

A. M. TARTAKOFF

Department of Pathology, University of Geneva, Switzerland.

How proteins are transported into mitochondria

Walter Neupert and Gottfried Schatz

Most mitochondrial polypeptides are synthesized outside the organelle as precursors which are usually larger than the 'mature' polypeptides found within mitochondria. The precursors are imported into the mitochondria by a process which is independent of protein synthesis but dependent on high-energy phosphate bonds inside the mitochondria. This mechanism is basically different from that which governs the movement of secretory polypeptides across the membrane of the endoplasmic reticulum.

How are mitochondria made? They contain their own genetic system that makes the mitochondrial RNAs and a few mitochondrial polypeptides such as the three largest subunits of cytochrome oxidase, the cytochrome b apoprotein, two subunits of the ATPase complex and one protein of the 30 S mitochondrial ribosomal subunit[1]. However, these polypeptides account for less than 10% of the total protein mass of a mitochondrion. All the other several hundred mitochondrial polypeptides are coded by nuclear genes, translated on cytoplasmic ribosomes and imported into the mitochondrion. This import process must be highly specific since each mitochondrial polypeptide occupies a precisely defined location within the organelle. Mitochondria (Fig. 1) consist of two distinct membranes: an outer membrane (whose function is still somewhat obscure) and a highly infolded inner membrane (which houses the machinery for oxidative phosphorylation). The space between the outer and inner membrane is the 'intermembrane space'; the space bounded by the inner membrane is the 'matrix space'.

The massive movement of polypeptides into the mitochondria raises several intriguing questions. Why is a newly made mitochondrial protein only imported into mitochondria and not into other organelles? How is an imported polypeptide steered to its proper intramitochondrial location? Why do the enzymes destined to be imported not become active in the cytosol? How are water-insoluble membrane proteins transported? How is the transport regulated?

Protein traffic in the eukaryotic cell

In addition to transporting proteins into mitochondria, a typical eukaryotic cell must also transport proteins across other intracellular membranes and proteins must be inserted into membranes such as the endoplasmic reticulum and the plasma membrane. One of the simplest hypotheses to account for ordered 'protein traffic' is that each protein carries in itself a specific structure that recognizes a complementary structure on the membrane across or into which it has to move. A similar principle might govern the assembly of oligomeric proteins.

Secretion of proteins is coupled to protein synthesis

The secretion of proteins by cells is closely related to intracellular protein traffic[2]. In eukaryotes, the synthesis of a protein destined to be secreted occurs on a ribosome which is attached to the cytoplasmic side of the endoplasmic reticulum:

protein synthesis and translocation are obligatorily linked. To explain the specificity of transport, it has been suggested that the amino terminal region of the nascent polypeptide recognizes some hypothetical receptor on the target membrane[3,4].

This process has two salient features: Firstly, transmembrane movement is 'co-translational'; it is coupled to elongation of the polypeptide chain. As a result, this translocation process is often termed 'vectorial translation'. Secondly, the protein is immediately sequestered on the opposite side of the membrane; the completed polypeptide chain is thus probably never present in the compartment where it is synthesized.

Mitochondrial protein import is not obligately coupled to protein synthesis

Do mitochondria import proteins by vectorial translation? Apparently not, since neither of the two criteria for vectorial translation is met. This was first indicated by studies *in vivo*[5,6] which established: (1) pulse-labelled mitochondrial proteins are initially found outside the mitochondria; (2) newly made mitochondrial proteins can be chased into mitochondria only after a lag; (3) this lag differs for different pulse-labelled proteins; (4) inhibition of protein synthesis by cycloheximide fails to prevent the appearance of pulse-labelled proteins inside the mitochondria.

Clearly, these observations do not support a typical cotranslational mechanism. Instead, they open the possibility that import is independent of concomitant translation and proceeds via extramitochondrial pools of mitochondrial proteins. Further evidence for a post-translational mechanism came from subsequent studies with cell-free systems. On the one hand, these experiments demonstrated clearly that the synthesis of a mitochondrial protein can be separated from its transport into the mitochondria in time as well as space. On the other hand, the cell-free systems showed that all imported mitochondrial proteins studied so far are initially made as precursors which are qualitatively different from their functional counterparts inside the mitochondria.

Precursors of mitochondrial proteins made in the cytoplasm

If cytoplasmically-made mitochondrial proteins are synthesized in heterologous cell-free systems and isolated by immunoprecipitation, most but not all of them display an apparent molecular weight which is several thousand daltons larger than that of the corresponding mature proteins*. The precursors which are larger than the mature polypeptides probably have additional sequences at the amino terminus. So far, several dozen mitochondrial precursors have been identified; a few representative examples are listed in Table I.

It has been reported that cytoplasmically-made subunits of oligomeric pro-

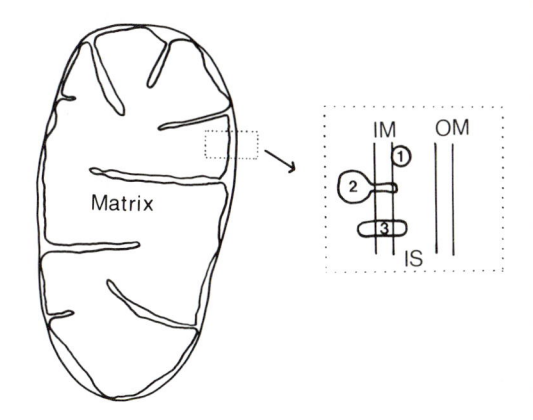

Fig. 1. The structure of a mitochondrion. IM = inner membrane; OM = outer membrane; IS = intermembrane space; 1 = cytochrome c; 2 = oligomycin-sensitive ATPase complex; 3 = ADP/ATP carrier.

* It is impossible to cite all reports on mitochondrial precursor polypeptides in this brief review. A few typical examples are listed in Table I.

TABLE I.
A few representative precursors to cytoplasmically-made mitochondrial proteins.

Polypeptide	Intra-mito-chondrial location	Organism	Apparent molecular weight Mature	Apparent molecular weight Precursor	Precursor found in vivo	Transport in vitro demonstrated	Reference
ATPase	Matrix	Yeast					7
α-subunit			58,000	64,000	+	+	
β-subunit			54,000	56,000	+	+	
γ-subunit			34,000	40,000	+	+	
Citrate synthase		Neurospora	45,000	47,000	+	−	8
ADP/ATP carrier	Inner membrane	Neurospora and yeast	32,000	32,000	−	+	9,10
Cytochrome c_1		Neurospora	31,500	37,000			
		and yeast	31,000	37,000	+	−	11,[a]
Cytochrome c	Intermembrane space	Neurospora and yeast	12,000	12,000	+	+	12,13,14[b]
Cytochrome c peroxidase		Yeast	33,500	39,500	+	+	15

[a] R. Zimmermann, H. Weiss and W. Neupert, unpublished work.
[b] S. Gasser and G. Schatz, unpublished work.

teins such as cytochrome oxidase are made as a polyprotein which is imported and cut to the mature subunits inside the mitochondria[16]. However, recent studies failed to confirm this; in fact, the individual cytoplasmically-made subunits of cytochrome oxidase and ATPase are made as discrete precursors which are only a few thousand daltons larger than the mature subunits[17,18,19]. In several cases the larger molecular weight precursors detected in the cell-free translation systems can also be found in pulse-labelled intact cells (Table I). The kinetic behaviour of these larger forms upon a chase was consistent with their role as precursors to the corresponding 'mature' polypeptides.

Import in vitro

Yeast or Neurospora mitochondria can incorporate mitochondrial precursors, synthesized in a cell free system, in the absence of protein synthesis[7,12−15]. This appears to be a specific process: First, it is accompanied by proteolytic removal of any extra sequences which had been present in the precursor. Second, the resulting mature polypeptide (but not any residual precursor) becomes resistant to externally added proteases. Third, non-mitochondrial proteins synthesized in vitro or radiolabelled mature mitochondrial proteins are neither altered nor rendered protease-resistant by isolated mitochondria. Figs. 2 and 3 illustrate import in vitro of precursors with or without amino terminal extensions.

Mitochondrial precursors are synthesized on free polysomes

The available evidence strongly suggests that precursors of mitochondrial proteins are predominantly synthesized on free polysomes (see Refs 9, 12, 20, also M. Suissa and G. Schatz, unpublished work). In one type of experiment, the different classes of cytoplasmic polysomes were first isolated from different cell types and allowed to complete their nascent polypeptide products in vitro ('read-out'); in the second type of experiment, the polysome-associated mRNA was isolated and translated in a reticulocyte lysate. In both instances, the radiolabelled translation products were checked for mitochondrial precursors by immunoprecipitation. Most of the mRNA for mitochondrial proteins was found to be associated with free polysomes, regardless of whether the mitochondrial protein tested for was a membrane protein or a matrix protein.

However, the situation may be more complex, at least in yeast. If yeast mitochondria are isolated under conditions preventing degradation of polysomes, they contain cytoplasmic polysomes bound to their surface[21]. Compared to free polysomes, these mitochondria-associated polysomes are enriched in mRNA for mitochondrial proteins (M. Suissa and G. Schatz, unpublished work). Nevertheless, more than 90% of the *total* mRNA for yeast mitochondrial proteins is associated with free polysomes, indicating that these polysomes make the bulk of the imported mitochondrial proteins. This result, together with the observations mentioned earlier, makes it extremely unlikely that the mitochondria-associated polysomes found in yeast represent the obligate port of entry for polypeptides into mitochondria. The biological significance of these mitochondria-bound polysomes remains unknown.

In what form do precursor proteins occur outside mitochondria?

Mitochondrial proteins synthesized in heterologous translation systems are always released into the post-ribosomal supernatant fraction. This is not only true for soluble matrix proteins but also for membrane proteins such as the ADP/ATP carrier or subunit V of cytochrome oxidase. The precursors discharged into the supernatant are accessible to added protease or antibody. Thus, precursors synthesized *in vitro* are apparently neither enclosed in

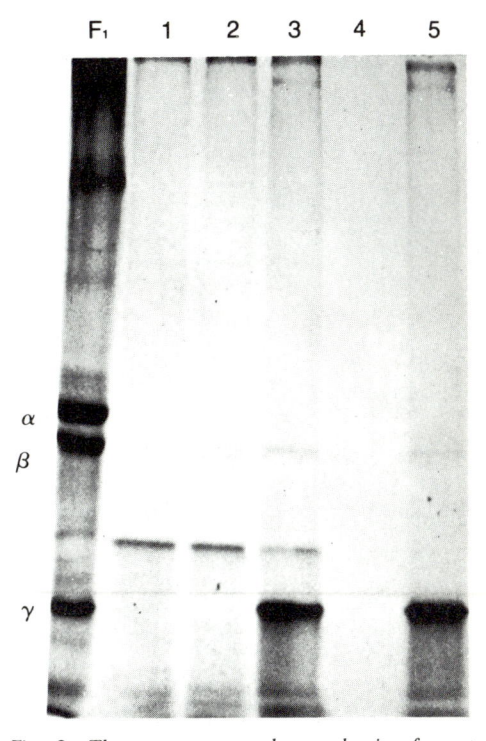

Fig. 2. The precursor to the γ-subunit of yeast mitochondrial F_1 ATPase is processed and imported into mitochondria in vitro. A radioautogram of an SDS-polyacrylamide gel is shown. The track labelled F_1 contains radiolabelled mature F_1-ATPase from yeast as a standard. Track 1: γ-subunit precursor synthesized in vitro. Track 2: after labelling in vitro, the reticulocyte lysate was inhibited with cycloheximide and incubated for 60 min at 29°C with isolated yeast mitochondria; the mitochondria were then spun down and the supernatant was subjected to immunoprecipitation with anti-γ-serum. Track 3: same as track 2 except that the mitochondrial pellet was analysed. Track 4: same as track 2 except that the lysate-mitochondria mixture was treated with trypsin and chymotrypsin before removal of the mitochondria; the mitochondria-free supernatant was analysed. Track 5: same as track 4 except that the mitochondrial pellet was analysed.

membrane vesicles nor deeply buried in a lipid phase. Similarly, after pulse-labelling intact yeast cells, at least 50% of the precursor to cytochrome oxidase subunit V is in the 'cytosolic' fraction (A. Lewin and G. Schatz, unpublished work). These observations suggest that mitochondrial protein precursors are initially released into the cytosol. However, it is difficult to exclude that the precursors are first discharged into some highly fragile vesicle which are disrupted when the cells are homogenized.

How can precursors to insoluble membrane proteins exist in the cytosol? We do not know, but a key to this question may be provided by observations from our laboratories that the precursors to the ADP/ATP carrier or to cytochrome oxidase subunit V do not exist as monomers but as larger aggregates. It is not known whether these aggregates are simply homo-oligomers of the precursor or complexes between precursor and other proteins. Preliminary observations suggest, moreover, that precursors to some mitochondrial membrane proteins are amphipathic. This raises the possibility that the precursor aggregates are protein micelles in which the hydrophobic domains are buried in the interior.

The molecular mechanism of import

What are the molecular events that transport a polypeptide into mitochondria? Is there a single mechanism for all proteins or are there different mechanisms for proteins transported into different locations within the organelle? We cannot yet answer these questions, but we can present some ideas as to how import might work.

Step 1: The polypeptide destined for import into mitochondria is made on free polysomes as a precursor which differs from the mature polypeptide in conformation and (in most cases) in molecular weight. The precursor is then discharged into a cytosolic pool whose size may vary depending on the metabolic conditions. The evidence for this is quite good and has already been discussed.

Step 2: The precursor binds to a 'receptor' on the mitochondrial surface. Most likely, each precursor polypeptide contains a domain which recognizes some structure on the cytoplasmic surface of the mitochondrial outer membrane. There is indeed some evidence that mitochondrial precursors can specifically bind to isolated mitochondria, that this binding is temperature- and energy-independent[22] and that it is abolished if the mitochondria are stripped of their outer membranes (D. L. Sabatini, personal communication). At present, there are no data proving the existence of distinct 'import receptors' on the mitochondrial surface, but we suspect that there exist at least two classes of such receptors: one class for proteins transported only across the outer membrane and the second class for proteins transported across both membranes. This second class of receptors might be located at those sites on the outer membrane where the inner and outer membrane are closely apposed (Fig. 1).

Step 3: Interaction between the precursor and its corresponding receptor opens up a pore which allows at least partial diffusion of the precursor across one or both membranes. This step requires high-energy phosphate bonds. The evidence for this is as follows: if ATP in the mitochondrial matrix of intact yeast cells is depleted by suitable mutations or inhibitors, import of proteins into mitochondria stops and unprocessed precursors accumulate in the cytosol[11]. The energy-dependent step is import, not proteolytic processing: on the one hand, energy-dependence of import was demonstrated for the precursor of ADP/ATP carrier[22] which has the same molecular weight as the mature protein; on the other hand, the processing protease (cf. below) does not require ATP. The energy-dependence of import by isolated Neurospora mitochondria has also been demonstrated *in vitro*: carbonyl cyanide *m*-chlorophenylhydrazone (CCCP), which apparently lowers the intramitochondrial ATP concentration, blocks import of mitochondrial protein precursors.

Step 4: The precursor is converted to the mature polypeptide on the opposite side of the membrane barrier. The resulting conformational change could then trap the polypeptide inside the mitochondria and thereby provide a driving force for unidirectional import. Almost certainly, step 4 differs for different precursors. This must be anticipated because not all precursors are proteolytically processed to the mature polypeptide. Differences might exist even among those precursors that are not proteolytically processed. For example: with cytochrome *c*, the heme-free precursor polypeptide is probably fitted with its heme group in the intermembrane space and thereby converted to the mature holocytochrome *c* which then attaches to the outer face of the mitochondrial inner membrane[12,13,14]. The precursor to the ADP/ATP carrier (which differs from the mature polypeptide with respect to solubility and chromatographic properties, but not molecular weight – see Table I and Fig. 3) might be converted to its mature conformation on being inserted into the hydrophobic core of the mitochondrial inner membrane and subsequent dimerization. Finally, precursors that carry amino-terminal extensions are proteolytically processed by an intramitochondrial protease. Removal of the extra amino acid sequences might trigger an essentially

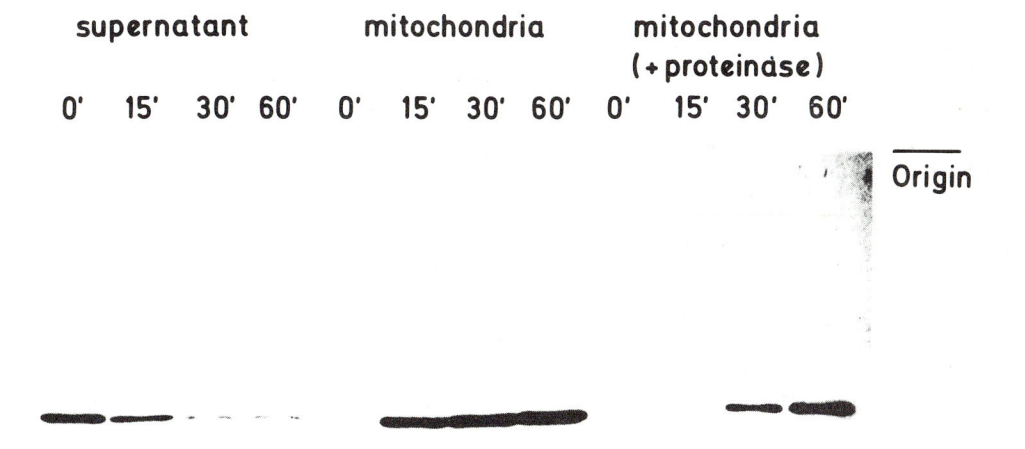

Fig. 3. The precursor to the ADP/ATP carrier is imported into Neurospora *mitochondria without proteolytic processing. A reticulocyte lysate containing radiolabelled precursor to the ADP/ATP carrier from* Neurospora *was incubated with isolated* Neurospora *mitochondria. Aliquots were withdrawn at the times indicated and freed from mitochondria by centrifugation. The supernatant was saved for analysis (cf. below). The mitochondrial pellets were resuspended and one half of each suspension was treated with proteinase K to degrade accessible proteins, followed by inactivation of the proteinase by phenyl methyl sulfonyl fluoride. Finally, all mitochondrial pellets were solubilized with Triton X-100 and the Triton extracts as well as the mitochondria-free supernatants (cf. above) were subjected to immunoprecipitation with an antiserum against the ADP/ATP carrier. The immunoprecipitates were resolved on SDS-polyacrylamide gels and radioautographed.*

irreversible conformational change which exposes sites necessary for catalytic function or assembly with partner polypeptides ('vectorial processing').

A mitochondrial protease that specifically processes yeast mitochondrial precursors has recently been identified and purified several hundredfold from yeast[23]. The protease is water-soluble, sensitive to o-phenanthroline and EDTA but insensitive to most inhibitors of serine proteases. It cleaves mitochondrial precursors to their mature size (but no further) and is inactive towards all non-mitochondrial proteins tested. Since it is localized in the matrix space, its role *in vivo* is probably directed at those precursors that are transported across both mitochondrial membranes. This implies that the intermembrane space, too, should contain its own distinct protease for processing proteins (such as cytochrome *c* peroxidase) which are transported into this space.

What next?

A beginning has been made but we do not yet really understand how the precursor polypeptides are translocated across the mitochondrial membrane(s). To know more we must isolate and study the molecules that participate in this translocation process. In addition to the processing proteases we have to isolate chemically significant amounts of precursors and the (still hypothetic) mitochondrial receptor molecules. In the long run, it should be possible to study the internalization of mitochondrial precursors with liposomes which contain a processing protease in their lumen and a mitochondrial 'import receptor' inserted into their phospholipid bilayer. Even if this could be achieved, there would still be some lingering doubt as to whether such experiments *in vitro* faithfully reflected the situation within a living cell. Such doubts could perhaps be silenced by the discovery of temperature-sensitive mutants in which specific steps of the import process are blocked. Since genetic studies are most conveniently performed with microbial systems, we suspect that *Neurospora* and yeast will continue to feature prominently in future efforts to understand how mitochondria import proteins from the cytoplasm.

References

1 Schatz, G. and Mason, T. L. (1974) *Ann. Rev. Biochem.* 43, 51–87
2 Palade, G. W. (1975) *Science* 189, 347–358
3 Milstein, C., Brownlee, G. G., Harrison, T. M. and Matthews, M. B. (1972) *Nature New Biology* 239, 117–120
4 Blobel, G. and Dobberstein, B. (1975) *J. Cell Biol.* 67, 835–851
5 Hallermayer, G., Zimmermann, R. and Neupert, W. (1977) *Eur. J. Biochem.* 81, 523–532
6 Schatz, G. (1979) *FEBS Letters* 103, 201–211
7 Maccechini, M.-L., Rudin, Y., Blobel, G. and Schatz, G. (1979) *Proc. Natl. Acad. Sci. U.S.A.* 76, 343–347
8 Harmey, M. A. and Neupert, W. (1979) *FEBS Letters* 108, 385–389
9 Zimmermann, R., Paluch, U., Sprinzl, M. and Neupert, W. (1979) *Eur. J. Biochem.* 99, 247–252
10 Nelson, N. and Schatz, G. (1979) in: *Membrane Bioenergetics* (Lee, C. P., Schatz, G. and Ernster, L., eds), pp. 133–152, Addison-Wesley, Reading
11 Nelson, N. and Schatz, G. (1979) *Proc. Natl. Acad. Sci. U.S.A.* 76, 4365–4369
12 Zimmermann, R., Paluch, U. and Neupert, W. (1979) *FEBS Letters* 108, 141–151
13 Korb, H. and Neupert, W. (1978) *Eur. J. Biochem.* 91, 609–620
14 Neher, E.-M., Harmey, M. A., Hennig, B., Zimmermann, R. and Neupert, W. (1980) in *Structure and expression of the mitochondrial genome* (Kroon, A. M. and Saccone, C., eds), 413–422, North-Holland, Amsterdam
15 Maccechini, M.-L., Rudin, Y. and Schatz, G. (1979) *J. Biol. Chem.* 254, 7468–7471
16 Poyton, R. O. and McKemmie, E. (1979) *J. Biol. Chem.* 254, 6763–6771
17 Lewin, A., Gregor, I., Mason, T. L., Nelson, N. and Schatz, G. (1980) *Proc. Natl. Acad. Sci. U.S.A.* 77, 3998–4002
18 Mihara, K. and Blobel, G. (1980) *Proc. Natl. Acad. Sci. U.S.A.* 77, 4160–4164
19 Schmelzer, E. and Heinrich, P. C. (1980) *J. Biol. Chem.* 255, 7503–7506
20 Raymond, Y. and Shore, G. S. (1979) *J. Biol.*

Chem. 254, 9335–9338
21 Kellems, R. E., Allison, V. F. and Butow, R. A. (1975) *J. Cell Biol.* 65, 1–14
22 Zimmermann, R. and Neupert, W. (1980) *Eur. J. Biochem.* 109, 217–229
23 Böhni, P., Gasser, S., Leaver, C. and Schatz, G. (1980) in *Structure and expression of the mitochondrial genome* (Kroon, A. M. and Saccone, C., eds), pp. 423–433, North-Holland, Amsterdam

Walter Neupert is at the Institute of Physiological Chemistry, University of Göttingen, German Federal Republic; Gottfried Schatz is at the Biocenter, University of Basel, Switzerland.

The signal hypothesis – a working model

David I. Meyer

For nascent secretory and membrane proteins to be inserted into the rough endoplasmic reticulum they must first be singled out from the proteins which are destined to remain in the cytoplasm. Moreover, they must contact the rough endoplasmic reticulum. The latest results elucidate these first steps, pointing out the role of specific receptors.

Although it may have been sufficient for the most primitive cells to translocate proteins *en bloc* across or into their membrane, an additional, highly efficient mechanism is used by complex cell types. It involves the transfer of nascent proteins across a membrane during their synthesis. Such a co-translational transfer offers distinct advantages: thermodynamically, only a small portion of the entire nascent chain is traversing the lipid bilayer at any given time; and a transfer concomitant with synthesis guards against an accumulation of (potentially harmful) proteins within the cytoplasm.

In 1971, a model was put forward postulating a series of molecular events comprising the co-translational transfer of secretory proteins across the membrane of the rough endoplasmic reticulum (RER)[1]. This 'signal hypothesis' depicted the way in which the RER selects, translocates and processes proteins during their synthesis (see Fig. 1). The hallmark of nascent secretory and membrane protein precursors is an N-terminal extension of some 15–30 amino acids, termed the signal sequence. It is now firmly established that only peptides which bear such signal sequences are capable of being co-translationally inserted into, or transferred across, the membrane of the RER. The removal of the signal sequence is usually concomitant with transfer and central to the signal hypothesis was the assumption that the microsomal membrane could select signal sequence-bearing nascent chains by means of a specific receptor[2]. Two separate pieces of evidence confirmed the existence of such a receptor: (a) synthetic signal sequences[3] and other secretory or membrane proteins[4,5] were shown to compete with proteins undergoing translocation; and (b) translocation could be prevented by inactivating the microsomal proteins (e.g. by proteases or alkylating agents)[6,7] or removing them from their microsomes (e.g. via salt extraction or proteolysis)[8–10]. This evidence brought into doubt the theory that nascent chains alone possessed all the features needed to ensure their transfer through a membrane[19–21].

Although attention continues to be focused on the nascent proteins *per se*, in the past 2 years the role played by the microsome in the interpretation of the signal sequence has become better understood.

The time element

Early studies indicated that microsomes had to recognize proteins which were to be translocated within a few minutes of their translation[11,12]. This period corresponded to the time (as judged from experiments *in vitro*) when the signal sequence emerged from the ribosome and became available as a ligand for its receptor. Later in translation, the protein begins to fold and the hydrophobic signal sequence probably becomes buried within the growing peptide chain. *In vivo*, where proteins are being synthesized at far higher rates, this period is

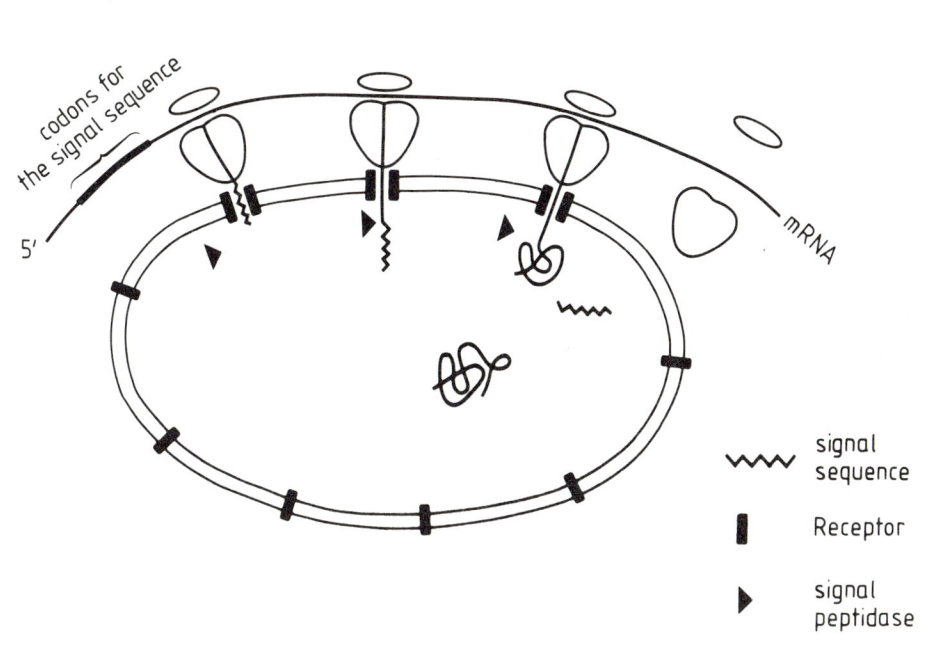

Fig. 1. Early models of the signal hypothesis suggested that the signal sequence, emerging from the large ribosomal subunit, mobilized membrane-bound receptors which formed a tunnel or pore in the ER membrane. The nascent chain would thus be able to be transferred through this tunnel to the lumen of the ER. Once the signal sequence has been transferred, it is subject to removal by the endoproteolytic action of the signal peptidase. Upon completion of translation, ribosomal subunits were free to commence translation elsewhere.

judged to be only a few seconds. As the synthesis of membrane and secretory, proteins can be initiated in the cytoplasm[12], the complex between the nascent chain and the ribosome would have to be fairly close to the correct membrane – and moving in the right direction – in order to ensure that recognition is accomplished during the time the signal sequence is accessible.

An elegant and efficient way of guaranteeing that all chains are translocated would be to eliminate this time element. If translation could be stopped once the signal sequence is exposed, every protein would have the time it needed to find the molecules mediating its membrane transfer. This is precisely what recent results lead one to conclude: translocation is a process which includes a transient cessation of protein synthesis.

Signal recognition protein (SRP)

When rough microsomes are washed with high salt concentrations they become unable to translocate nascent proteins[8]. The factor removed in this way is a protein complex of Mr = 250,000[13,14]. It has a low affinity for ribosomes alone but a very high affinity for ribosomes synthesizing secretory and membrane proteins. This high affinity is expressed, when a signal sequence forms part of the nascent chain. Addition of this 250,000 mol. wt complex to a wheat germ cell-free system blocked translation as soon as the nascent chain was long enough to be recognized as a putative membrane or secretory protein (i.e., when the signal sequence was exposed. The block remained in effect until salt-washed rough microsomes were added to the system. At this point translation continued and translocation commenced. For these reasons it has been given the name Signal Recognition Protein (SRP)[15]. Although it was originally purified from salt extracts of rough microsomes, recent results indicate that it is also present in the cytoplasm[16] and may be already attached to ribosomes.

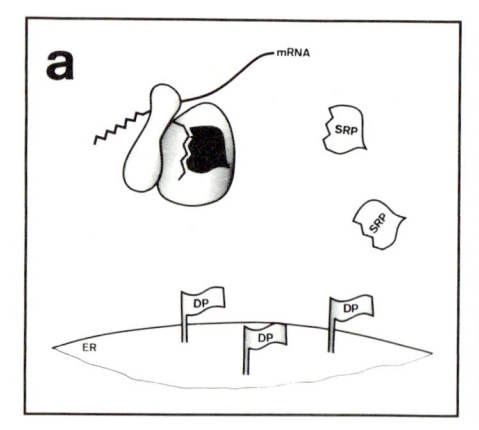

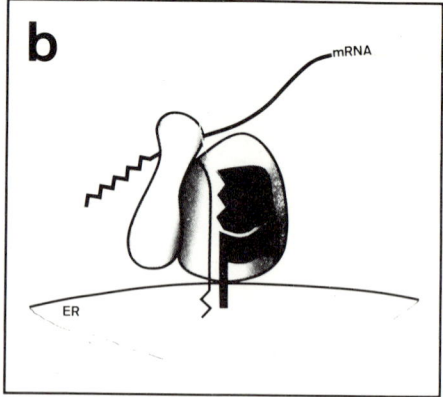

Fig. 2. (a) An up-to-date version of the initial events occurring in vectorial transfer indicates the participation of two cytoplasmically-disposed proteins. The signal recognition protein (SRP) binds to the ribosome which bears a signal sequence. Translation is stopped immediately. (b) This block persists until the ribosomal complex contacts the ER-specific SRP receptor, the docking protein (DP). By virtue of this SRP–DP interaction, the block is removed, translation resumes and translocation commences.

With the signal receptor located in the cytoplasm, nascent peptides would have a virtually unlimited amount of time to contact the membrane. As a nascent secretory or membrane protein reveals itself through the exposure of its signal sequence, cytoplasmic SRP will recognize it and shut down its further translation. This block persists until the nascent chain–SRP–ribosomal complex reaches and binds to the RER. Only then can translation continue and transfer through the membrane commence.

The docking protein

Just as a specific molecule, SRP, blocks translation, the next step in transfer must involve a system for removing the block and allowing transfer to proceed. Such a system must have the following characteristics: it must be capable of overcoming the block, and it must be ER-specific. In the past 2 years a membrane protein fulfilling these requirements has been isolated and characterized.

A component required for the translocation of nascent secretory proteins was isolated from rough microsomes using mild proteolysis in conjunction with high salt[10]. Subsequent purification yielded an active peptide with a Mr = 60,000[17]. Antibodies raised against this 60,000 mol. wt fragment revealed that it represented the cytoplasmically-disposed domain of a membrane protein of Mr = 72,000[18]. Using indirect immunofluorescence it was also possible to establish the molecule as ER-specific[18]. Most importantly, the addition of this protein to cell-free translation systems removed the SRP-induced block of secretory protein synthesis[16]. On the basis of this data (depicted diagrammatically in Fig. 2) the protein was assumed to be the membrane SRP receptor required for docking of the SRP-blocked ribosome with the rough ER. It was therefore dubbed the 'Docking Protein'.

The unanswered question

This series of steps indicates the way in which the cell discriminates between proteins to be translocated across the ER and those which must remain cytoplasmic. Consistent with the signal hypothesis is the fact that the information contained within signal peptides is interpreted by the sequential interaction of specific proteins. Once a nascent protein has traversed the membrane, it is known that another component postulated in the signal hypothesis, the signal peptidase, removes the signal sequence. Questions regarding the specific mechanism by which nascent chains cross the lipid bilayer of the RER have not been answered. Here again the signal hypothesis suggests that specific proteins are involved. Overall, those predictions of the signal hypothesis which have been tested are correct. Whether proteins mediate the transfer through the bilayer remains to be seen.

References

1 Blobel, G. and Sabatini, D. D. (1971) in *Biomembranes Vol. 2* (Manson, L. A., ed.), pp. 193–195, Plenum Press, New York
2 Blobel, G. and Dobberstein, B. (1975) *J. Cell Biol.* 67, 835–851
3 Majzoub, J. A. *et al.* (1980) *J. Biol. Chem.* 255, 11478–11483
4 Lingappa, V. R., Katz, F. N., Lodish, H. F. and Blobel, G. (1978) *J. Biol. Chem.* 253, 8667–8670
5 Prehn, S., Nürnberg, P. and Rapoport, T. A. (1981) *FEBS Lett.* 123, 79–84
6 Jackson, R. C., Walter, P. and Blobel, G. (1980) *Nature (London)* 286, 174–176
7 Prehn, S., Tsamaloukas, A. and Rapoport, T. A. (1980) *Eur. J. Biochem.* 107, 185–195
8 Warren, G. and Dobberstein, B. (1978) *Nature (London)* 273, 569–571
9 Walter, P. *et al.* (1979) *Proc. Natl Acad. Sci. U.S.A.* 76, 1795–1799
10 Meyer, D. I. and Dobberstein, B. (1980) *J. Cell Biol.* 87, 498–502
11 Blobel, G. and Dobberstein, B. (1975) *J. Cell Biol.* 76, 852–862
12 Rothman, J. E. and Lodish, H. F. (1977) *Nature (London)* 269, 775–780
13 Dobberstein, B. (1978) *Hoppe-Seyler's Z. Physiol. Chem.* 359, 1469–1470
14 Walter, P. and Blobel, G. (1980) *Proc. Natl Acad. Sci. U.S.A.* 77, 7112–7116
15 Walter, P. and Blobel, G. (1981) *J. Cell Biol.* 91, 545–561
16 Meyer, D. I., Krause, E. and Dobberstein, B. *Nature (London)* 297, 647–650
17 Meyer, D. I. and Dobberstein, B. (1980) *J. Cell Biol.* 87, 503–508
18 Meyer, D. I., Louvard, D. and Dobberstein, B. (1982) *J. Cell Biol.* 92, 579–583
19 Wickner, W. (1979) *Ann. Rev. Biochem.* 48, 23–45
20 von Heijne, G. and Blomberg, C. (1979) *Eur. J. Biochem.* 97, 175–181
21 Engelman, D. M. and Steitz, T. A. (1981) *Cell* 23, 411–422

David I. Meyer is at the European Molecular Biology Laboratory, 6900 Heidelberg, F.R.G.

The secretory pathway in yeast

R. Schekman

Yeast mutants that are temperature-sensitive for secretion and growth accumulate secretory organelles which contain glycoprotein intermediates. The secretory pathway, defined by these mutants, is also responsible for plasma membrane and vacuole assembly.

Protein secretion occurs in almost all cell types. Despite the wide range of activity that this implies, the stages in the secretory process are quite similar in all organisms. In eukaryotes, the sequence ER → Golgi → vesicle → cell surface is the generally accepted mode of transport for soluble and membrane proteins and, although prokaryotes do not have specialized secretory organelles, polypeptides penetrate the bacterial cytoplasmic membrane via processes quite analogous to those employed in protein import to the endoplasmic reticulum and mitochondrion.

In addition to a role in cell surface assembly, the secretory process may contribute to the assembly of the lysosome, mitochondrion, and nucleus. Lysosomal glycoprotein precursors are translocated into ER membranes by the same system used for secretory and plasma membrane proteins[1]. Furthermore, both in histochemical and in organelle fractionation studies, lysosomal enzymes are detected in ER and Golgi cisternae[2]. Mitochondrial membranes are assembled with lipids synthesized in the ER and although many mitochondrial enzymes derive from soluble cytoplasmic precursors, the ER and mitochondrial outer membranes appear to share a number of integral proteins[3]. Finally, the nuclear envelope is continuous with the ER. This is most apparent during interphase, when the nuclear envelope is reconstructed by outgrowth from the ER. Certain soluble nuclear components may also be derived from secretory organelles. The observation that chromatin-associated high mobility group proteins (HMGs) contain N-glycosidically linked, complex oligosaccharides suggests that these proteins may gain access to the nucleus via the Golgi body[4]. Clearly, secretory organelles play a major role in cell architecture and metabolism.

While a considerable amount is known about the gross features of secretory organelles, and about the structure and covalent modifications of molecules that are transported through the organelles, it has been much more difficult to define the cellular functions involved in protein transport. New approaches involving genetic and biochemical techniques are essential. In this regard, recent advances in identifying proteins involved in the penetration of secretory polypeptides across a membrane[5-9], and in the discrimination of lysosomal and secretory proteins[10-12] have provided important examples of these cellular functions.

A genetic approach can be used to identify the full range of functions required for protein transport. For this and other reasons, we have studied the secretory process in the yeast *Saccharomyces cerevisiae*. Although much less is known about the secretory process in yeast, and there are special technical difficulties in the use of yeast, the potential for a combined genetic and biochemical approach may prove a crucial advantage.

Organization of the yeast cell surface

The yeast cell surface consists of at least three layers: the cell wall which contains

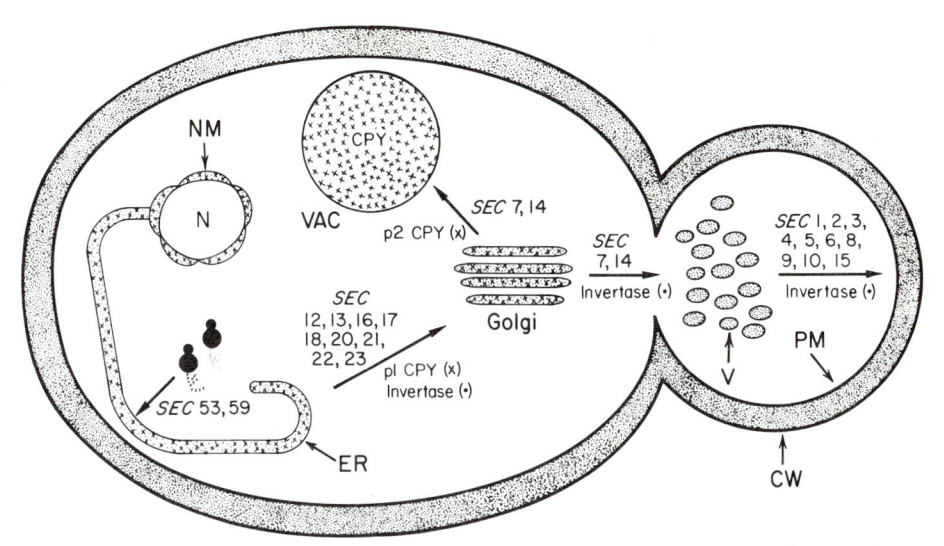

Fig. 1. Secretory and vacuolar protein transport pathways in yeast. N: nucleus; NM: nuclear membrane; ER: endoplasmic reticulum; SEC: wild-type gene; VAC: vacuole; V: vesicle; PM: plasma membrane; CW: cell wall; CPY: 61 Kd mature carboxypeptidase Y; p1 CPY: a 67 Kd pro-enzyme form of CPY; p2 CPY: a 69 Kd pro-enzyme form of CPY.

mannoproteins and structural polysaccharides (β-1,3 and β-1,6-linked glucan), a periplasm that contains mannoproteins and a plasma membrane. Most of the soluble secreted enzymes, such as invertase and acid phosphatase, are located in the periplasm or in the cell wall where they are accessible to low molecular weight substrates. Certain smaller non-glycosylated proteins, such as α-factor and killer toxin, are secreted through the cell wall into the culture medium.

Secretion is correlated topologically to the region of cell surface growth. Invertase and acid phosphatase are secreted into the bud portion of a growing cell which corresponds to the point of cell surface addition during most of the division cycle[13,14]. The correlation between secretion and budding is best accounted for by an exocytic mechanism of surface growth. Secretory vesicles may fuse with the inner surface of the bud and deliver mannoproteins to the periplasm and membrane precursors to the plasma membrane. The available cytologic evidence strongly supports this notion. Electron microscopic thin section and freeze fracture views show 50–100 nm vesicles which fuse with the bud plasma membrane[15] and an acid phosphatase-specific stain has been taken up by bud-localized vesicles, the ER and a Golgi-like organelle[16].

Although the yeast secretory process appears to resemble the mechanism used by plant and animal cells, one striking difference is the low level of secretory organelles revealed by standard EM thin section analysis. This low level is consistent with a rapid transit time for the export of invertase[17] and a low level of invertase export precursors[18]. The small internal pool of secretory precursors provides a sensitive experimental system for the evaluation of mutants that block the secretory pathway and cause an accumulation of secretory enzymes and organelles.

Isolation and characterization of secretory mutants

Given the possibility that the secretory process contributes generally to yeast cell surface growth, Peter Novick assumed that secretory mutants would be lethal. To get around this problem, Novick screened a collection of temperature-sensitive growth mutants for ones that failed to export active invertase and acid phosphatase at the non-permissive temperature (37°C), but which performed normally at the permissive temperature (25°C). Mutants representing two complementation groups (*sec1*, *sec2*) were found which accumulated secretory enzymes in an intracellular pool[18]. A large number of additional *sec* mutants have been isolated based on the observation that *sec1* cells become dense at 37°C. Susan Henry showed that during inositol starvation of an auxotrophic strain, net cell surface growth stopped while cell mass increased[19]. Starved cells could be resolved from normal cells on a Ludox density gradient. Similarly, *sec* mutants cells can be enriched from a mutagenized culture by incubation at 37°C followed by Ludox density gradient sedimentation.

The *sec* mutants are of two types. Class A *sec* mutants (192 total) are like *sec1* and *sec2*, in that active secretory enzymes accumulate in an intracellular pool[20]. Class B *sec* mutants (23 in total) do not secrete or accumulate active secretory enzymes, yet protein synthesis continues at a near normal rate for several hours at 37°C. Complementation analysis has revealed 23 *sec* loci in the A class and 4 *sec* loci in the B class. The distribution of mutant alleles suggests that more of both classes could be found.

The initial characterization of the *sec* mutants suggested a general block in secretion and cell-surface growth at the non-permissive temperature. In addition to an immediate halt in bud growth, export of a number of secreted proteins (invertase, acid phosphatase, L-asparaginase, α-galactosidase, α-factor, killer toxin) and plasma membrane permease activities (SO_4^{2-} permease, arginine permease, proline-specific permease) are blocked. A more general probe of surface assembly that was used to examine the export and turnover of macrophage plasma membrane proteins[21] has now been adapted by Novick to examine surface assembly in yeast[22]. Modification of cell surface amino groups with trinitrobenzensulfonate (TNBS) followed by precipitation with TNP antibody allows the analysis of newly-exported proteins. In this procedure, wild-type and mutant cells are labelled with protein synthesis precursors at 37°C and then tagged with TNBS at 0°C. Under these conditions, TNBS does not penetrate into the cell. Both secreted and plasma membrane surface proteins are tagged in this procedure and can be examined separately: secreted proteins are released when cells are converted to spheroplasts, and tagged membrane proteins are recovered in a sedimented fraction from lysed spheroplasts. Wild-type cells export distinct sets of membrane and secreted proteins are revealed by the SDS-gel electrophoresis of TNP-antibody precipitates. The major proteins in both fractions are not exported in *sec* mutant cells at 37°C, but are exported at 24°C. These results suggest that the secretory process is responsible for the localization of most cell surface proteins in yeast.

Many of the Class A *sec* mutants secrete a large fraction of the invertase that accumulates at 37°C when cells are returned to the permissive temperature. In most cases the secretion of accumulated invertase is insensitive to cycloheximide. This implies that the affected gene product is reversibly inactivated by the temperature shift. The result demonstrates that ongoing protein synthesis is not essential for post-translational transit of secretory enzymes, and thus excludes the possibility that newly-synthesized secretory protein forces the flow of the export process. Mutations in one gene (*sec7*) allow thermoreversible secretion only in growth medium containing a low concentration of glucose.

Perhaps the most dramatic feature of the class A *sec* mutants is that they accumulate or exaggerate specific secretory organelles. Mutations in ten groups produce 80–100 nm vesicles at 37°C that are distributed throughout the cytoplasm, unlike the bud-localized vesicles seen in wild-type cells. Mutations in another nine genes produce exaggerated endoplasmic reticulum. In these mutants the ER lines the inner surface of the plasma membrane and winds through the cytoplasm where multiple connections with the nuclear envelope are seen. The lumen of both the ER and the nuclear envelope is wider than the corresponding wild-type structure. However, the high density of ribosomes in the background has made it impossible to determine if the exaggerated ER is in the rough or smooth form. A third class of mutant, represented by two genes, produces a different organelle depending on the growth medium in which the cells are incubated at 37°C. Mutations in the *sec7* gene cause the accumulation of typical Golgi-like structures when mutant cells are incubated at 37°C in medium containing 0.1% glucose. The same mutant, when incubated at 37°C in medium with 2% glucose, accumulates cup- and toroid-shaped organelles that we have called Berkeley bodies. This change in organelle morphology correlates with the effect of glucose on thermoreversible secretion mentioned earlier, and suggests that the Golgi-body structure is a more natural intermediate. In each case where reversible secretion is observed, a return to the permissive temperature allows the accumulated organelle to diminish in abundance. Histochemical staining of mutants representing each of the distinct cytologic types has shown that the secreted enzyme, acid phosphatase, is contained in the lumen of the accumulated organelle[23].

Susan Ferro-Novick has found that the class B *sec* mutants produce enzymatically inactive forms of invertase and acid phosphatase[24]. In two of the mutants (*sec53* and *sec59*), immunoreactive forms of invertase are produced at 37°C which appear to remain embedded in the ER membrane[25]. Perhaps as a result of this aberrant accumulation, the ER in these mutants appears fragmented in contrast to the smooth, thin ER envelope seen in wild-type cells. These mutants also show greatly reduced incorporation of [³H]mannose into a total glycoprotein fraction at 37°C, although oligosaccharide synthesis is not directly affected. Reduced mannose incorporation appears to be due to a defect in translocation of nascent polypeptide chains to the lumenal surface of the ER membrane where oligosaccharides are transferred to protein. Surprisingly, protease protection experiments have indicated that a significant portion of the invertase polypeptide (mol. wt 11,000 of a 60,000 mature polypeptide length) is inserted into and across the ER membrane in mutant cells at 37°C[26]. Furthermore, as with the class A *sec* mutants, *sec53* and *sec59* are thermoreversible. Upon return to 25°C, in the presence of cycloheximide, the transmembrane form of invertase is transferred into the lumen of the ER, glycosylated, and transported to the cell surface through the normal pathway. Thus, the *sec53* and *sec59* gene products define functions required for completion, but not initiation of protein penetration across the ER membrane.

Order of events in the pathway

A simple technique exists for the ordering of events along a linear irreversible pathway in which distinct intermediates accumulate in different mutants. In this circumstance a double mutant will accumulate the intermediate prior to the first block that is encountered. This analysis has been performed with mutants representing each of the four stages that are identified by the *sec* mutations. Double mutants containing *sec53* or *sec59* together with any of the class A *sec* mutants fail to accumulate active invertase[25]. Thus, these B *sec* mutants are epistatic to the other mutants. Among the class A *sec* mutants, the ER-accumulating phenotype is epistatic to the Golgi body- and vesicle-accumulating phenotypes, and a Golgi body accumulating mutant is epistatic to all of the vesicle-blocked mutants[17]. The order of events determined by this analysis is shown in Fig. 1.

An independent line of evidence supports this order of events. Brent Esmon has analysed the extent of glycosylation of invertase accumulated at 37°C in the *sec* strains and identified at least two stages in oligosaccharide assembly[23]. The invertase polypeptide produced in *sec53* has little or no carbohydrate. During translocation into the ER, invertase acquires 9–10 N-glycosidically linked core oligosaccharides which have a composition identical to the mammalian 'high-mannose' oligosaccharide (Man$_9$GlcNAc$_2$)[27]. Secreted invertase contains a core portion of 11–14 mannose residues and an outer chain portion that has as many as 150 mannose residues[28]. When movement beyond the ER is blocked, invertase and most other glycoproteins accumulate with Man$_9$GlcNAc$_2$ N-linked oligosaccharides. Under these conditions about 50% of the total mannan synthesized is in the O-glycosidic linkage, and instead of the usual mannotriose and -tetraose, the O-linked sugar is mannose and mannobiose. Golgi- and vesicle-blocked mutants accumulate mature N- and O-linked oligosaccharides. These results suggest that oligosaccharide assembly is compartmentalized as indicated in Fig. 2.

A role for the pathway in vacuole assembly

Although profiles of Golgi bodies have not been convincingly seen in wild-type *Saccharomyces* cells, the organelle that accumulates in *sec7* cells at 37°C (in medium containing 0.1% glucose) looks and functions just as a mammalian Golgi body in oligosaccharide maturation. An additional role for the yeast Golgi body has been realized by Tom Stevens in experiments on maturation and localization of the vacuolar glycoprotein, carboxypeptidase Y[29].

The yeast vacuole is a lysosome-like organelle that has a set of hydrolytic enzymes distinct from the secreted glycoproteins. Among the vacuolar enzymes, carboxypeptidase Y (CPY) has been studied in some detail. CPY is made as a 69,000 mol. wt precursor containing four N-glycosidically-linked oligosaccharides[30]. The amino-terminal 8000 mol. wt propeptide portion is cleaved under the direction of at least two gene products (PEP4, PEP17) that are required for the maturation of several other vacuolar enzymes[31]. A *pep4* mutant accumulates enzymatically inactive proCPY in the vacuole, suggesting that maturation occurs after the precursor is localized[29]. If vacuolar protein require the secretory pathway for transport, proenzyme forms will not be localized and processed at 37°C in the *sec* mutants. Stevens has shown that mutants blocked in movement from the ER or from the Golgi body accumulate proCPY in some place other than the vacuole, presumably in the accumulated organelle. Upon return to the permissive temperature the accumulated proenzyme forms become processed normally. Mutants that block after the Golgi step have no effect on CPY localization. These results suggest that vacuolar and secretory proteins travel together from the ER to the Golgi body where sorting may occur (Fig. 1). The results also rule out a secretion–recapture mechanism of localization such as has been suggested in mammalian lysosomal enzyme studies[10].

A mannose-6-P determinant on N-linked oligosaccharides has been implicated in the targeting of lysosomal enzymes in human fibroblasts. Although sorting of lysosomal and secretory proteins in the Golgi body may rely on a carbohydrate structure, the ultimate source of discrimination lies in an amino acid sequence or structural feature of the targeted protein. Carbohydrate does not serve this role in yeast because at least two vacuolar proteins, CPY and an alkaline phosphatase, are synthesized and activated normally in the absence of oligosaccharide synthesis[30,32]. Mutant alleles of CPY that result in misdirection of an otherwise normal proenzyme may reveal the signal responsible for normal localization.

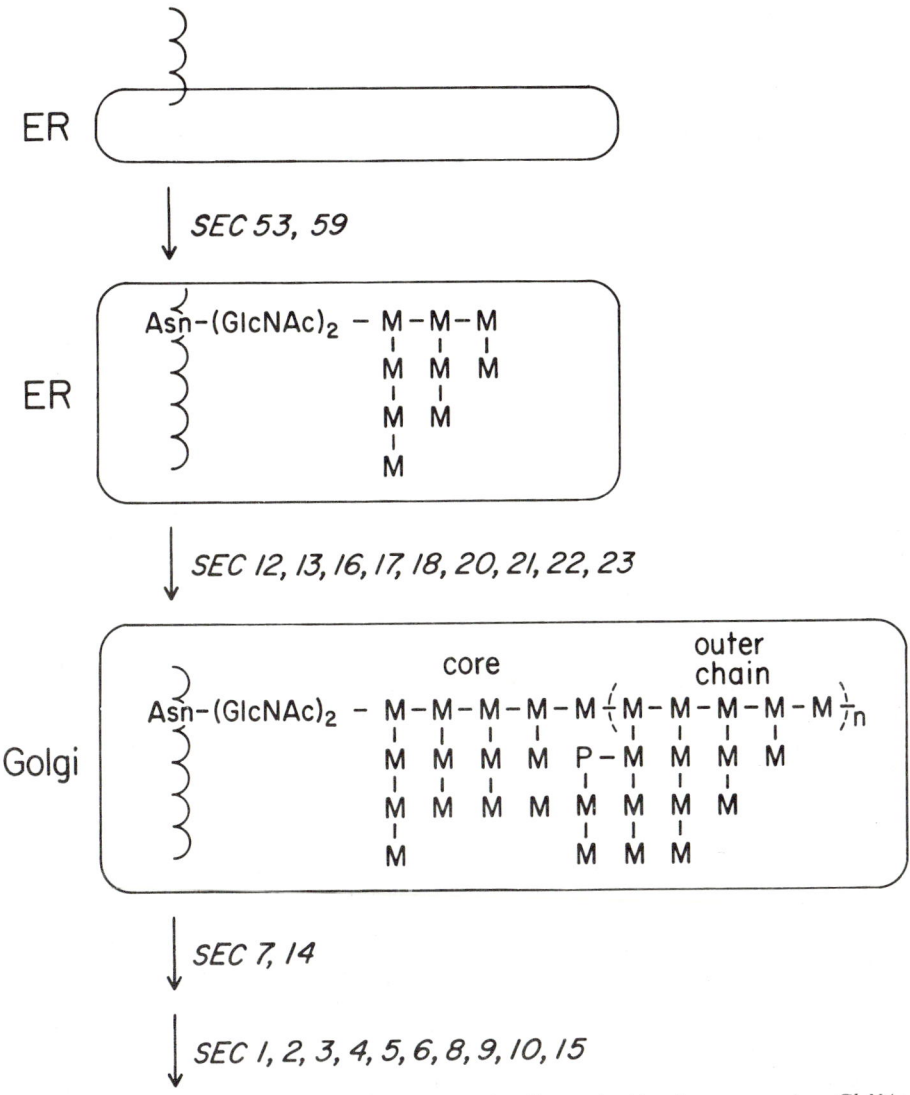

Fig. 2. *Compartmentalized assembly of mannoprotein oligosaccharide. Asn: asparagine; GlcNAc, N-acetylglucosamine; M: mannose;* ∫ *: polypeptide.*

Conclusions

The transport of cell surface macromolecules requires a large number of cellular functions. Lesions in these essential functions lead to an interruption of plasma membrane and secretory protein export at one of four stages in a linear pathway (Fig. 1). Mutants blocked early in the pathway have shown that glycoprotein carbohydrate synthesis is compartmentalized: core oligosaccharides are added in the ER, and the outer chain structure is extended in the Golgi body (Fig. 2). These same mutants have revealed that part of the pathway is responsible for localization of at least one vacuolar glycoprotein.

The *sec* mutants have provided a new method for tracing the pathway of protein localization in a eukaryote, and have in a limited sense defined the cellular functions required for transport. *In vitro* reactions that require the *sec* gene products and thus reflect portions of the secretory pathway will be needed to understand the mechanism of transport.

Acknowledgements

I gratefully acknowledge R. Roon, R. Buckholz, D. Julius, H. Bussey, P. Call, and W. Courchesne for their communication of preliminary results concerning the behavior of *sec* mutants. The research conducted in my laboratory is supported by grants from the NIH (National Institute of General Medical Sciences and the National Institute of Environmental Health Science) and the NSF.

References

1 Erikson, A. H. and Blobel, G. (1979) *J. Biol. Chem.* 254, 11771–11774
2 Rome, L. H., Garvin, A. J., Allietta, M. M. and Neufeld, E. F. (1979) *Cell* 17, 143–153
3 Shore, G. C. (1979) *J. Cell Sci.* 38, 137–153
4 Reeves, R., Chang, D. and Chung, S.-C (1981) *Proc. Natl Acad. Sci. U.S.A.* 78, 6704–6708
5 Walter, P., Ibrahimi, I. and Blobel, G. (1981) *J. Cell Biol.* 92, 545–550
6 Walter, P., Ibrahmi, I. and Blobel, G. (1981) *J. Cell Biol.* 91, 551–556
7 Walter, P., Ibrahimi, I. and Blobel, G. (1981) *J. Cell Biol.* 91, 557–561
8 Meyer, D. I., Louvard, D. and Dobberstein, B. (1982) *J. Cell Biol.* 92, 579–583
9 Zwizinski, C. and Wickner, W. (1981) *J. Biol. Chem.* 255, 7973–7977
10 Hickman, S. and Neufeld, E. F. (1972) *Biochem. Biophys. Res. Commun.* 49, 992–999
11 Kaplan, A., Achord, D. T. and Sly, W. S. (1977) *Proc. Natl Acad. Sci. U.S.A.* 74, 2026–2030
12 Reitman, M. L. and Kornfeld, S. (1981) *J. Biol. Chem.* 256, 11977–11980
13 Tkacz, J. S. and Lampen, J. O. (1973) *J. Bacteriol.* 113, 1073–1075
14 Field, C. and Schekman, R. (1980) *J. Cell Biol.* 86, 123–128
15 Moor, H. (1967) *Arch. Mikrobiol.* 57, 135–146
16 Linnemans, W. A. M., Boer, P. and Elbers, P. F. (1977) *J. Bacteriol.* 131, 638–644
17 Novick, P., Ferro, S. and Schekman, R. (1981) *Cell* 25, 461–469
18 Novick, P. and Shekman, R. (1979) *Proc. Natl Acad. Sci. U.S.A.* 76, 1858–1862
19 Henry, S. A., Atkinson, K. D., Kolat, A. I. and Culbertson, M. R. (1977) *J. Bacteriol.* 130, 472–484
20 Novick, P., Field, C. and Schekman, R. (1980) *Cell* 21, 205–215
21 Kaplan, G., Unkeless, J. C. and Cohn, Z. A. (1979) *Proc. Natl Acad. Sci. U.S.A.* 76, 3824–3828
22 Novick, P. and Schekman, R. (in preparation)
23 Esmon, B., Novick, P. and Schekman, R. (1981) *Cell* 25, 451–460
24 Ferro-Novick, S., Novick, P., Field, C. and Schekman, R. (in preparation)
25 Ferro-Novick, S., Hansen, W. and Schekman, R. (in preparation)
26 Ferro-Novick, S., Stevens, T. and Schekman, R. (in preparation)
27 Esmon, B. and Schekman, R. (in preparation)
28 Lehle, L., Cohen, R. E. and Ballou, C. E. (1979) *J. Biol. Chem.* 254, 12209–12218
29 Stevens, T., Esmon, B. and Schekman, R. (in preparation)
30 Hasilik, A. and Tanner, W. (1978) *Eur. J. Biochem.* 91, 567–575
31 Hemmings, B. A., Zubenko, G. S., Hasilik, A. and Jones, E. W. (1981) *Proc. Natl Acad. Sci. U.S.A.* 78, 435–439
32 Onishi, H. R., Tkacz, J. S. and Lampen, J. O. (1979) *J. Biol. Chem.* 254, 11943–11952

R. Schekman is at the Department of Biochemistry, University of California, Berkeley, CA 94720, U.S.A.

Index

Accuracy
 in DNA synthesis 226
 in translation 219, 216
 of splicing 127
ACTH-β-lipotropin precursor 130
Actin genes 138
Adenovirus
 DNA 31
 RNA, splicing of 171
 gene expression 162, 171
ADP/ATP carrier, mitochondria 270
alaS promoter 238
Allelic exclusion 148
Alternative RNA processing 143
Alu family of DNA sequences 119
Amanitin, inhibition of RNA
  polymerases 156, 196
Aminoacyl-tRNA synthetases
 binding to DNA 239
 specificity 226, 233
 structure 233
Amplification
 of chorion genes 137
 and evolution of collagen genes 127
Aphidicolin 31
AT spacers 80
Attenuation 18

Bacterial conjugation 48
Bacteriophage lambda 58
Bacteriophage mu 56

Capping of RNA 196, 209
'CAT' box 129
Catabolic repression 50
Chorion genes 137
Chromatin 51
Chromosomes 112
 initiation of DNA replication in
   eukaryotes 23
Collagen gene 125
Control *see* Regulation
Core particles of nucleosomes 51
Corticotropin-β-lipotropin precursor 130
Cruciform DNA 42
Cytochrome *c*, synthesis 271

DNA
 binding of aminoacyl-tRNA
   synthetases 238
 complexes with RNA polymerase 11
 cruciform structures 42
 gyrase 41, 43
 helicases 43
 histone interactions 52

methylation 117
mitochondrial 67–107
polymerase, eukaryotic 30
polymerases, evolution 33
polymerase, exonuclease activity 230
polymerase inhibitors 30
recombination 59
repeated sequences 119
replication 23, 230
satellite 111, 117
single-stranded, binding proteins 46
supercoiled 37, 49
synthesis, accuracy in 226
synthesis, initiation 23
synthesis reactions 32
transposition 56
unwinding 13, 43
Drosophila 28, 34, 114, 137

Editing mechanisms 226, 231
Elongation factors 213, 219
Endoplasmic reticulum 257, 265, 274
Enhancer sequences 181
Errors in DNA and protein synthesis 226
Eukaryotic
 chromosomes, DNA replication 23
 DNA polymerase 30
 RNA polymerases 155
Evolution 97
 Alu family 119
 collagen gene 125
 DNA polymerases 33
 mitochondrial genome 74, 85, 92, 99
 peptide hormones 130
 5S RNA 246
 16S RNA 252
 satellite DNA 111
Exons
 collagen gene 126
 corticotropin-β-lipotropin precursor 130
Exonuclease activity, of DNA
  polymerase 230
5'-untranslated region 133

'Foot printing' 11

Gene
 amplification 137
 actin 138
 collagen 125
 families 139
 immunoglobulin 139–152, 183
 mitochondrial 67–107
 peptide hormones 130
 translocation rearrangement 146

Genetic code, mitochondrial 86
Genome-regulatory proteins 3
Glycoproteins, in yeast secretory mutants 278
Golgi complex 258, 278

HeLa cells, mitochondrial DNA 87, 92
Helicases 43
Helix destabilization 32
Heterochromatin 111
his operon 20, 22
histones 28, 52
HMG proteins 55

ilv operon 22
Immunoglobulin
  genes 139–152
  RNA processing 183
Influenza virus RNA 196
Initiation
  DNA synthesis 23
  of mitochondrial genes 94
  protein synthesis 205
  of replication 24
Initiator tRNA 206
IS elements 56
Isotopic exclusion 148
λ integrase 41

lac promoter 12
leu operon 22
Linking number 37
Long terminal repeats 123

m⁷Gppp 173, 196, 209
Membrane-bound ribosomes 257
Membrane protein synthesis 261, 274
Messenger RNA
  control of translation 240
  influenza 196
  splicing 171, 183
  translation 205–225
  mitochondrial 92
'Minus 35' region 12, 15
Mitochondrial DNA 67–107
Mitochondrial
  genome, yeast 67, 76, 101
  genome, mammalian 85, 92
  genome, plant 99
  genome, evolution 74, 85, 92, 99
  genome, excision 76
  genome, replication 76
  polypeptide synthesis 266
  rRNA and mRNA 92
Multigene families 139
Mutation rates 226

Nuclear RNA polymerases 155
Nucleosome 51

Operators, interaction with repressors 4
Operons 18, 22
Origins of replication 78
Ovalbumin synthesis 262

Peptide hormones 130
Peptidyl transferase 216, 223
Peptidyl-tRNA hydrolysis 222
Petite mutants 67, 76
Phage λ integrase 41
phe operon 20, 22
Plant male sterility 99
Plant mitochondrial genes 99
Poly A
  addition 186
  mitochondrial 92
Polyoma 25
Polypeptide synthesis 213
  termination of 220
Pribnow box 12, 15
Prolactin synthesis 265
Promoters
  adenovirus 162
  collagen gene 128
  E. coli 10
  eukaryotic 181
  sequences 12
  tRNA synthetases 238
Proof reading mechanisms 226
Protein-nucleic acid interactions 3–55, 213, 248
Protein secretion 185, 257
Protein synthesis 205–225
  accuracy in 226, 257
  elongation 213
  initiation 205
  release factors 220
Protein traffic in cells 257–284

Regulation
  feedback in *E. coli* 240
  of adenovirus genes 162
  of collagen gene 125
  of expression, mitochondrial genome 85, 92
  of immunoglobulin genes 139
  of protein synthesis 211
  of transcription 3, 55, 238
  RNA processing 143
  through enhancer sequences 181
Release factors in protein synthesis 220
rep protein 43
Repeated DNA sequences 119
Replication of DNA 23, 30, 47, 230
  and transposition 59
  mitochondrial 76, 85, 92

Repression, of aminoacyl-tRNA synthetases 238
Repressor
  lac 3
  intersegment transfer 4
  sliding 4
Ribophorin 265
Ribosomal protein synthesis 240
Ribosomes
  association with mRNA 206
  membrane-bound 257
  structure 246, 252
RNA
  5S 246
  16S 252
  polymerase, E. coli 10
  polymerase, eukaryotic 155
  polymerase, interaction with promoters 8, 10
  processing in immunoglobulin synthesis 183
  mitochondrial ribosomes 92
  splicing, adenovirus 162
  splicing of tRNA precursors 188
  synthesis and processing 155–202
  tumor viruses 57

Satellite DNA 111, 117
Secretory protein synthesis 257, 266, 274, 278
Selfish DNA 119
Signal peptides 258, 265, 266, 274
Signal recognition proteins 264, 265, 275
Spacers, AT rich in mitochondrial genomes 72
Splicing 183
  accuracy of 127
  of adenovirus RNA 171
  of Alu family 119
  of collagen gene 125
  of immunoglobulin genes 145
  of peptide hormones 130

of tRNA precursors 188
sites 179
Supercoiling of DNA 37, 38, 49
SV 40 25, 181

Tautomerism, of pyrimidines 230
Termination, of polypeptide synthesis 220
Terminator 20
Topoisomerases 41, 42, 49
Topoisomers of DNA 38
Transcription
  Alu sequences 122
  E. coli 10
  enhancer sequences 181
  influenza virus 196
  mitochondrial DNA 87
  role of topoisomerases 49
  termination 18
Transfer RNA
tRNA
  mitochondrial 87
  precursors, splicing of 188
Translation, accuracy in 219
  elongation 213
  initiation 205
  termination 220
Translational regulation 244
Transposable elements 56, 123
Transposition, role of topoisomerases 49
Transposons 56
Tryptophan operon 18
Ty l elements 57
Ty l transposon of yeast 123

Xenopus egg
  DNA replication 25
  gene amplification 137
Yeast, petite mutants 67, 76
Yeast, synthesis of secretory proteins 278

Z-DNA 42

The articles in this book have been reprinted
from the monthly review journal

TRENDS IN BIOCHEMICAL SCIENCES

This journal is available on subscription from
  Elsevier Biomedical Press (Cambridge)
  Department DRP
  68 Hills Road
  Cambridge CB2 1LA
  United Kingdom

Write for details of the personal edition and library edition of *Trends in Biochemical Sciences* and our special reduced prices for students.